PAPERS OF THE MICHIGAN ACADEMY OF SCIENCE, ARTS, AND LETTERS

VOLUME XLV

I. *Natural Science*
II. *Social Science*
III. *Philosophy, Literature, and Art*

Papers of the
MICHIGAN ACADEMY
OF SCIENCE, ARTS, AND LETTERS

VOLUME XLV

Editor
SHERIDAN BAKER

ANN ARBOR · THE UNIVERSITY OF MICHIGAN PRESS
1960

Library of Congress Catalog Card No. 23-11028

Printed and bound by CPI Group (UK) Ltd, Croydon, CR0 4YY

Paperback ISBN: 978-0-472-75013-9

OFFICERS OF THE ACADEMY
1958-1959

PRESIDENT: J. Sutherland Frame, Michigan State University
PRESIDENT-ELECT: Claude W. Hibbard, The University of Michigan
VICE-PRESIDENT: Conrad R. Lam, Henry Ford Hospital
SECRETARY: Donald A. Ringe, The University of Michigan
TREASURER: Volney H. Jones, The University of Michigan
EDITOR: Sheridan Baker, The University of Michigan
LIBRARIAN: Frederick H. Wagman, The University of Michigan
COORDINATOR OF JUNIOR ACADEMY: Lincoln C. Pettit, Michigan State University
COORDINATOR OF JUNIOR ACADEMY FOR NORTHERN PENINSULA: Clarence Bjork, Northern Michigan College

SECTION CHAIRMEN

AMERICAN STUDIES: Richard D. Miles, Wayne State University
ANTHROPOLOGY: Moreau S. Maxwell, Michigan State University
BOTANY: Jack C. Elliott, Michigan State University
ECONOMICS: Thomas Mayer, Michigan State University
FINE ARTS: C. Howard Church, Michigan State University
FISHERIES-WILDLIFE: Eugene W. Roelofs, Michigan State University
FOLKLORE: Martha Curtis, Eastern Michigan College
FORESTRY: John Carow, The University of Michigan
GEOGRAPHY: Robert W. McIntosh, Michigan State University
GEOLOGY-MINERALOGY: James H. Fisher, Michigan State University
HISTORY-POLITICAL SCIENCE: Carl O. Smith, Wayne State University
LANDSCAPE ARCHITECTURE: Laurence M. Cook, 911 Cleo St., Lansing
LANGUAGE-LITERATURE: John Hepler, Central Michigan College
MATHEMATICS: G. Y. Rainich, University of Notre Dame
MEDICAL SCIENCES: Lester F. Wolterink, Michigan State University
PHILOSOPHY: Richard Rudner, Michigan State University
PSYCHOLOGY: Marvin Hyman, VA Regional Office, Detroit
SOCIOLOGY: Frank Hartung, Wayne State University
ZOOLOGY: John Shaver, Michigan State University

CONTENTS

Part I. Natural Science

MEDICINE

ZOOLOGY

BOTANY

FORESTRY

PALEONTOLOGY

GEOLOGY

Part II. Social Science

FOLKLORE

GEOGRAPHY

ECONOMICS

HISTORY

POLITICAL SCIENCE

SOCIOLOGY

PSYCHOLOGY

Part III. Philosophy, Literature, and Art

PHILOSOPHY

LITERATURE

ART

PART I

NATURAL SCIENCE

Papers of the Michigan Academy of Science, Arts, and Letters
Vol. XLV, 1960 (1959 Meeting)

MEDICINE

BLOCKING OF VIRAL HEMAGGLUTININS BY SERA FROM CORONARY THROMBOSIS PATIENTS[1]

L. TUNSTALL AND L. H. MATTMAN

Wayne State University

It is established that large numbers of patients who are hospitalized with coronary thrombosis have a history of antecedent or concurrent upper respiratory infections. A significant gap in our knowledge exists in the relationship between these two pathologic conditions. Medical opinion in this regard is fragmentary and contradictory. The basis for divergent views is essentially that it is virtually impossible for the clinician to make an accurate diagnosis of influenza or other viral respiratory infections from the symptomatology alone (Silber, 1958).

A fortunate discovery by Hirst in 1941 has made possible a remarkable progress in the study of the relation of certain animal viruses to animal cells. In harvesting the allantoic fluid of eggs infected with influenza virus, Hirst observed that if blood was shed into virus-containing fluid, the erythrocytes quickly became agglutinated. No such agglutination occurred in uninfected fluids. The agent responsible for this "hemagglutination" is the virus particle itself. The particles are taken up by the red blood cell; this can be shown either by titration of free virus or by observation with the electron microscope of virus adsorbed on the red blood cells. Antivirus serum inhibits this hemagglutination. The inhibitory titer of a serum is related, although not strictly proportional, to its virus-neutralizing titer.

The study of viruses shows that many possess hemagglutinating power. In the present study, however, we employed only those viruses that resemble influenza virus. These agglutinate red blood cells

[1] Publication No. 37 of the Biology Department of Wayne State University. Supported by Michigan Heart Grant No. 32-403-3034 and by an American Medical Association Equipment grant.

directly, the virus particles serving as bridges that hold the cells together.

The following viruses are included:

Influenza Type A strains
- PR 8
- FM 1
- Denver
- Asian

Influenza Type B strains
- Lee
- G. L.

Influenza Type C strain J. J.

Mumps, Enders strain

Newcastle (Parke-Davis)

We used the standard source for virus, that is, the allantoic fluid from egg embryos inoculated two days previously with a small amount of the strains.

Hemagglutination is detected either by an increase in the rate of sedimentation of a suspension of red blood cells measured by the residual turbidity of the supernatant or by Salk's so-called "pattern test" (1944). Nonagglutinated cells sediment into a round solid pellet at the bottom of a test tube, whereas agglutinated cells settle in a characteristic, scattered pattern of sedimentation. A hemagglutination test can be read within an hour or less. The result of a typical method of virus hemagglutination titration and a hemagglutination inhibitor titration is given in Tables I and II.

TABLE I

METHOD OF VIRUS ANTIGEN TITRATION*

Dilution of virus	1:32	1:64	1:128	1:256	1:512	1:1024	1:2048	Saline Control
Degree of hemagglutination	4	4	4	4	3	2	0	0

* The titer of the antigen is based on the last tube showing 50 per cent (2+) agglutination.

TABLE II

METHOD OF HEMAGGLUTINATION INHIBITOR TITRATION*

Dilution of serum	1:32	1:64	1:128	1:256	1:512	1:1024	1:2048	1:4096	Saline Con- trol
Degree of he- magglutina- tion	0	0	0	0	0	2	3	4	4

* The titer of the serum is the dilution in which 2+ hemagglutination occurred. The usual inhibitor is antibody.

The development of the immunology of influenza has followed classic lines. First was demonstrated that sera from the patients convalescent from the diseases contained specific antibodies which combine with and neutralize the infectivity of the viral agent. Presuming our coronary sera had antibodies of some type, we tested it against all the stated viruses and found that the inhibition of hemagglutination did occur.

During the course of studies on the inhibition of hemagglutination by immune sera, Francis (1947) observed that when titrations of sera were made, some sera nonspecifically gave blocking unless live virus was used.

Normal serum contains other nonspecific inhibitors of hemagglutination besides the Francis factor. Gottschalk (1951) and Sampaio and Isaacs (1953) have shown that the use of trypsin degrades those mucoprotein inhibitors and avoids the competition between cells and the inhibitor particles for the virus, a reaction which is similar to that of antibody.

In order to obtain specificity in our hemagglutination inhibition, we treated our serum with trypsin and inactivated it at 56 degrees C. for thirty minutes. We also used both inactivated and live virus and were able to demonstrate the ability of coronary sera to prevent viral hemagglutination of the influenza-Newcastle group of viruses. The sera used was from sixty-three coronary cases compared with sera from seventeen patients hospitalized for other conditions. The sera represents cases from three successive years, hence from epidemic and nonepidemic winters for influenza. Some sera were tested after stor-

TABLE III

CORONARY SERUM BLOCKING OF PR 8 VIRUS HEMAGGLUTINATION

Dilution of Serum	1:32	1:64	1:128	1:256	1:512	1:1024	1:2048	1:4096	Saline Control
Hemagglutination:									
with inactivated commercial virus	0	0	0	0	0	0	2	4	4
with inactivated cultured virus	0	0	0	2	3	4	4	4	4
with live virus	0	0	2	2	3	4	4	4	4

age in the frozen state; and some were tested on the days of withdrawal.

Controls did not give an inhibition at a serum dilution of 1:32 or with higher dilutions.

The coronary patients' sera consistently showed some inhibition, and in some cases to a serum dilution of 1:1024. Trypsinizing the serum, which should remove most nonspecific inhibitors, did not, however, prevent virus inhibition. By using the PR 8 strain as the virus source, the presence of inhibitory substance from coronary serum is demonstrated in Table III.

The results of our tests do give weight to the presumption of a relationship between upper respiratory infection and coronary thrombosis. The evidence that post-coronary syndrome might be related to antibody was presented by Dressler in 1956. We must make more studies to determine whether the hemagglutination inhibitor is a normal serum constituent, merely present in an increased amount, or a complement-fixing antibody. The presence of any substance in sera which is constant and definitive may aid in the diagnosis of coronary occlusion.

LITERATURE CITED

DRESSLER, M. 1956. Post-myocardial Infarction Syndrome. Preliminary Report of Complications Resembling Idiopathic, Recurrent, Benign Pericarditis. J.A.M.A., 160: 1379–1383.

FRANCIS, T., JR., J. E. SALK, AND J. J. QUILLIGAN, JR. 1947. Experience with Vaccination Against Influenza in the Spring of 1947. Am. Journ. Pub. Health, 37: 1013–1016.

GOTTSCHALK, A. 1951. N-substituted Isoglucosamine Released from Mucoproteins by the Influenza Virus Enzyme. Nature, 167: 845–847.

HIRST, G. K. 1941. The Agglutination of Red Cells by Allantoic Fluid of Chick Embryos Infected with Influenza Virus. Science, 94: 22–23.

SALK, J. E. 1944. A Simplified Procedure for Titrating Hemagglutinating Capacity of Influenza Virus and Corresponding Antibody. Journ. Imm., 49: 87–98.

SILBER, E. N. 1958. Respiratory Viruses and Heart Disease. Ann. Int. Med., 48: 228–242.

PAPERS OF THE MICHIGAN ACADEMY OF SCIENCE, ARTS, AND LETTERS
VOL. XLV, 1960 (1959 MEETING)

ZOOLOGY

THE EFFECT OF COLORATION ON GREAT BARRIER REEF ANIMALS[1]

JOSEPH G. ENGEMANN
Michigan State University

MANY brilliant-colored fish are normally associated with coral. The explanation of this coloration has been ascribed to concealing coloration by many authors (Dakin, 1950: 108). The main purpose here is to consider the selective advantage of the pattern of a strikingly colored fish, *Chaetodontoplus personifer* (McCulloch) Fraser-Brunner (1933). *C. personifer* is a member of the Chaetodontidae, which Jordan (1908: 613) considers to be one of the most characteristic families of coral reef fishes.

THE ROLE OF NATURAL SELECTION

Darwin, in his *Origin of Species* (1859: 198), indicates that color can be subject to natural selection. Wallace (1889: 265–266) thought conspicuous colors of tropical fish could either aid concealment among brilliant coral reef organisms or, in some instances, be warning colors. Fisher (1958) has calculated the survival probability for characteristics giving slight selective advantage. Very few introductions of an advantageous trait are needed to assure its continuance. McAtee (1932) concluded that protective adaptations and natural selection were unimportant because predation was in proportion to population and availability. His tables of bird stomach analyses, however, showed that chaetodont fishes were not included in the fish consumed by birds. Poulton (1932) points out the fallacy of assuming no value for protective coloration just because some who have it are consumed.

[1] These observations were made possible by the U. S. Government, which provided a Fulbright Scholarship to Australia, by the Great Barrier Reef Committee, which allowed me to use its facilities on Heron Island, and by Mr. Richard Slack-Smith, who identified the fish and extended many other courtesies. I am indebted to Dr. T. W. Porter for reviewing this paper. Contribution from the Department of Zoology, The University of Michigan.

Since the part sexual selection has played in color and pattern does not readily permit examination, I find it desirable to omit discussion of it by following the same invalid assumption others have used, namely, when male and female have a close correspondence in color and pattern, sexual selection has had little effect. I further assume that brilliant color is not necessarily indicative of a more vigorous physiology and thus selected for on an intraspecific level. The main protective adaptations of coloration in animals are discussed by Cott (1957) as providing concealment, warnings, or mimicry. Mimicry in operation resolves itself into one of the other two categories. I will consider the mode of protection afforded by the color pattern of *C. personifer.*

The Visual Effect of Color

Hue or color is less important in visual apprehension and perception than luminosity contrast (Tinker, 1932). Objects with greater contrast are judged closer than those with less contrast as compared to the background (Mount, et al., 1956). The evidence in both cases is from humans. Proving that fish see exactly what humans do is impossible. But studies like Reighard's (1908) indicate some similarity, probably even color vision. Furthermore, colors may be vivid and contrasting but aid concealment because the pattern blends with the background. Friedmann (1942) discusses such methods of camouflage.

Chaetodontoplus personifer has a blue head with orange spots, a black body, and a yellow caudal fin (Fig. 1). This pattern constitutes warning coloration in coral reef fishes by Bristol's definition (1903). But Thayer (1909) maintains the brilliant colors of reef fishes present an average background picture and thus are concealing colors. Many corals of Heron Island contain algal cells giving them a color similar to the blue head of *C. personifer.* Thus if *C. personifer* were feeding on small invertebrates near the base of such a coral it might look something like Figure 2, with its blue head against the blue algal cells, its black body in the shadow, and its yellow tail superimposed on the bright coral sand. Behavior patterns of other fish suggest that *C. personifer* might take a similar position during periods of danger. The position described would tend to conceal the fish from casual observation.

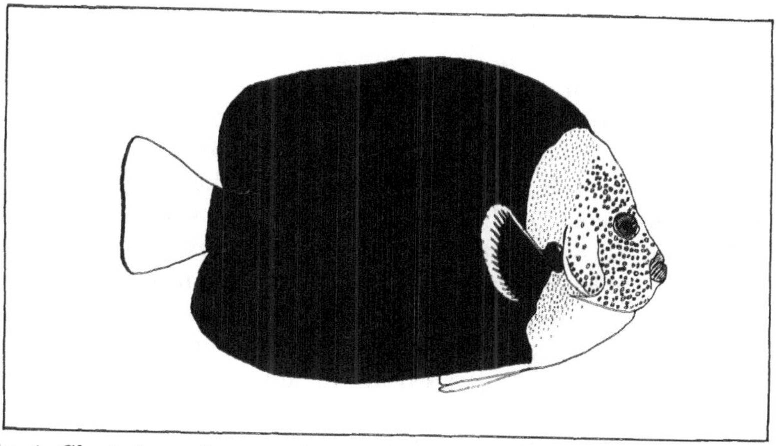

FIG. 1. *Chaetodontoplus personifer*, an angel fish. Length, 19 cm. Coloration: clear area, yellow; black area, black; finely stippled area, pale blue; spotted area, yellowish-orange spots on a blue background.

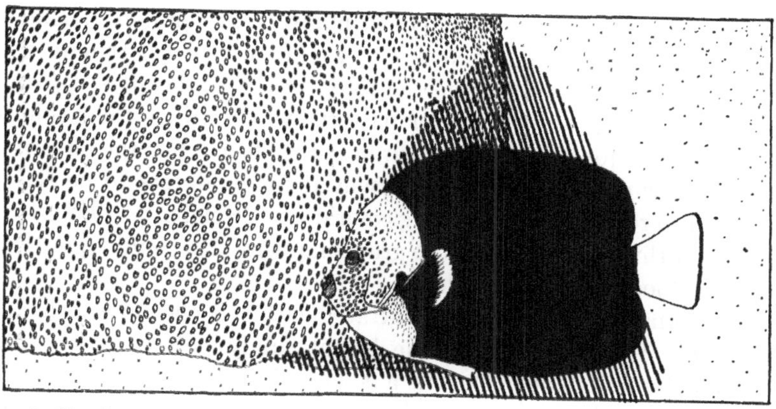

FIG. 2. The hypothetical position of *Chaetodontoplus personifer* in the shadow of a coral colony.

BEHAVIOR

Rabbitfishes (*Siganus* spp.) on Heron Island, when pursued, concealed themselves by taking positions close beside pieces of coral rock. I have observed guppies, pursued by larger aquarium fish, press themselves into the corner of the aquarium. I saw a carp in shallow water in Lake Huron obtain some degree of concealment by stopping next to an isolated obstruction of smaller width than itself. It remained motionless in that position until actually touched. Apparently, alignment with another object is a frequent position of retreat for fish. In the three cases cited the fish tended to blend with the background or object where they stopped. Breder and Rasquin (1951) observed fish taking positions near background or bottom matching their color. They found the movements of the fish to be rapid and erratic over contrasting bottom. Longley (1918) was able to induce color changes that matched the environment by enticing fish from one environment to another with food. Odiorne (1948) discusses the physiological basis for rapid color change in fish. Color changes in fish are not restricted to providing concealment. Braddock and Braddock (1955) have found a correlation of color with phases of the aggressive behavior pattern in *Betta splendens*. Physiological changes seem unimportant in *C. personifer*, since Rasquin (1958) was not able to produce appreciable change in body color of chaetodont fishes by injecting them with substances capable of altering the pattern of representatives of many other families. Longley (1914) found that bright red fish of the Tortugas reef remained in hiding during the day. From the foregoing examples it is apparent that many fish possess behavior patterns which utilize their coloration. The major predators of fish are other fish. For the effectiveness of warning coloration it is essential for predators to possess a mechanism responsive to warning coloration. Reighard (1908) demonstrated this mechanism in the gray snapper. He could train them to avoid food of a chosen color.

Two kinds of small, dorsoventrally flattened elasmobranchs are commonly found at rest on the bright sandy bottom of the reef around Heron Island. One is a small, flat-headed shark with a light color closely matching the bottom. The other is a ray of much darker color. When the ray settles down on the bottom, however, sufficient sand swirls up from the bottom and settles on the ray to reduce its contrast markedly. Thus its behavior results in partial concealment by sand.

Invertebrate Animals

The invertebrates of many phyla associated with the crinoids of a coral reef resemble the crinoids in color (Potts, 1915). Mollusks are abundant. Most pelecypod shells are similar to the coral sand in color. *Tridacna*, the giant clam, has a mantle conspicuously colored by algae. The clam closed as quickly when a shadow was passed over it as when it was touched. Closing was less vigorous and less complete when I made vibrations in its vicinity by stomping my feet. The gastropod shells tend to have their brighter colors inside where they are concealed by the body of the animal in life.

The more colorful invertebrates are found on the underside of or among coral rocks. Two major exceptions are the corals and some echinoderms. The corals derive their protection from their stinging cells and sharp skeletons. Conspicuous blue starfish are occasionally found in the open. So is a red species. They are probably protected by their spines as is the more dangerous needle-spined sea urchin.

An interesting puzzle is presented by two types of common large holothuroideans. Frequently several of each are associated with each group of coral colonies. One is a conspicuous black and often is extended on the coral sand. The other is a mottled brown, which blends well with the coral rock, but in the daytime it stays hidden under the rock and at night is as abundant on the coral sand as the black species. Thus it appears that the one possessing concealing coloration does not utilize it during the day.

Discussion and Conclusion

Most authors have considered warning colors distinct from concealing colors. This is not necessarily so, since it is unlikely that distastefulness or other defensive qualities will be repulsive to all predators. Thus a concealing color pattern in a species distasteful to a predator would function as a warning coloration when recognized by the predator. This would probably occur at a closer range than in species lacking effective concealing coloration. I interpret Reighard's (1908) work as indicating that recognition is the essential factor in response to a coloration that experience has revealed to be repellent. Selection for a concealing coloration would be slower in animals having an undesirable attribute capable of recognition as a warning, and not at all when it would increase the chance of their being destroyed or eaten by accident. Obviously, many fish have concealing coloration. I have suggested here that a gaudy color pattern could provide concealment in a restricted situation in the environment of *C. personifer*, and that natural selection in an environment with constant physical conditions would eventually result in all organisms acquiring coloration to conceal themselves from any enemies with vision they may have. The full extent of concealing coloration among reef animals will not be recognized until we are acquainted with the position, behavior, and environment of each of them.

Summary

The gaudy coloration of *Chaetodontoplus personifer* (McCulloch) is considered in terms of concealing coloration and behavior patterns needed for its effectiveness. Most reef animals have concealing coloration. Natural selection should evolve concealing coloration for most warningly colored forms.

LITERATURE CITED

Braddock, J. C., and Z. I. Braddock. 1955. Aggressive Behavior Among Females of the Siamese Fighting Fish, *Betta splendens*. Physiol. Zool., 28(2): 152–172.

Breder, C. M., and P. Rasquin. 1951. A Further Note on Protective Behavior in Fishes in Reference to Background. Copeia, 1951: 95–96.

Bristol, C. L. 1903. On the Color Patterns of Certain Bermuda Fishes. Science, 17(430): 492.

Cott, H. B. 1957. Adaptive Coloration in Animals. London: Methuen. xxxii + 508 pp., 48 pls.

Dakin, W. J. 1950. Great Barrier Reef. Melbourne: Australian Nat. Travel Assn. 135 pp.

Darwin, C. 1859. The Origin of Species. London: John Murray. 502 pp.

Fisher, R. A. 1958. The Genetical Theory of Natural Selection. 2d rev. ed. New York: Dover. 291 pp.

Fraser-Brunner, A. 1933. A Revision of the Chaetodont Fishes of the Subfamily Pomacanthinae. Zool. Soc. London, Proc., 1933(3): 543–599, 1 pl.

Friedmann, H. 1942. The Natural-History Background of Camouflage. Smithsonian Institution Publication 3700. 17 pp., 16 pls.

Jordan, D. S. 1908. Fishes. New York: Holt. xv + 789 pp., 18 pls.

Longley, W. H. 1914. Report upon Color of Fishes of the Tortugas Reefs. Carnegie Inst. Wash. Yearbook, 13(1914): 207–208.

———. 1918. Marine Camoufleurs and Their Camouflage: The Present and Prospective Significance of Facts Regarding the Coloration of Tropical Fishes. Smithsonian Institution. Ann. Rep., 1918: 475–485, 5 pls.

McAtee, W. L. 1932. Effectiveness in Nature of the So-Called Protective Adaptations in the Animal Kingdom, Chiefly as Illustrated by the Food Habits of Nearctic Birds. Smithsonian Misc. Coll., 85(7): 1–201.

Mount, G. E., H. W. Case, J. W. Sanderson, and R. Brenner. 1956. Distance Judgement of Colored Objects. Jour. Gen. Psychol., 55: 207–214.

Odiorne, J. M. 1948. Morphological Color Changes in Vertebrates. Spec. Publ. New York Acad. Sci., 4: 288–308.

Potts, F. A. 1915. The Fauna Associated with the Crinoids of a Tropical Coral Reef: With Especial Reference to Its Color Variations. Carnegie Inst. Wash., 8(212): 71–96.

Poulton, E. B. 1932. 'Protective' Adaptations of Animals. Nature, 130: 848.

Rasquin, P. 1958. Studies in the Control of Pigment Cells and Light Reactions in Recent Teleost Fishes. Bull. Am. Mus. Nat. Hist., 115(1): 1–68, 10 pls.

Reighard, J. E. 1908. An Experimental Field-Study of Warning Coloration in Coral-Reef Fishes. Carnegie Inst. Wash., 2(103): 257–325, 5 pls.

Thayer, G. H. 1909. Concealing-Coloration in the Animal Kingdom. New York: Macmillan Co. xix + 260 pp., 72 pls.

Tinker, M. A. 1932. The Effect of Color on Visual Apprehension and Perception. Genetic Psychol. Mono., 11: 61–136.

Wallace, A. R. 1889. Darwinism. London: Macmillan & Co. xvi + 494 pp.

Papers of the Michigan Academy of Science, Arts, and Letters
Vol. XLV, 1960 (1959 Meeting)

THE GENUS *ANTICHAETA* HALIDAY, WITH SPECIAL REFERENCE TO THE AMERICAN SPECIES (DIPTERA, SCIOMYZIDAE)

GEORGE C. STEYSKAL
Grosse Ile, Michigan

The genus *Antichaeta* is one of the less common groups of the family Sciomyzidae. It consists of five Palaearctic species and six Nearctic species, including one new species, *Antichaeta fulva,* described below. Only one of the Palaearctic forms has been recorded from this continent, and that record must be considered very doubtful. Only one of the species, *Antichaeta melanosoma,* is fairly common; the others may be considered rare. Of the immature stages, only fragmentary notes on the puparium of *Antichaeta analis* have been recorded, but it is likely that all of the species, like all others in the family that have been observed, are parasitic or predaceous upon fresh-water or land snails.

Antichaeta belongs in the subfamily Tetanocerinae and may be distinguished from other genera by the following characters: propleural bristle not developed; anal cell of wing truncate; vallar bristles lacking; ocellar bristles well developed; hind tibiae with two strong preapical bristles; arista with black hairs; lunule covered; midfrontal stripe broad, shining, complete, sometimes divided down the middle; wing veins *Sc* and R_1 ending well basad of *ta*; only one pair of *dc* bristles present.

The closest related genus is *Hemitelopteryx* Cresson, which is unknown in America, although one of the species of *Antichaeta, johnsoni* Cresson, was described in that genus. *Hemitelopteryx* differs from *Antichaeta* in the last two of the above characters, having wing veins *Sc* and R_1 close together, with R_1 ending well basad of *ta*; and having only one pair of *dc.*

Genus *Antichaeta* Haliday

1839, Ann. Nat. Hist., **2**: **187**, with one species, *Tetanocera vittata* Hal. (= *Sciomyza analis* Meigen, *teste* Hendel, **1902**); Hendel, **1902**, Zool.-Bot. Ges. Wien, Abhandl., **2**: **78**; Melander, **1920**, Entom. Soc. Am., Ann., **13**: **317**;

Hendel, 1923, Konowia, 2: 209; Curran, 1934, Fam. Gen. No. Am. Dipt.: 369 (in key, position transposed with *Renocera*); Séguy, 1934, Faune de France, 28: 274; Enderlein, 1939, Deut. Kolon.- u. Uebersee-Mus. Bremen, Veröffentl., 2: 201; Sack, 1939, *in* Lindner, Die Fliegen d. pal. Reg., 5, fasc. 37: 45; Verbeke, 1948, Mus. Roy. Hist. Nat. Belg., Bull., 24 (3): 18.

Paranticheta Enderlein, 1936, *in* Brohmer, Ehrmann, und Ulmer, Die Tierwelt Mitteleuropas, 6: 145, with one species, *Antichaeta bisetosa* Hendel; 1939, Deut. Kolon.- u. Uebersee-Mus. Bremen, Veröffentl., 2: 205. *Subgenus, new status.*

Lioya Enderlein, 1939, Deut. Kolon.- u. Uebersee-Mus. Bremen, Veröffentl., 2: 205. *New synonym.*

KEY TO KNOWN SPECIES OF THE GENUS *ANTICHAETA*

1 (4). Scutellum with 2 bristles; 1 pair of *dc* present ..... Subgenus *Paranticheta* Enderlein

2 (3). Thorax yellowish, 2 broad stripes and anterior margin black; fore coxae silvery white (North America) ..... *A. johnsoni* (Cresson)

3 (2). Thorax wholly black; fore coxae blackish with silvery sheen (Europe) ..... *A. bisetosa* Hendel

4 (1). Scutellum with 4 bristles; 2 pairs of *dc* present ..... Subgenus *Antichaeta* s.s.

5 (8). Shining interfrontal stripe divided medianly; black species (Europe).

6 (7). One pair of *fo*; parafrontalia dull black; eyes banded in life; legs yellowish red ..... *A. atriseta* (Loew)

7 (6). Two pairs of *fo*; parafrontalia reddish yellow; eyes not banded; legs black ..... *A. nigra* Karl

8 (5). Shining interfrontal stripe not divided, simple and more or less parallel-sided.

9 (16). Palpi black; upper pleura shining.

10 (11). Antennae with third segment black; reddish yellow species (Europe) ..... *A. analis* (Meigen)

11 (10). Antennae with third segment wholly yellowish, no more than upper edge darkened; head, thorax, and abdomen black.

12 (13). With 1 pair of *fo*; front wholly black; *ta* beyond middle of discal cell, at tip of stigma; male postabdomen as in Figs. 1 and 9 (North America) ..... *A. melanosoma* Melander

13 (12). With 2 pairs of *fo*, the anterior sometimes much the smaller; front yellowish, at least anteriorly.

14 (15). Front wholly yellow; anterior tarsi blackish, the apical segment yellowish (Europe) ..... *A. obliviosa* Enderlein

15 (14). Front blackish, anterior margin more or less yellow; *ta* basad of middle of discal cell and well basad of tip of stigma; tarsi wholly blackish; male epandrium as in Fig. 8 (North America) ..... *A. canadensis* (Curran)

16 (9). Palpi yellowish; pleura more or less pruinose above; with 2 pairs of *fo*; largely yellowish species (North America).

17 (18). Antennae with third segment black in apical half; notum not vittate; male postabdomen as in Figs. 2–4 ..... *A. fulva* Steyskal, sp. nov.

18 (17). Antennae wholly yellowish, third segment at most a little darkened above.

19 (20). Notum wholly yellowish, not vittate; wings uniformly yellowish; male epandrium as in Figs. 5–6. . . . . *A. testacea* Melander

20 (19). Notum trivittate with dark gray and yellow; wings almost hyaline, with yellowish anterior border; male epandrium as in Fig. 10. *A. robiginosa* Melander

Subgenus *Parantichaeta* Enderlein

Antichaeta bisetosa Hendel

Antichaeta bisetosa Hendel, 1902, Zool.-Bot. Ges. Wien, Abhandl., 2: 78; Sack, 1939, *in* Lindner, Die Fliegen d. pal. Reg., 5, fasc. 37: 46.

Parantichaeta bisetosa (Hendel) Enderlein, 1936, *in* Brohmer, Ehrmann, and Ulmer, Die Tierwelt Mitteleuropas, 6: 145; 1939, Deut. Kolon.- u. Uebersee-Mus. Bremen, Veröffentl., 2: 205.

Apparently known only from the type locality in the vicinity of Vienna, Austria.

Antichaeta johnsoni (Cresson), comb. nov.

Hemitelopteryx johnsoni Cresson, 1920, Am. Entom. Soc., Trans., 46: 51; Johnson, 1925, Boston Soc. Nat. Hist., Occas. Pap., 7: 250.

I have seen the type from Franconia, New Hampshire, and find that it must be referred here rather than to *Hemitelopteryx*, which genus has not been found in North America. Cresson also noted a female specimen from North Adams, Massachusetts, and Johnson added a record from Briggsville, in the same State. I have also seen two females, one from Hamburg, New York (United States National Museum) and one from Sand Lake, Ontario (Canadian National Collection).

Subgenus *Antichaeta* s.s.

Antichaeta analis (Meigen)

Sciomyza analis Meigen, 1830, Syst. Beschr., 6: 15.

Antichaeta analis (Mg.) Hendel, 1902, Zool.-Bot. Ges. Wien, Abhandl., 2: 79; Séguy, 1934, Faune de France, 28: 274; Verbeke, 1948, Mus. Roy. Hist. Nat. Belg., Bull., 24 (3): 18.

Anthichaeta [*sic.*] *analis* (Mg.) Lundbeck, 1923, Dansk. naturh. Fören., Vidensk. Medd., 76: 101.

Tetanocera vittata Haliday, 1833, Ent. Mag., 1: 168; Curtis, 1837, Guide Arrang. Brit. Ins.: 371; Hendel, 1900, Zool.-Bot. Ges. Wien, Verhandl., 1900: 358.

Sciomyza vittata (Hal.) Rondani, 1868, Soc. Ital. Sci. Nat. Milano, Atti, 115: 229; Soc. Nat. Modena, Ann., 47: 10.
Antichaeta vittata (Hal.) Haliday, 1838, Ann. Nat. Hist., 2: 187; Hendel, 1902, Zool.-Bot. Ges. Wien, Abhandl., 2: 79.

This species appears to be rather rare in Europe, known from Sweden, Germany, Galicia, Ireland, England, and Belgium. I have not seen examples. The only American record is one mentioned by Osten-Sacken of a specimen from Massachusetts in the Loew collection, but Aldrich (1905, Cat.: 578) stated that he could not trace the specimen.

Antichaeta atriseta (Loew)

Sciomyza atriseta Loew, 1849, Stettin. Entom. Ztg., 10: 340; Schiner, 1862, Fauna Austr., 2: 48.
Antichaeta atriseta (Loew) Hendel, 1902, Zool.-Bot. Ges. Wien, Abhandl., 2: 80; Séguy, 1934, Faune de France, 28: 274; Sack, *in* Lindner, Die Fliegen d. pal. Reg., 5, fasc. 37: 46.
Lioya atriseta (Loew) Enderlein, 1939, Deut. Kolon.- u. Uebersee-Mus. Bremen, Veröffentl., 2: 209.

Loew did not cite a locality, but it has since been recorded from Germany, Austria, and Livonia. I have not seen examples.

Antichaeta canadensis (Curran), comb. nov.

(Fig. 8)

Pteromicra canadensis Curran, 1923, Canad. Entom., 55: 277; Steyskal, 1954, Mich. Acad. Sci., Arts, Letters, Pap., 39: 269.

Guy E. Shewell first recognized that this species rightfully belongs in *Antichaeta*. I concurred when I saw the types from Banff, Alberta. I have also seen material from Saskatchewan (Saskatoon), Manitoba (Churchill; Wabowden), Ontario (Miners Bay; Moose Factory), Wisconsin (Dane and Washburn Cos.), and Michigan (Sanilac Co., in alder swamp).

Antichaeta fulva Steyskal, new species

(Figs. 2–4)

Male.—Length of wing, 3.5–3.7 mm. Color tawny, only apical half of third segment of antennae, arista, tip of fore tibiae, and preabdomen (segments 1–5), blackish. Bristles and hairs, including those of arista, black; only squamal ciliae yellow. Wings hyaline, broadly seamed with brownish yellow along veins.

Head: Midfrontal shining stripe simple, a little wider anteriorly; 2 *fo*, anterior pair about half the length of the posterior pair and placed close to them; hairs of arista neither dense nor long.

Thoracic dorsum moderately yellowish-pruinose, with a pair of light brownish median stripes and rather indistinct brownish sublateral areas. Pleura heavily white-pruinose below, thinly so above; wings with *ta* a little beyond middle of discal cell and only slightly before tip of stigma.

Abdomen with terminal part as in Figs. 2–4; 5th sternite with simple, bluntly pointed apical lobe underhanging the postabdomen; epandrium with long, tongue-shaped lobe on dextral side and two short protuberances on sinistral posteroventral margin; dextral surstylus roughly keystone-shaped in lateral view; hypandrium and aedeagus small and simple.

Female.—Length of wing, 3.8 mm. Similar to male; abdomen with segments 1–4 blackish, remainder tawny; fore femora largely pitchy; fore tibiae in apical half and all of fore tarsi black.

Holotype (male) and *allotype*, Ringwood, Dryden, Tompkins County, New York, May 29, 1955 (C. O. Berg and G. Steyskal); male *paratype*, same locality, June 1, 1957 (G. Steyskal), all in my collection.

The distinctions cited in the key will suffice to separate this species from its relatives.

Antichaeta melanosoma Melander

(Figs. 1, 9)

Antichaeta melanosoma Melander, 1920, Entom. Soc. Am., Ann., 13: 318; Leonard, 1928, Cornell Univ. Agric. Exper. Sta., Mem., 101: 846; Ouellet, 1941, Nat. Canad., 68: 135.

This seems to be the most abundant species in the genus, ranging from Ontario (Moose Factory; Ottawa; Point Pelee; Simcoe), to Quebec (Aylmer; Berthierville), Wisconsin (Dane Co.; Polk Co., type), Michigan (Clinton, Gratiot, Huron, Iosco, Keweenaw, Kent, Leelanau, Livingston, Midland, Newaygo, Roscommon, Saginaw, Sanilac, Schoolcraft, Tuscola, and Wayne Cos.), Ohio (Delaware), and New York (Beaver Kill; Ithaca).

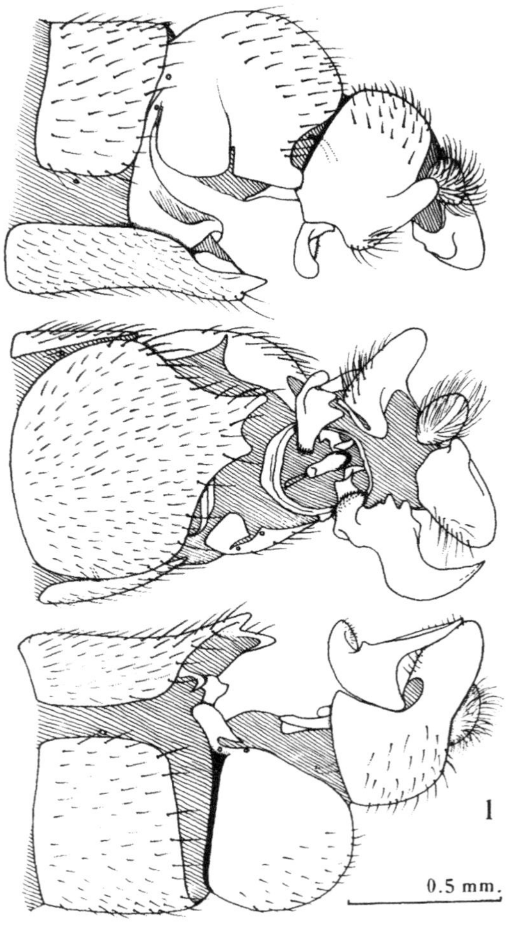

Antichaeta melanosoma Mel., male, Grosse Ile, Michigan

Fig. 1. Apical part of abdomen (fifth segment and postabdomen), sinistral, ventral, and dextral views.

Antichaeta nigra Karl

Antichaeta nigra Karl, 1921, Stettin. Entom. Ztg., 82: 125.
Antichaeta nigroaenea Frey, 1935, Notulae Entom., 15: 97; Sack, 1939, *in* Lindner, Die Fliegen d. pal. Reg., 5, fasc. 37: 85.

A rare European form, described from northeastern Germany, and recorded by Frey from Finland.

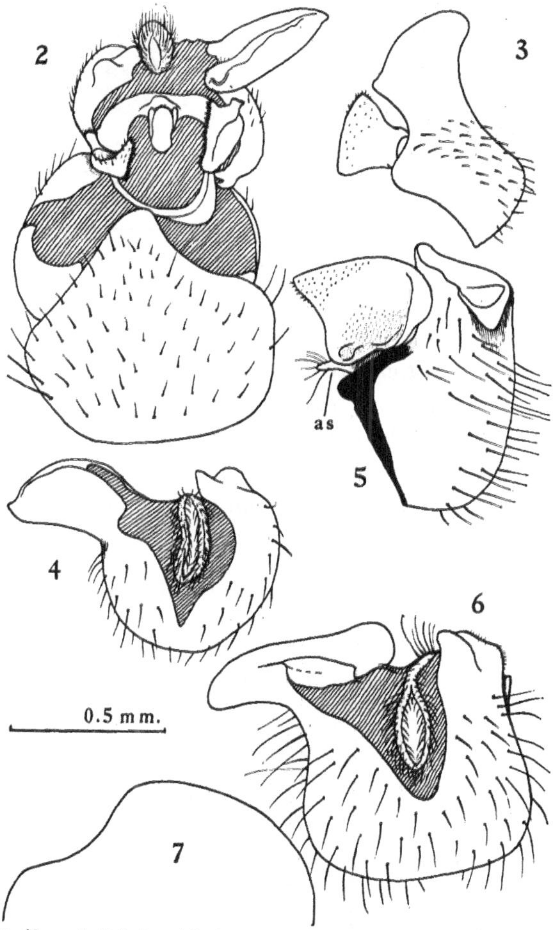

Details of Male Abdomen of *Antichaeta* Species

FIGS. 2–4. *Antichaeta fulva*, sp. nov., paratype: 2, ventral view of fifth segment and postabdomen; 3, dextral view of epandrium and surstylus; 4, posterior view of epandrium and proctiger.

FIGS. 5–7. *A. testacea* Mel., Spearfish, South Dakota: 5, dextral view of epandrium and surstyli (*as*—anterior surstylus); 6, posterior view of epandrium and proctiger; 7, outline of posterior half of 5th ventrite.

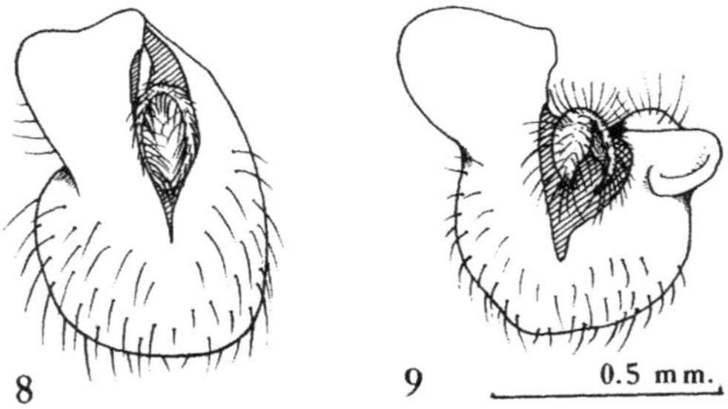

FIG. 8. *Antichaeta canadensis* (Curran), Sanilac County, Michigan: posterior view of epandrium.
FIG. 9. *A. melanosoma* Mel., Grosse Ile, Michigan: posterior view of epandrium.

Antichaeta obliviosa Enderlein

Antichaeta obliviosa Enderlein, 1939, Deut. Kolon.- u. Uebersee-Mus. Bremen, Veröffentl., 2: 209; Verbeke, 1948, Mus. Roy. Hist. Nat. Belg., Bull., 24 (3): 18.

Another rare form, described from Berlin, Germany, and cited by Verbeke from Belgium.

Antichaeta robiginosa Melander

(Figs. 10–11)

Antichaeta robiginosa Melander, 1920, Entom. Soc. Am., Ann., 13: 317.

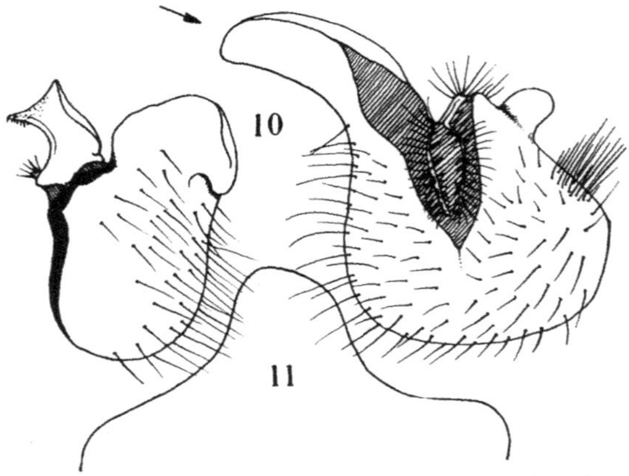

FIGS. 10–11. *A. robiginosa* Mel., Roberts, Oregon: 10, posterior view of epandrium with view in direction of arrow, showing surstyli of dextral side; 11, outline of posterior half of 5th ventrite.

This species was described from a female from Three Forks, Montana. I have seen one additional female specimen, somewhat paler than described, from Truro, Nova Scotia, in the Cornell University collections. What is evidently the male of this species is a specimen from Roberts, Marion County, Oregon, March 5, 1940 (R. E. Rieder), from the Oregon State College collections. The figures have been drawn from this specimen.

Antichaeta testacea Melander

(Figs. 5–7)

Antichaeta testacea Melander, 1920, Entom. Soc. Am., Ann., 13: 318.

This species was described from Gallatin County, Montana. I have seen material that agrees well with the description from Spearfish, South Dakota, and Cimarron Canyon, Colfax County, New Mexico.

Phylogeny as Indicated by Postabdominal Structure

The five American species whose postabdomens I have been able to examine exhibit an interesting series of characters which seem to indicate the phylogenetic position of the species. The genus is one of the more highly specialized (apomorphic) of the Sciomyzidae, as shown by the strongly asymmetrical epandrium, the reduced hypandrium and aedeagus, and the extension of the 5th sternite under the postabdomen.

The three yellow forms, *Antichaeta testacea, A. robiginosa,* and *A. fulva* are apparently the least specialized (most plesiomorphic). The asymmetry is least developed, the 5th sternite the least extended, the hypandrium and aedeagus the least reduced. *Antichaeta testacea* and *A. robiginosa* are the only species in which a rudiment of the anterior surstylus (Fig. 5, *as*) is present on the dextral side only. *Antichaeta fulva* is somewhat more specialized than either *A. testacea* or *A. robiginosa*; its hypandrium and aedeagus are more reduced and the anterior surstylus wholly lacking.

Antichaeta canadensis and *A. melanosoma* are closely related and more specialized than either of the preceding three species. Both are extensively black in color, have a well-extended fifth sternite ending in two points, and have a strongly reduced aedeagus and hypandrium. *Antichaeta melanosoma* would appear to be somewhat more apomorphic than *A. canadensis*, but lack of a prepared postabdomen of the latter forces us to rely on other characters for this conclusion—*A. melanosoma* has developed a wholly black front and has lost the anterior (second) *fo* bristle. However, the development of protuberances of the epandrium in *A. melanosoma* seems to have gone somewhat further than in *A. canadensis.*

The disposition of the spiracles and the development of the sclerites of the 6th and 7th somites in all five species is similar.

When data on the postabdomen of the other species becomes available, an unusually good basis for phylogenetic reconstruction may be presented.

PAPERS OF THE MICHIGAN ACADEMY OF SCIENCE, ARTS, AND LETTERS
VOL. XLV, 1960 (1959 MEETING)

BOTANY

TAXONOMIC PROBLEMS IN POISON IVY[1]

WILLIAM T. GILLIS
Michigan State University

IN recent years, there has been much study of poison ivy from a non-botanical standpoint. Strauss, Langs, et al. have developed a drug, "Aqua Ivy," as a reasonably certain preventive from the dermatitis (Passenger, Spain, and Strauss, 1956; Gaillard, 1956; Langs and Strauss, 1958; Langs, Strauss, and Fuchs, 1958). Kligman (1958) has written an excellent summary of his own work and that of others on poison ivy dermatitis, its clinicobiology, treatment, and prophylaxis. Epstein and Claiborne (1957) have studied human racial and environmental factors in susceptibility to dermatitis from members of the poison ivy complex. In recent months, the entrance into the United States of Haitian voodoo dolls used as cocktail stirrers has disturbed public health officials, inasmuch as the dolls' heads were made of the toxic cashew nut, a relative of poison ivy, and the eyes of jequirity nuts (Leguminosae), probably more deadly. Charles Dawson has published extensively during the last half decade on the chemistry of the poisons in poison ivy (1954, 1956a, 1956b; Dawson, Wasserman, and Keil, 1946). It has been over two decades since Barkley (1937) published his monograph on the *Rhus* complex and, even yet, his nomenclature is little heeded. The time is thus ripe for a review of poison ivy from a botanical standpoint.

Poison ivy is the northernmost species found in the seven genera of the *Rhus* complex in the Anacardiaceae occurring north of Guatemala—*Actinocheita, Cotinus, Malosma, Metopium, Rhus, Schmaltzia,* and *Toxicodendron* (Barkley, 1937, 1940). It has been recorded as far north as Grand Rapids in Manitoba near Lake Winnipeg (Scoggan, 1957), and as far south as Huehuetenango department in Guatemala (Standley and Steyermark, 1949). It occurs in the Gaspé Peninsula and extends to Vancouver Island (Scoggan, 1950). Outside

[1] Publication No. 59-7, Department of Botany and Plant Pathology, Michigan State University.

of continental North America, it is found in Bermuda and in the Bahamas. It occurs in every state in the Union, except California, in the taxon accepted as *Toxicodendron radicans*. Several references have been noted to a supposedly native poison ivy in Japan (Greene, 1905; Murakoshi, 1934; Ohwi, 1956). This plant, its taxonomy, and its presence in Japan all need further study. I shall say more about the distribution and habitat preferences of poison ivy when I have delimited the species more clearly.

Barkley (1937) felt that there were sufficiently strong characters consistent within the genus *Toxicodendron* to warrant separating it from *Rhus*. Further, Heimsch (1940, 1942) corroborated this generic distinction on the basis of wood anatomy and pollen morphology. These characters hold for the Asiatic species as well. In addition to the stated characters of nonglandular pubescence of the fruit coat, waxy mesocarp, consistently poisonous resin, and paniculate inflorescence given by Barkley, I have also noted that root hair color is an additional character difference. North American Toxicodendrons have a distinctly brown root with dark brown root hairs. The root hair color of *Rhus*, as opposed to *Toxicodendron*, is generally tawny pink or lighter. The reddish fruit coat of *Rhus* is a fairly consistent character, separating it from the whitish or dun-colored fruit coat of *Toxicodendron* (Barkley, 1957). Poison ivy thus belongs to the genus *Toxicodendron*.

It is interesting to note that previously this plant has been placed by botanists in four genera (not to mention the genus *Edera* [Barkley and Barkley, 1938], where it was first placed by botanists before Linnaeus mistaking it for English ivy, or a close relative). One proposal was a compounded genus, *Rhus-Toxicodendron*, an evident attempt to circumvent the whole problem.

The species epithet is another question entirely. Barkley (1937) lists fifty-three synonyms that have been validly published over the years, including an even baker's dozen published by Greene (1905)! Others, usually referred to by nonbotanists, are often found in dermatological, medical, chemical, or popular journals. One such combination, fortunately attributed to no author, even intimates that poison ivy, poison oak, and poison sumac all form one polytypic species. Whether Barkley's four subspecies are in reality good separations, based on physiological and ecological tests, is a question still to be answered. In any event, the correct name for poison ivy is *Toxicodendron radicans* (L.) Kuntze (Kuntze, 1891).

The common name poison ivy is usually written as two words, although Harlow (1949) makes a strong case for introducing the one-word spelling. Because of confusion between a poisonous ivy and a name originating from its kinship to English ivy, Harlow suggests the use of "poison-ivy." Fowler (1958) indicates that a hyphen is used to make two words into one: therefore, the more correct form would be a hyphenated word, poison-ivy. Due to common usage and the fact that the literature is saturated with the name spelled as two words, however, it is probably best to regard "poison ivy" as a *nomen conservandum* in a popular sense.

Barkley has also recognized two plants with the common name poison oak (1937). One is *Toxicodendron diversiloba* (Torrey & Gray) Greene, restricted to Washington, Oregon, California, Lower California, and two inland states in Mexico. The second poison oak is *Toxicodendron quercifolia* (Michx.) Greene, found south of an arc from southern New Jersey to Texas. The two poison oaks are allopatric, but they are, except in California, sympatric with *T. radicans.* Hybrids among these forms have been hypothesized by McNair (1925), but no specimens have been cited, nor any artificial hybrids made. The southeastern poison oak and poison ivy have flowering periods that generally do not overlap, whereas the western poison oak and poison ivy have flowering periods which generally do overlap, McNair reports. Since they are all insect-pollinated, there is reason to suspect occasional exchange of genetic material.

Sandy, oak-pine forest is the habitat in which *T. quercifolia* is most likely to occur (Biederman, 1938). On the other hand, poison ivy is not normally abundant in dry, sandy, oak-pine forests although found among pines and other trees in the mountains of Mexico. It is rare in the jack pine forests of Michigan (not listed in the flora by Zimmerman, 1956), and seldom encountered except at bog margins in the New Jersey Pitch Pine Barrens. Martin's studies (1959) at Island Beach, New Jersey, indicate that poison ivy is virtually omnipresent in all habitats but the salt marsh and pine forest type. This suggests that poison oak may be an ecotype of poison ivy, which has, in certain climatic, geographic areas, been able to adapt to the environment of the pine forest type, with accompanying morphological changes.

The western poison oak enjoys a greater variety of soil conditions than any other California shrub, ranging from sandy flats to blue adobe and from gravels to salt marshes. Although it does best on moist hillsides, it, like its eastern counterpart, will tolerate the most sterile soils (Jepson, 1936). Shelmire (1941) grew specimens of the two poison oaks and poison ivy from Wyoming, Nebraska, Texas, and California in his backyard in Texas under "identical conditions of soil, sunshine, . . . and moisture." No constant variation was found in pubescence, tendency to climb, or definite lobation. He concluded from his limited studies that "this variable plant seems best treated taxonomically as a single pleomorphic species with several more or less consistent geographic variations."

Neither of the two recognized poison oak taxa occur in the North Central States, although a "poison oak" is often said to occur there. The name poison oak is frequently applied by laymen to the shrub form and poison ivy to the climbing or herbaceous form, although neither separation is valid in all analyses. Laymen and physicians often diagnose dermatitis venenata from the size of blisters whether poison ivy or poison oak has been the culprit. It is for this reason that many persons believe that poison oak occurs in such states as Illinois and Michigan. "Oak-shaped" leaves have been found on otherwise typical poison ivy in Michigan. Such collections (*Gillis No. 2623*, Ingham Co., Mich., M.S.U. campus, 30 July 1958; *Gillis No. 2846*, Ingham Co., Mich., Baker Woods, M.S.U. campus. Prostrate vine, 26 September 1958) are deposited in the herbarium of Michigan State University. The poison ivy-oak complex is one that demands further investigation to determine the constancy of the variation upon which the species separations are made.

Haack (1940) and Dawson and his co-workers have clearly identified the four poisonous substances in poison ivy as variations in saturation of the alkyl side chain in 3-*n*-pentadecylcatechol (1956a). No work has been done to determine if there is any difference in the chemical nature of the poisons of poison oak, nor, for that matter, of poison sumac (*Toxicodendron vernix*). The Japanese lac tree (*T. vernicifera*) differs in poison chemistry in only one of the four poisons, having two conjugated double bonds in the side chain, a situation not present in the poisons of poison ivy. Inasmuch as I am not susceptible to the poisons of poison ivy and have, indeed, been desensitized by the use of Aqua Ivy as a precautionary measure, yet find myself "allergic" to poison sumac, there is some evidence that the chemistry of the poisons in poison sumac is different from that of poison ivy.

Hybrids between poison ivy and poison sumac have even been postulated. Such a find would strengthen the concept of their close relation within the same distinct genus. Several plants on the Proud Lake Recreation Area of Oakland County, Michigan, have given evidence of being hybrids between these two species. One plant shows the especially weird combination of three to nine leaflets and a tendency to climb. I am inclined to consider this a juvenile form of poison sumac, but plan further study of transplanted individuals and others in a nearby "hybrid" population.

A five-leaflet poison ivy is known. It has been studied by Barkley (1936) in Texas and grows within the city limits of Houston. Normally the quinquefoliolate leaves are produced as fall approaches. They are pinnately compound, whereas the leaves of the normal, trifoliolate variety are palmately compound. (It is interesting to consider here the difference between a palmately trifoliolate leaf, as in poison ivy, and a pinnately trifoliolate one, as in clover.) A dumping area near Syracuse, New York, is also known to have a quinquefoliolate poison ivy plant (William M. Harlow and John Andresen, personal communication). A five-leaflet poison oak is also described by McNair (1936) as a vigor-response to cutting back in an effort to kill the plant.

Habitat variations probably led to many of the species segregations made over the years. The shrub form is found frequently in open, sunny, dry sites along roads and on sand dunes. The climbing form frequents fencerows and woodlands. There is a "perennial herb" form in which the plant rarely attains a height of 3 to 4 dm. This form predominates farther north, the area around Itasca Park, Minnesota, having only the "perennial herb" form. According to Dr. E. G. Voss in his studies for the publication of the Michigan Flora (personal communication), his specimen *No. 7105,* Manistee Co., Michigan, was collected about as far north as the climbing form occurs in the state of Michigan. The trailing vine form, common in deep shade of mesic woods, is another growth form. The trailing stems may branch extensively and cover several square meters. Occasionally, leaves may be spaced several meters apart on the same trailing stem. Neither the "perennial herb" (subshrub) form nor the trailing vine form flowers regularly.

Geographical variation has some interesting aspects. As one progresses south, he notices little apparent difference in populations of the typical *Toxicodendron radicans*. Plants from southern Michigan may look exactly like those collected in Central Mexico. The fruits in specimens examined from Mexico, however, have a more roughened, brownish appearance than those farther north. In the Southeast, the stems take on a more reddish hue. Vines from Bermuda have distinctly red twigs and petioles, with villous pubescence and a distinct red border on the leaf margins, especially of the young leaves. (Living specimens from Bermuda were sent through the courtesy of I. W. Hughes, Paget East, Bermuda.)

Furthermore, microscopic studies have revealed resin canals in stems, petioles, and even bracts, sepals, and petals (usually three canals in each of the perianth parts). Leaflet size is variable; I have seen variations from 2 to almost 30 cm. for the terminal leaflet. The leaflets tend to be larger on climbing forms, and larger on floodplain sites than elsewhere. The leaves are larger in the sun, contrary to the usual concept of sun and shade leaves. I found an interesting morphological variation at Island Beach State Park, New Jersey, in 1958 (*Gillis No. 2709*, 14 August 1958; *Gillis No. 2791*, 31 August 1958, both from secondary foredune thicket), with feathery structures in the axils of several leaves. A first glance suggested that clusters of opposite, simple leaves comprised the "pinnae" of the feather. Closer examination indicated that these leaves were enlarged primary and secondary bracts, whose flowers failed to develop. In these aborted flowers, the bracts grew to several times their normal length (up to 1.4 cm.) and were arranged like the flowers, occasionally opposite and occasionally alternate.

Vegetatively, poison ivy may occasionally be confused with box elder (*Acer negundo*),[2] virgin's bower (*Clematis virginiana*), fragrant sumac (*Schmaltzia crenata—Rhus aromatica* of Fernald, 1950), strawberry (*Fragaria spp.*), hoptree (*Ptelea trifoliata*), bladdernut (*Staphylea trifolia*), Virginia creeper (*Parthenocissus quinquefolia*), and buckbean (*Menyanthes trifoliata*). The presence of the characteristic fruits and winter buds is the best identifying character in the winter. The angles of leaf venation (lateral veins at 60 degrees from the midvein), absence of attenuate bases on the terminal leaflet, al-

[2] All binomials agree with authors given in Fernald (1950) except those of the Anacardiaceae, which agree with authors given in Barkley (1937, 1940).

ternate leaves, and twig characters are probably the best means of identifying the plant in the growing season. Leaf shape is too variable to be of help.

It can be seen that poison ivy is a problem from the standpoint of nomenclature and clear delimitation of species. The confusions arise in part because of the extreme variation in the species. It is confusing even to the professional botanist, because he may not know it in all its variations and can easily mistake it for other trifoliolate plants. Thus, pity the poor layman, whose error in identification may result in physical torture.

LITERATURE CITED

BARKLEY, FRED A. 1936. "Five-Leaved" Poison Ivy. Biologist, 17: 122.

——. 1937. A Monographic Study of *Rhus* and Its Immediate Allies in North and Central America, Including the West Indies. Ann. Mo. Bot. Gard., 24: 265–498.

——. 1940. *Schmaltzia*. Am. Midl. Nat., 24: 647–665.

——. 1957. Generic Key to the Sumac Family (Anacardiaceae). Lloydia, 20: 255–265.

—— AND ELIZABETH D. 1938. A Short History of *Rhus* to the Time of Linnaeus. Am. Midl. Nat., 19: 265–333.

BIEDERMAN, JOSEPH B. 1938. Observations on the Relation of Poison Ivy and Poison Oak. N.E. Jour. Med., 219: 117–118.

DAWSON, CHARLES R. 1954. The Toxic Principle of Poison Ivy and Related Plants. Rec. Chem. Prog., 15: 39–53.

——. 1956a. The Toxic Principle of Poison Ivy. Jour. Chem. Ed., 33: 94–96.

——. 1956b. The Chemistry of Poison Ivy. Trans. N.Y. Acad. Sci., 18: 427–443.

——, DAVID WASSERMAN, AND HARRY KEIL. 1946. 3-*n*-Pentadecylcatechol. Jour. Am. Chem. Soc., 68: 534.

EPSTEIN, ERVIN, AND EARL CLAIBORNE. 1957. Racial and Environmental Factors in Susceptibility to *Rhus*. A.M.A. Arch. Dermat., 75: 197–201.

FERNALD, M. L. 1950. Gray's Manual of Botany. 8th ed., New York: American Book Company.

FOWLER, H. W. 1958. A Dictionary of Modern English Usage. London: Oxford University Press.

GAILLARD, G. EVERETT. 1956. The Modern Treatment of Poison Ivy. N.Y. State Jour. Med., 56: 2255–2259.

GREENE, EDWARD L. 1905. Leaflets of Botanical Observation and Criticism, 1: 114–144. Washington, D.C.

HAACK, N. H. 1940. Vesicant Constituents of the Milk-Sap of Some Anacardiaceae. Thesis (in Dutch). J. B. Wolters' University-Company, Ltd., Groningen, Batavia, Netherlands.

HARLOW, WILLIAM M. 1949. Poisonivy and Poisonsumac. N. Y. State College For. Publ., 22: 1–27.

HEIMSCH, CHARLES. 1940. Wood Anatomy and Pollen Morphology of *Rhus* and Allied Genera. Jour. Arn. Arb., 21: 279–291.

——. 1942. The "Gruinales" and "Terebinthales." Lilloa, 8: 83–198.

KLIGMAN, ALBERT M. 1958. Poison Ivy (*Rhus*) Dermatitis. A.M.A. Arch. Dermat., 77: 149–180.

KUNTZE, OTTO. 1891. Revisio Generum Plantarum, 1: 153–154. Leipzig.

LANGS, ROBERT J., AND MARGARET B. STRAUSS. Paper read at the A.M.A. meeting, Dermat. Sec.; San Francisco, Calif., June 25, 1958.

——, MARGARET B. STRAUSS, AND ABNER M. FUCHS. Paper read at the A.M.A. Clinical Meeting, Minneapolis, Minn., December 5, 1958.

MARTIN, WILLIAM E. 1959. The Vegetation of Island Beach State Park, New Jersey. Ecol. Mono., 29: 1–46.

McNAIR, JAMES B. 1925. Taxonomy and Range of Poison Ivy. Science, 61: 589.

——. 1936. Five-Leaflet Poison Oak. Bull. Torr. Bot. Club, 63: 473–476.

MURAKOSHI, MISAO. 1934. New Illustrated Flora of Japan (in Japanese). Tokyo.

OHWI, JISABURO. 1956. Flora of Japan (in Japanese). Shibundo, Tokyo.

PASSENGER, RICHARD E., WILL COOK SPAIN, AND MARGARET B. STRAUSS. 1956. Aqua Ivy. Jour. Allergy, 27: 409–423.

SCOGGAN, H. J. 1950. The Flora of Bic and the Gaspé Peninsula, Quebec. Natl. Mus. Can. Bull., 115 (Biol. Series No. 39).

——. 1957. Flora of Manitoba. Natl. Mus. Can. Bull., 140 (Biol. Series No. 47). Ottawa.

SHELMIRE, BEDFORD. 1941. The Poison Ivy Plant and Its Oleoresin. Jour. Invest. Dermat., 4: 337–348.

STANDLEY, PAUL C., AND JULIAN STEYERMARK. 1949. Flora of Guatemala, Part VI. Fieldiana, 24: 186–187.

ZIMMERMAN, DALE A. 1956. The Jack Pine Association in the Lower Peninsula of Michigan: Its Structure and Composition. Thesis, University of Michigan, Ann Arbor, Mich.

Papers of the Michigan Academy of Science, Arts, and Letters
Vol. XLV, 1960 (1959 Meeting)

THE RELATION OF ANTHOCYANINS TO *GAILLARDIA* TAXONOMY[1]

Warren P. Stoutamire
Cranbrook Institute of Science

The rapid development of chromatographic procedures during the past decade has provided a valuable method for separating and identifying organic compounds. This technique has stimulated work on the chemical relationships of species and on the biochemistry of gene action. It is an easily used procedure which may have valuable application in studies of plant and animal evolution. Several papers based on this concept have been published with this end in view.

The interest in chemical differences between species began long before the development of chromatography. Baker and Smith in 1902 and 1920 (Gibbs, 1958) correlated the essential oils of over 100 species of *Eucalyptus* with the morphology of these species and produced an evolutionary sequence based on this relationship. Paper chromatography was not used at this time, but most of the recent work on the subject has been based on this technique. Among the several papers on animal relationships investigated by means of paper chromatography are those of Buzzati-Traverso and Rechnitzer (1953) with fish, Micks (1954) with mosquitoes, and Kirk, Main, and Beyer (1954) with land snails. These workers, analyzing amino-acids and related compounds, found that species differences could be detected in the chromatograms, and that intraspecific races generally gave the same chromatographic patterns. Few papers have appeared on plant chemical relationships, although Roberts and Wood (1958), working with the *Thea* section of *Camellia*, found that two ill-defined species consist of three chemical entities and that the non-*Thea* Camellias are quite distinct from them. The plant investigations of antigen-antibody relationships are more numerous and involve entirely different techniques—not considered here.

The present investigation includes eleven species of *Gaillardia* whose flowers contain anthocyanin. Of these, *G. pulchella* Foug. is

[1] Part of an investigation on the evolution of the genus *Gaillardia*; supported by National Science Foundation Grant G-3943.

one of the best-known species and is widespread in southwestern United States. Its numerous geographic races differ in ray color as well as in morphological characters. This ray color varies from deep red to pink, and differences persist when plants are grown in Michigan. Several other species of the genus produce ray and disk flowers in shades of red and maroon which do not occur in *Gaillardia pulchella.* I suspected that different anthocyanin pigments were the cause of these color variations. Therefore, I have analyzed the colors chromatographically and compared them with the morphological relationships between species.

Methods

Anthocyanins and other compounds were removed from freshly collected rays and disk flowers by boiling in 95 per cent ethyl alcohol which contained 1 per cent HCl. The extract was filtered, evaporated to a dark red liquid, and applied as spots to Whatman No. 1 filter paper strips. The ascending chromatograms were developed in glass cylinders, using a mixture of ethyl acetate:t-butanol:water: glacial acetic acid (5:4:3:1) as the developing solution. All strips were placed in an atmosphere saturated with HCl vapor for two minutes before being run. In most cases two strong anthocyanin bands separated during development in addition to the hydrolyzed cyanidin which traveled near the solvent front. The determination of cyanidin in *Gaillardia* was made by Forsyth and Symmonds (1954). Comparison of the R_f values of several extracts on the same chromatogram indicates which races have anthocyanins in common and which are not identical. Ray and disk extracts of each race must be handled separately, because they contain different pigment mixtures.

Results

All races of *Gaillardia pulchella* tested produced the same chromatographic pattern (Fig. 1) no matter what their geographic origin. Ray extracts separated into two compounds with R_f values of approximately .22 and .28. Disk extracts produced two compounds with R_f values of approximately .22 and .25. It should be noted that these values varied for each chromatogram, and the only accurate way of comparing different extracts was to compare each with a standard, in this case a ray extract of race 1291 from Galveston, Texas. Three anthocyanins were involved in producing the red and orange

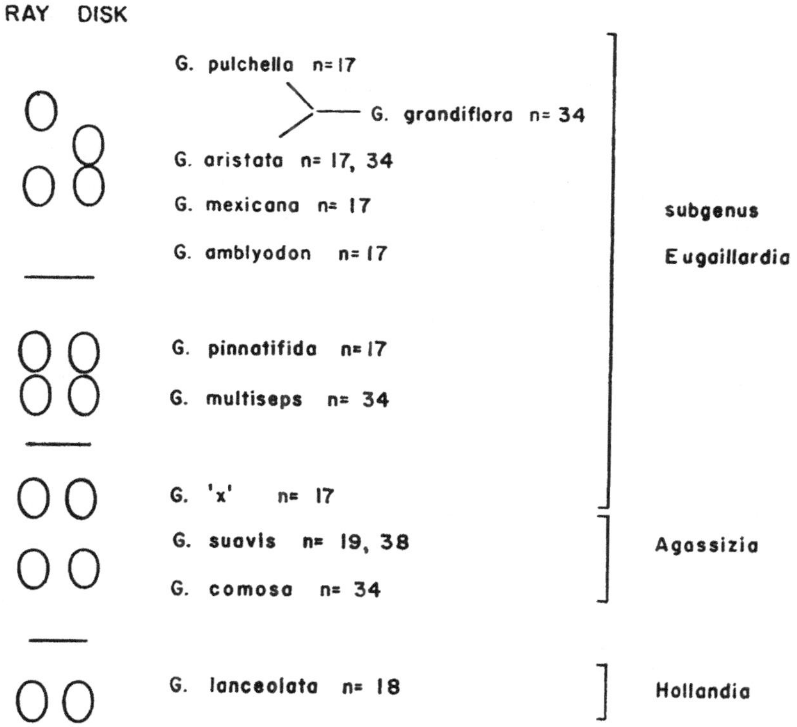

FIG. 1. Relative positions of anthocyanins from ray and disk extracts of eleven *Gaillardia* species.

colors of this species, only one of which was common to both the ray and disk flowers. The color variations in geographic races of *G. pulchella* were apparently due to quantitative differences in these three pigments.

Four other species—*Gaillardia aristata* Pursh., *G. grandiflora* Van Houtte, *G. mexicana* Gray, and *G. amblyodon* Gay—produced pigments at locations identical with those of *G. pulchella* when their extracts were run on the same chromatogram. These species are closely related morphologically, and *G. pulchella* has been experimentally hybridized with each, indicating fairly close relationship.

Races of *Gaillardia pinnatifida* Torr. and *G. multiceps* Greene have similar compounds in both ray and disk flowers, and these correspond to the pigments in *G. pulchella* disk flowers. All species considered so far belong to the morphological subgenus *Eugaillardia* (Biddulph, 1944).

Ray and disk flowers of *Gaillardia suavis* (Gray and Engelm.) Britt. and Rusby, *G. comosa* Gray, and an undescribed species "X" contain the same pigments, and these correspond to the ray pigments of *G. pulchella*. The first two species are included in the subgenus *Agassizia*. The undescribed species belongs on morphological grounds to subgenus *Eugaillardia* but its anthocyanins correspond to those of *G. suavis* and *G. comosa*.

The third subgenus, *Hollandia*, includes the species *Gaillardia lanceolata* Michx., whose ray and disk flowers contain only one anthocyanin with a R_f of .22, and this corresponds to one of the pigments in the other subgenera.

In order to determine whether these species contained the cyanin molecule, the separate bands were clipped out of the chromatogram, eluted with ethyl alcohol, evaporated to dryness, and hydrolyzed five minutes at 100 degrees C. in 50 per cent HCl. These preparations were evaporated, applied to a chromatogram, and run in the previously described solvent. Cyanidin appeared in all cases, and the species differences are therefore due to different sugars or related substances in the side chains of the molecule.

Conclusions

Gaillardia species with red rays can be separated into four groups on the basis of the anthocyanin mixtures present in ray and disk flowers, and these groups contain species closely related on the basis of their morphology. The only departure from this close correlation is the undescribed species which is morphologically closest to the *G. pinnatifida–G. multiceps* group of subgenus *Eugaillardia*, but whose anthocyanin complex is like that of the *Agassizia* subgenus. The chemical differences in ray color closely parallel the morphological differences and provide another tool for analyzing species relationships. A study of other classes of organic compounds would probably yield considerably more information on species relationships.

LITERATURE CITED

Biddulph, Susan F. 1944. A Revision of the Genus *Gaillardia*. Res. Studies State Coll. Wash., 12(4): 195–256.

Buzzati-Traverso, A. A., and A. B. Rechnitzer. 1953. Paper Partition Chromatography in Taxonomic Studies. Science, 117: 58–59.

FORSYTH, W. G. C., AND N. W. SIMMONDS. 1954. A Survey of the Anthocyanins of Some Tropical Plants. Proc. Roy. Soc. (London), B.142: 549–564.

GIBBS, R. D. 1958. Chemical Evolution in Plants. Jour. Linn. Soc. London, Bot., 56(365): 49–57.

KIRK, R. L., A. R. MAIN, AND F. G. BEYER. 1954. The Use of Paper Partition Chromatography for Taxonomic Studies of Land Snails. Biochem. Jour., 57: 440–442.

MICKS, DON W. 1954. Paper Chromatography as a Tool for Mosquito Taxonomy: the *Culex pipiens* Complex. Nature (London), 174(4422): 217–218.

ROBERTS, E. A. H., AND D. J. WOOD. 1958. Paper Chromatography as an Aid to the Taxonomy of *Thea* Camellias. New Phytol., 57: 211–225.

PAPERS OF THE MICHIGAN ACADEMY OF SCIENCE, ARTS, AND LETTERS
VOL. XLV, 1960 (1959 MEETING)

POLLEN SPECTRA FROM BRYOPHYTIC POLSTERS, INVERNESS MUD LAKE BOG, CHEBOYGAN COUNTY, MICHIGAN[1]

WILLIAM S. BENNINGHOFF
The University of Michigan

EVER since the application of pollen analysis to the study of postglacial vegetation history we have been concerned about improvement of our interpretations of fossil pollen records. We are well aware of the wide variation in production of pollen and spores by different species and of the vagaries of transport of these minute bodies by wind and water (Cain, 1939). But the proven usefulness of the pollen analytical method in paleoecology, especially in the postglacial sediments of northwestern Europe and eastern North America, demonstrates that pollen spectra do reflect significant features of the vegetation of local areas when carefully interpreted. One approach to improving our interpretation of spectra of fossil pollen is through correlation of modern pollen rain with existing vegetation that contributes the pollen and spores.

A common method for sampling the modern pollen rain for specific sites is analysis of pollen and spores lodged in bryophytic polsters (Carroll, 1943; Potzger, Sylvio, and Hueber, 1956). Moss polsters, at least in constantly moist habitats, provide surfaces on which pollen lodges and is more or less protected against decay. Polsters on bog surfaces presumably hold the pollen that will eventually be incorporated in accumulated peat as microfossils. The term of life of moss polsters undoubtedly varies with the moss species and the habitat, but polsters of living stems of *Sphagnum* in bogs probably contain pollen from no more than the past ten years.

In this study I used bryophytic polsters to analyze the modern pollen rain at seven stations in Inverness Mud Lake Bog, Cheboygan County (Secs. 7 and 18, Inverness Tp., T. 37 N., R 2 W.), in the north-

[1] The field surveys and collections were made while the author was on the faculty of the University of Michigan Biological Station, and the laboratory work was performed in part through research assistance supported by the Horace H. Rackham Fund of the University of Michigan.

ern part of the Lower Peninsula of Michigan. Detailed phytosociological lists of the vegetation in the vicinities of the bryophytic polster localities permit detailed comparison of the spectra with the very local flora; and pollen profiles from the peat deposit in Inverness Mud Lake Bog (Sears, 1942) and from the bottom sediments in Douglas Lake (Wilson and Potzger, 1943) in southwestern Cheboygan County permit comparisons with the youngest fossil spectra that have been described for this region.

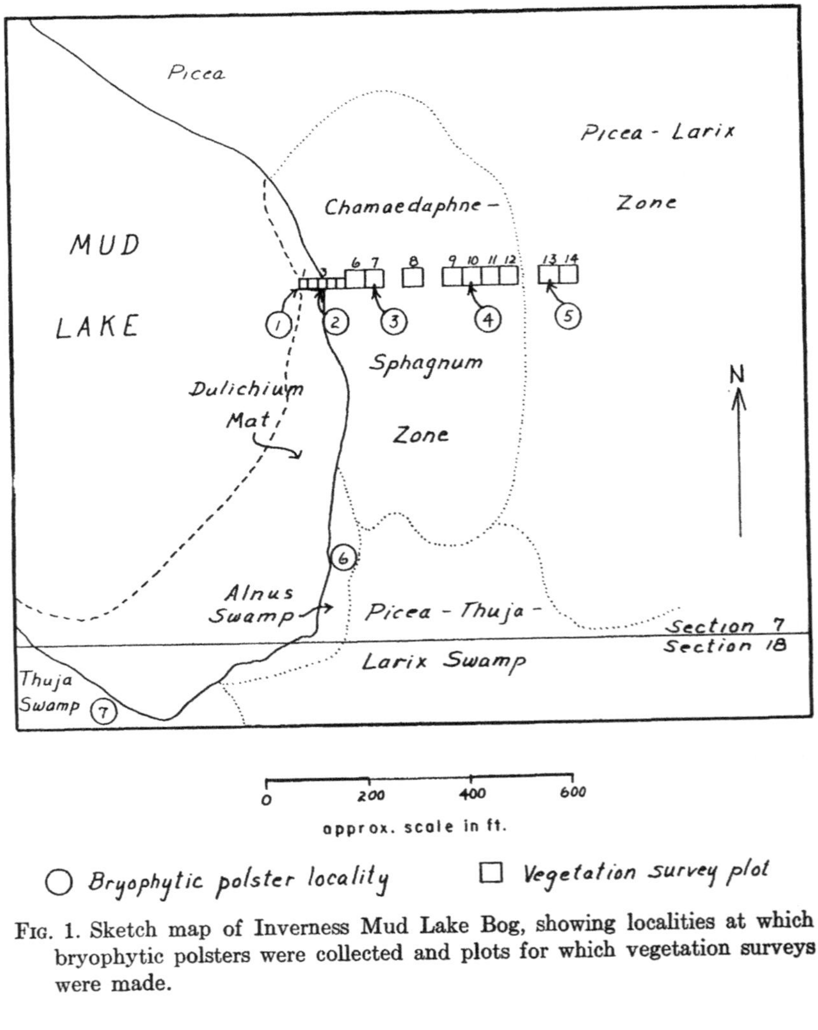

FIG. 1. Sketch map of Inverness Mud Lake Bog, showing localities at which bryophytic polsters were collected and plots for which vegetation surveys were made.

Methods

The sketch map of Inverness Mud Lake Bog, Figure 1, shows the locations from which bryophytic polsters were collected after a survey of fourteen vegetation plots by the students in my ecology class from the University of Michigan Biological Station on 13 July 1957. The phytosociological lists from these plots are presented in Tables I through V. I am indebted to Frederick B. Bevis for phytosociological lists which he made in 1958 in the vicinity of bryophytic polster stations 6 (*Alnus* Swamp, Table VI) and 7 (*Thuja* Swamp, Table VII) (Bevis, 1959). Mr. Bevis also furnished identifications of some of the lichens and mosses listed in Tables I through V.

TABLE I

PHYTOSOCIOLOGICAL LISTS FROM VEGETATION SURVEY PLOTS 1 AND 2, *DULICHIUM* MAT ZONE*

Polster No. 1, 13 July 1957

Survey by	Anderson	Gebben
Plot No.	1	2
Plot size (m)	3 × 5	7 × 5
Herb stratum		
Dulichium arundinaceum	3.5	3.4
Eleocharis olivacea	1.5	—
Nymphaea odorata	3.4	—
Lysimachia terrestris	+.1	—
Carex lasiocarpa	1.3	3.4
Hypericum virginicum	+.1	2.3
Lycopus virginicus	1.3	2.4--
Eleocharis acuminata	1.2	+.2
Carex comosa	+.2	+.2
Sium suave	—	+.1
Potentilla palustris	—	+.2
Campanula aparinoides	—	+.1
Galium trifidum	—	+.1
Epilobium strictum	—	+.1
Viola cf. pallens	—	+.3
Ground stratum		
Calliergon stramineum	+.3	2.3

* Addition of -- beneath the sociability value 3, 4, or 5 indicates that the patch, colony, or stand is interrupted or penetrated by other species locally.

TABLE II

Phytosociological Lists from Vegetation Survey Plots 3, 4, and 5, *Chamaedaphne* Zone Adjacent to *Dulichium* Mat*
Polster No. 2, 13 July 1957

Survey by	Heatwole	LaRoi	Farmer
Plot No.	3	4	5
Plot size (m)	3 × 5	5 × 5	5 × 5
Tree stratum			
Picea mariana	—	2.1	3.1
Larix laricina	—	2.2	1.1
Shrub stratum			
Chamaedaphne calyculata	4.5	4.4	3.3
Andromeda glaucophylla	+.3	+.2	2.2
Alnus rugosa	+.2	—	—
Amelanchier sp.	—	+.1	—
Vaccinium myrtilloides	—	r.1	—
Kalmia polifolia	—	+.2	2.3
Herb stratum			
Acer rubrum (seedling)	—	r.1	—
Thuja occidentalis (seedling)	—	r.1	—
Lythrum salicaria	+.1	—	—
Carex lasiocarpa	2.4	—	—
Lysimachia terrestris	+.1	—	—
Lycopus virginicus	+.1	—	—
Carex comosa	+.2	—	—
Sium suave	+.2	—	—
Campanula aparinoides	+.1	—	—
Epilobium strictum	+.1	—	—
Viola cf. pallens	+.2	—	—
Scirpus cf. cyperinus	+.2	—	—
Poa sp.	+.2	—	—
Calla palustris	+.4	r.1	—
Carex disperma	—	1.2	—
Carex trisperma	—	+.2	—
Calamogrostis sp.	—	r.2	—
Lycopus uniflorus	—	r.1	—
Rhynchospora alba	—	1.3	2.4
Eriophorum virginicum	—	+.2	1.2
Calopogon pulchellus	—	+.1	+.1
Sarracenia purpurea	—	r.1	2.2
Cypripedium acaule	—	+.1	+.1
Habenaria blephariglottis	—	—	+.1
Carex pauciflora	—	—	+.2
Menyanthes trifoliata	—	—	2.2
Ground stratum			
Vaccinium oxycoccus	—	1.2	1.3
Sphagnum magellanicum, S. warnstorfianum, S. spp.	—	4.4	5.5
Undet. mosses	—	+.3	—
Cladonia sp.	—	+.2	—

* *Ibid.*

TABLE III

Phytosociological Lists from Vegetation Survey Plots 6, 7, and 8, *Chamaedaphne–Sphagnum* Zone*

Polster No. 3, 13 July 1957

Survey by	Marks	Quaynor	Seda del Toro
Plot No.	6	7	8
Plot size (m)	10 × 10	10 × 10	10 × 10
Tree stratum			
Picea mariana	2.4	—	3.3
Larix laricina	+.2	—	2.3
Shrub stratum			
Chamaedaphne calyculata	3.5	4.3	3.4
Andromeda glaucophylla	2.3	3.3	2.3
Gaylussacia baccata	—	1.2	—
Vaccinium myrtilloides	—	+.2	—
Kalmia polifolia	—	—	2.3
Herb stratum			
Picea mariana (young)	—	2.1	—
Larix laricina	—	2.1	—
Eriophorum virginicum	1.1	1.2	1.2
Sarracenia purpurea	1.2	2.1	1.2
Habenaria blephariglottis	1.1	+.1	1.1
Cypripedium sp.	1.2	+.1	—
Pogonia ophioglossoides	1.2	—	—
Calopogon pulchellus	+.1	1.1	—
Scheuchzeria palustris	+.1	—	—
Carex pauciflora	+.3	—	—
Menyanthes trifoliata	1.3	—	—
Rhynchospora alba	2.3	—	2.3
Eriophorum spissum	—	—	+.2
Ground stratum			
Gaultheria hispidula	—	1.2	—
Vaccinium macrocarpon	2.4	1.2	2.3
Vaccinium oxycoccus	1.3	1.4	—
Drosera rotundifolia	1.3	—	1.3
Sphagnum magellanicum, S. warnstorfianum, S. spp.	4.4	4.5	4.5
Epiphytes			
Parmelia physodes	+.1	r.1	+
Cetraria pinastre	+.3	—	+
Evernia sp.	+.2	+	—
Alectoria sp.	+.3	—	—

* *Ibid.*

TABLE IV

PHYTOSOCIOLOGICAL LISTS FROM VEGETATION SURVEY PLOTS 9, 10, AND 11, *CHAMAEDAPHNE–SPHAGNUM* ZONE*

Polster No. 4, 13 July 1957

Survey by	Mensah	Wunderlich	Franklin
Plot No.	9	10	11
Plot size (m)	10 × 10	10 × 10	10 × 10
Tree stratum			
Picea mariana	—	—	3.3
Larix laricina	—	—	1.1
Shrub stratum			
Chamaedaphne calyculata	3.4	4.4	4.4
Andromeda glaucophylla	1.3	1.2	2.3
Kalmia polifolia	2.4	—	1.3
Ledum groenlandicum	+.2	r.2	+.2
Herb stratum			
Picea mariana (young)	2.1	1.3	—
Larix laricina (young)	—	1.1	—
Rhynchospora alba	2.3	+.2	2.2
Sarracenia purpurea	2.3	+.3	2.2
Habenaria blephariglottis	1.1	+.1	1.1
Scheuchzeria palustris	+.2	+.1	—
Carex pauciflora	—	+.2	—
Eriophorum spissum	—	+.3	+.3
Cypripedium sp.	—	—	+.1
Ground stratum			
Gaultheria hispidula	r.2	+.3	—
Vaccinium macrocarpon	2.3	+.2	—
Vaccinium oxycoccus	—	—	2.2
Drosera rotundifolia	1.2	+.3	+.3
Sphagnum magellanicum, S. warnstorfianum, S. spp.	4.4	5.5	5.5
Epiphytes			
Evernia mesomorpha	—	—	+
Cetraria pinastre	—	—	+
Cetraria sepincola	—	—	+
Alectoria nidulifera	—	—	+
Alectoria jubata	—	—	+
Parmelia physodes	—	—	+

* *Ibid.*

TABLE V

Phytosociological Lists from Vegetation Survey Plots 12, 13, and 14, *Picea–Larix* Zone*

Polster No. 5, 13 July 1957

Survey by	Hodson	Kepner	Bell
Plot No.	12	13	14
Plot size (m)	10 × 10	10 × 10	10 × 10
Tree stratum			
Picea mariana	3.2	3.2	3.2
Larix laricina	2.1	1.1	—
Shrub stratum			
Chamaedaphne calyculata	3.2	2.2	4.4
Ledum groenlandicum	2.3	3.2	1.1
Kalmia polifolia	1.1	1.1	1.1
Vaccinium myrtilloides	1.1	1.1	+.1
Andromeda glaucophylla	1.1	1.1	—
Herb stratum			
Eriophorum spissum	1.2	+.2	+.1
Eriophorum virginicum	1.1	—	—
Sarracenia purpurea	+.1	+.1	+.1
Calopogon pulchellus	r.1	r.1	r.1
Habenaria blephariglottis	—	r.1	+.1
Carex trisperma	—	r.2	+.1
Scirpus sp.	—	+.1	—
Cypripedium acaule	—	—	+.1
Ground stratum			
Gaultheria hispidula	1.3	1.1	+.1
Vaccinium macrocarpon	—	1.1	+.1
Drosera rotundifolia	—	r.1	+.2
Sphagnum magellanicum, S. warnstorfianum, and undetermined spp.	4.4	5.5	4.4
Calliergonella schreberi	2.4	—	2.3
Dicranum bergerii	+.3	—	+.3
Cladonia spp.	1.3	+.2	+.3
Epiphytes			
Arceuthobium pusillum	—	—	+
Evernia mesomorpha	+	+	+
Cetraria pinastre	+	+	+
Cetraria sepincola	+	+	+
Alectoria nidulifera	+	+	+
Alectoria jubata	+	+	+
Parmelia physodes	+	+	+

* *Ibid.*

TABLE VI

PHYTOSOCIOLOGICAL LIST FROM *ALNUS* SWAMP
(FROM BEVIS, 1959)

Polster No. 6, July 1958

Survey by	Bevis
Plot size (m)	2 × 2
Shrub stratum	
Alnus rugosa	4.3
Nemopanthus mucronata	1.2
Salix spp.	+.4

The vegetation surveys were made by means of phytosociological lists (*relevés*) according to the Braun-Blanquet method of notation (Braun-Blanquet, 1932; Ellenberg, 1956). This method requires listing of all observable plant species, and evaluation of cover abundance (first symbol or number in the notation) and "sociability" or gregariousness (second number in the notation). The scale of values for cover abundance is as follows:

*Value**	*Condition of abundance or cover for the species*
5	Covering more than 75 per cent of the sampled area
4	Covering 50 to 75 per cent of the sampled area
3	Covering 25 to 50 per cent of the sampled area
2	With any number of individuals covering 5 to 25 per cent of the sampled area, or very numerous individuals but covering less than 5 per cent of the area
1	Numerous, but covering less than 5 per cent of the sampled area, or fairly sparse but with greater cover value
+	Sparse and covering only a little of the sampled area
r	Very rare and covering only a very little of the sampled area (usually only one example)

The scale of values for sociability, which indicates the pattern of distribution of the individuals, is as follows:

*Value**	*Condition of "sociability" (gregariousness) for the species*
5	In large solid stands; very dense populations
4	In small colonies or larger mats; rather dense populations
3	In small patches or polsters; distinct groups
2	In small groups or clusters
1	Growing singly

* Addition of -- beneath the sociability value 3, 4, or 5 indicates that the patch, colony, or stand is interrupted or penetrated by other species locally.

TABLE VII

PHYTOSOCIOLOGICAL LIST FROM *THUJA* SWAMP
(FROM BEVIS, 1959)

July 1958

Survey by	Bevis
Plot size (m^2)	120
Tree stratum	
Thuja occidentalis	4.1
Abies balsamea	1.1
Acer rubrum	1.1
Betula lutea	r.1
Picea mariana	r.1
Shrub stratum	
Taxus canadensis	2.3
Acer spicatum	1.1
Amelanchier sp.	r.1
Herb stratum	
Coptis groenlandica	2.3
Rubus pubescens	1.2
Aralia nudicaulis	1.1
Galium aparine	1.2
Maianthemum canadense	1.1
Linnaea borealis	+.1
Trientalis borealis	+.1
Clintonia borealis	+.1
Athyrium filix-femina	+.3
Goodyera repens	r.1
Pyrola rotundifolia	r.1
Mitchella repens	r.1
Habenaria hyperborea	r.1
Dryopteris spinulosa	r.1
Erythronium americanum	r.1
Dryopteris marginalis	r.1
Ground stratum	
Mnium punctatum var. elatum	1.5
Mnium punctatum	+.4
Thuidium recognitum	+.5
Sphagnum magellanicum	+.5
Sphagnum wulfianum	+.5
Bazzania trilobata	+.5
Radula complanata	+.4
Heterophyllum haldnianum	r.1
Cladonia cristatella	r.1
Epiphytes	
Parmelia sulcata	r.1
Parmelia rudecta	r.1
Parmelia physodes	r.1
Bacidia muscorum	r.1
Usnea florida	r.1
Cetraria sepincola	r.1

The bryophytic polsters collected were all of *Sphagnum* except No. 1, which was a polster, or rather a mass of the living mat, of *Calliergon stramineum.* Only the uppermost and living portions of each polster were taken in order to avoid contamination from fossil pollen in underlying peat. The polsters were immediately sealed in individual waxed-paper bags, in which they were dried without heat for a period of five days before storing. The collecting bags were later opened in the laboratory, and the polsters were individually stirred up in large beakers with a solution of trisodium phosphate (0.5 per cent by weight) in distilled water (Benninghoff, 1947) to free the pollen and spores from the larger moss fragments. The sediment dislodged from each polster was then centrifuged, boiled three minutes in 5 per cent KOH solution, washed with glacial acetic acid, subjected to the Erdtman acetolysis procedure (Erdtman, 1943), washed with distilled water, and mounted in glycerin jelly on microscope slides.

Pollen counts were made by traverses back and forth across the slide preparations at precise settings of the calibrated mechanical stage under a 40× objective with 10× oculars. A reference point was marked with India ink in the lower left-hand corner of the cover glass of each preparation, and its position was recorded in terms of the mechanical stage calibration. In this way, the mechanical stage coordinates for any grain of interest could be used for relocating that grain on any microscope employing the millimeter scale on its mechanical stage. Counting of a preparation was continued until at least 100 grains of tree pollen were counted. With nonarboreal pollen and spores and undetermined grains the total count was 250 to 300 for most preparations. After more than 100 tree grains had been counted, two slide preparations were thoroughly searched for additional types of pollen and spores; new types found in this search were assigned "trace" (tr) values. Pollen analysts differ in their practices in regard to the minimum numbers of grains counted, depending in part upon the nature and geographic origin of the material. With bryophytic polster material I have found that the above-described practice results in a reliable sample, as further counts have failed to increase the "trace" categories to more than 1 per cent or to significantly alter the values for grains present in larger proportions.

Percentage values for the categories of pollen and spores were calculated on the basis of total tree pollen (ΣAP) as 100 per cent. Fungus spores, although counted under the microscope, are not presented

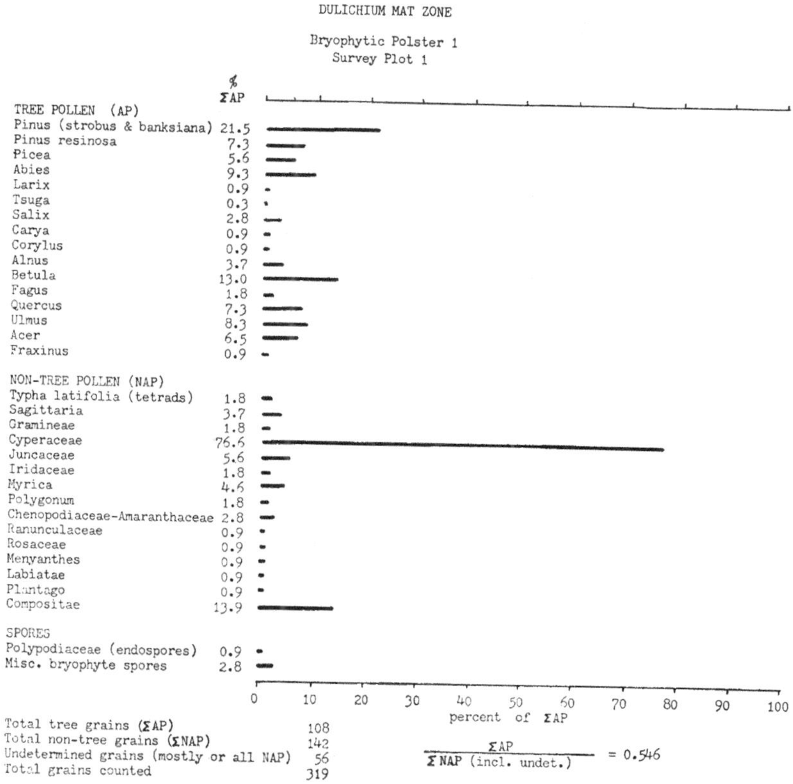

FIG. 2. Pollen spectrum of Bryophytic Polster No. 1 collected on Vegetation Survey Plot 1 in the *Dulichium* Mat Zone.

in the tabulated results (Figs. 2 through 8), as their identities could not be determined and their significance was therefore left in doubt. The pollen and spore categories shown in Figures 2 through 8 are at the limit of determination possible with the reference slide collections and literature available to me. Thus, "Sparganium-Typha" represents single grains like those of *Typha angustifolia* and of several species of *Sparganium*, but I could not find characters by which to further resolve the determination. The "Carya" could be any one of several species in the genus, but the *Juglans cinerea* grain is distinct from *J. nigra* and could be resolved to the species. *Corylus* and *Alnus*, usually placed in the nonarboreal pollen (NAP) category in

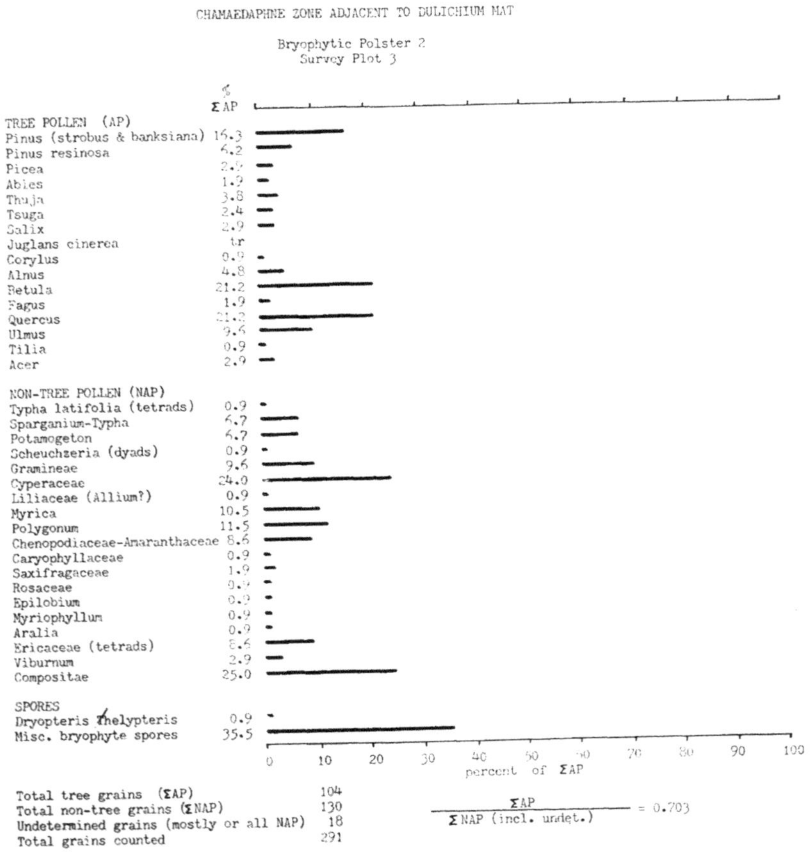

FIG. 3. Pollen spectrum of Bryophytic Polster No. 2 collected on Vegetation Survey Plot 3 in the *Chamaedaphne* Zone adjacent to the *Dulichium* Mat.

forested region studies, were included with the arboreal pollen (AP) because their pollen is commonly distributed as widely as that of the heavily producing forest trees, in contrast to the short-range distribution of most nonarboreal pollen (Gramineae and Compositae excepted). *Pinus resinosa* pollen was distinguished from the other pine species of the region (*P. strobus* and *P. banksiana*) by the presence of a thickened, scalloped fringe around the margin of the cap on the body of the bisaccate grain. With fossil pollen, of course, one could not so safely assume that only certain species have contributed the pollen,

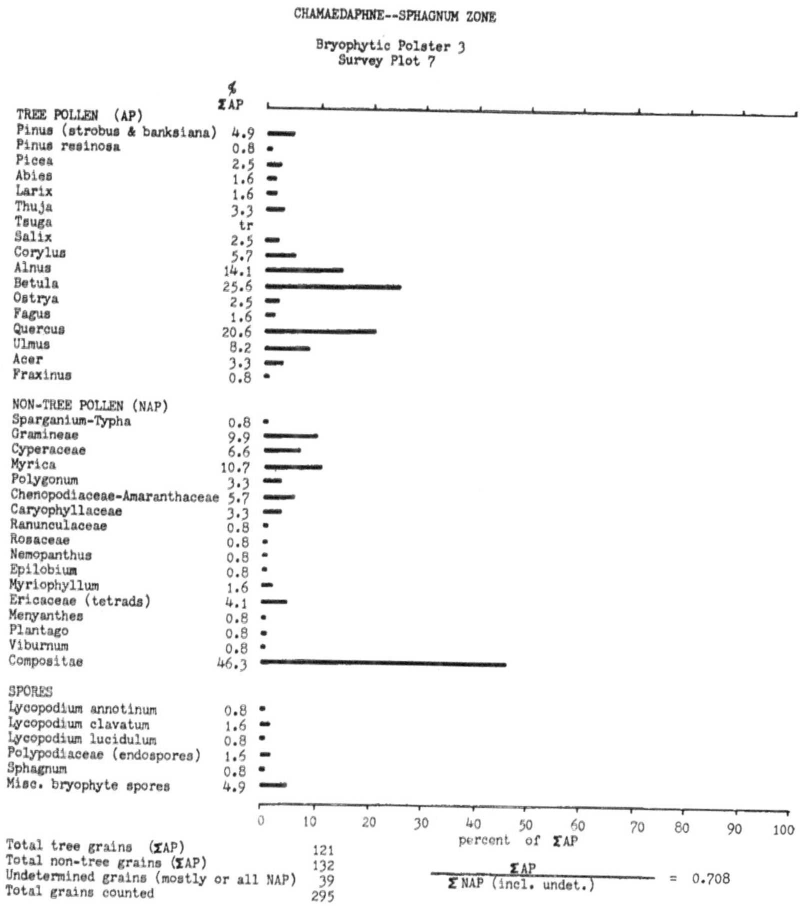

FIG. 4. Pollen spectrum of Bryophytic Polster No. 3 collected on Vegetation Survey Plot 7 in the *Chamaedaphne-Sphagnum* Zone.

and this treatment might not be recommended. The ratio of AP: NAP was calculated for each preparation and is presented with the spectra in Figures 2 through 8. This ratio is often useful in pollen statistics for evaluation of the relative prevalence of forest over non-forest vegetation. In these samples this ratio varied from 0.457 to 0.894, a reflection of the relative contribution of local AP or NAP component.

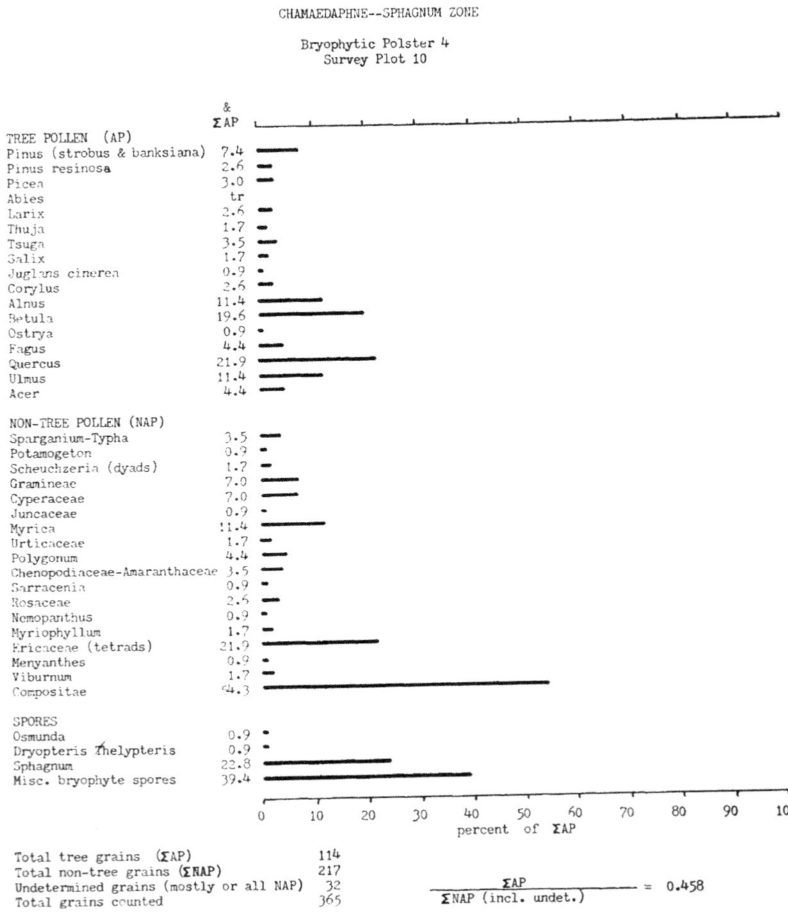

FIG. 5. Pollen spectrum of Bryophytic Polster No. 4 collected on Vegetation Survey Plot 10 in the *Chamaedaphne-Sphagnum* Zone.

DISCUSSION

Evaluation of the pollen spectra must take into consideration the fact that certain plants, primarily the anemophilous species, produce much greater amounts of pollen than others. *Quercus, Pinus, Alnus, Betula, Corylus, Ulmus,* and the Gramineae are such outstanding producers, as are also certain of the Compositae, especially those that are anemophilous. The pollen of trees is more likely to be distributed

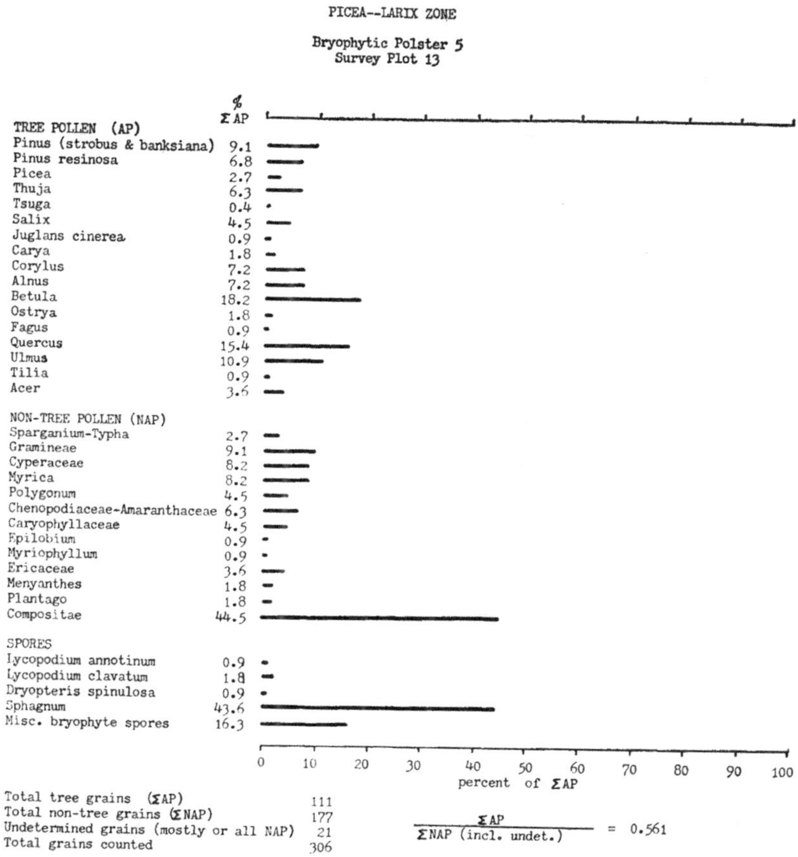

FIG. 6. Pollen spectrum of Bryophytic Polster No. 5 collected on Vegetation Survey Plot 13 in the *Picea-Larix* Zone.

greater distances by reason of being launched higher above the ground and in faster air currents. The lower the stature of the plant, the less opportunity its pollen will have for long-range transport by wind. Therefore, the tree component of the pollen rain on a given site tends to represent a more extensive sample of vegetation than the shrub and herb components.

The pollen spectra shown in Figures 2 through 8 may be compared with the respective phytosociological lists in Tables I through VII to evaluate the representativeness of the pollen rain for a given station

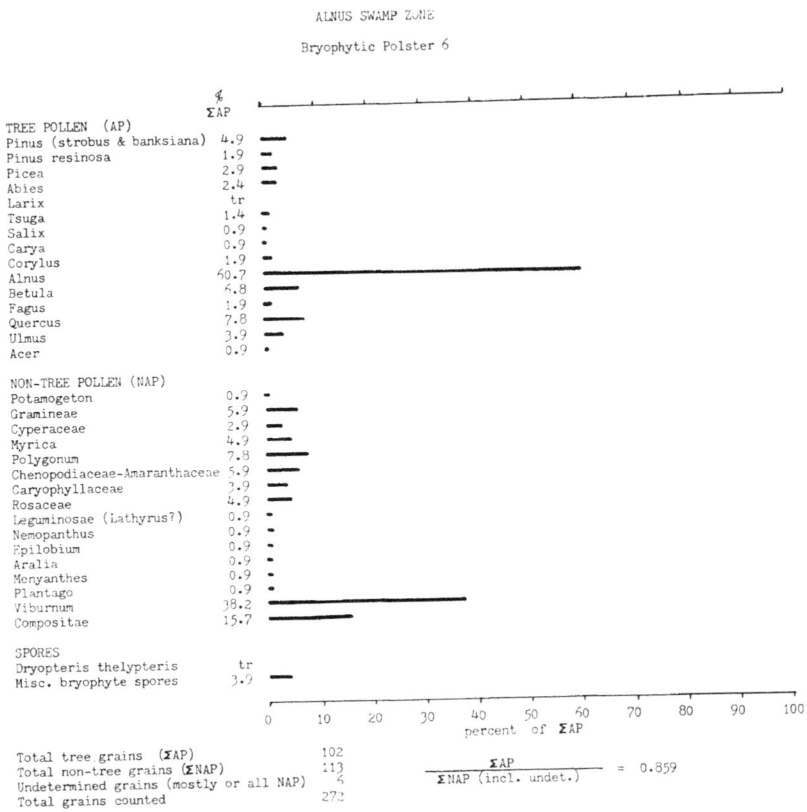

FIG. 7. Pollen spectrum of Bryophytic Polster No. 6 collected in the *Alnus* Swamp.

The pollen values for *Pinus, Tsuga, Ostrya, Fagus, Quercus, Ulmus, Tilia,* and *Fraxinus* undoubtedly represent the regional pollen rain (i.e., an area with a radius of several miles from the site sampled). Pollen of *Picea, Abies, Alnus, Betula,* and *Acer* is probably in part the regional component, but certainly in part from trees on the bog and in the *Thuja* Swamp. *Juglans cinerea* and *Carya* pollen may represent elements of "long-range drift" in the pollen rain, as these trees are not recorded for Emmet and Cheboygan counties in the natural vegetation (Voss, 1954). On the other hand, these two pollen types may have come from the few trees of these genera present as house plantings in the general area (E. G. Voss, personal communication).

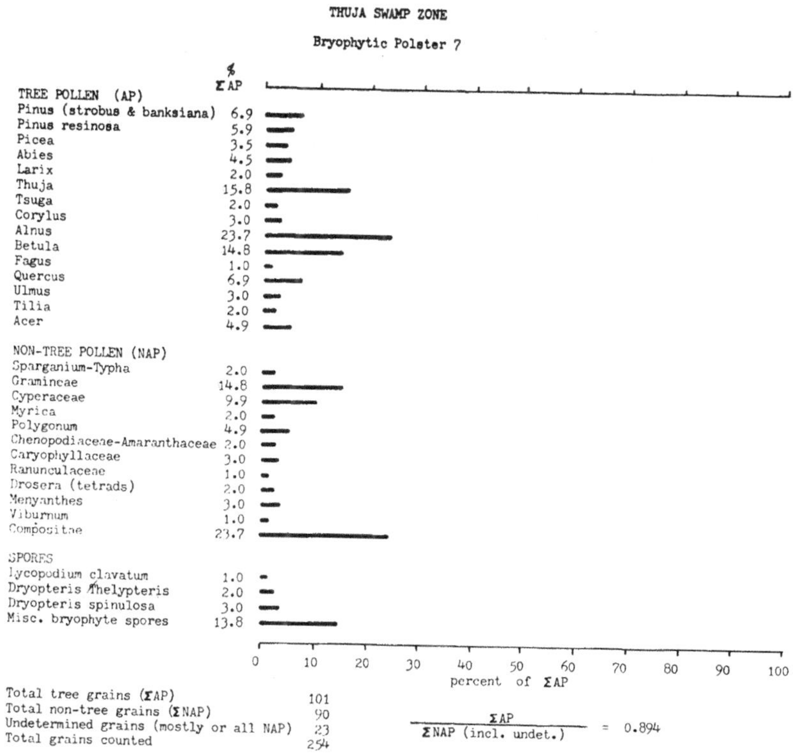

FIG. 8. Pollen spectrum of Bryophytic Polster No. 7 collected in the *Thuja* Swamp.

The *Larix* pollen is probably derived exclusively from trees on the bog, and the *Thuja* pollen exclusively from trees in the *Thuja* Swamp. A characteristic of the regional pollen components (e.g., *Quercus* and *Fagus*) is that the percentages vary more or less as a group from one polster sample to another, whereas the pollen types that are of local origin (e.g., *Thuja*) vary with the location of the sample over small distances. Pollen types with both regional and local components in the pollen rain (e.g., *Picea* and *Alnus*) have a relatively constant minimum percentage from one station to another, but show locally marked increases. Long-range drift components and regional flora elements that are present in small numbers or are small producers of pollen are erratic in their appearance in pollen spectra. *Carya* and *Juglans cinerea* produce moderate quantities of pollen, but it is difficult to determine in this instance what the source of the pollen might have been. *Tilia* is a relatively smaller producer of pollen and is commonly underrepresented in pollen spectra.

The Gramineae and Compositae are virtually absent from the bog and the *Thuja* Swamp. Their large representation must be entirely from regional pollen rain. The region has extensive areas of fields, pastures, and fallow ground, as well as weedy roadsides that are rich in grasses and composites. Almost half of the pollen grains of the Gramineae were found to be relatively large (more than 40 μ in diameter) like those of cultivated cereal grasses. The Chenopodiaceae-Amaranthaceae pollen probably comes from weed species on cultivated or disturbed ground.

The Cyperaceae is represented with relative accuracy in proportion to its very local occurrence. The Ericaceae, however, is greatly underrepresented in most of the spectra. The extreme underrepresentation in the spectrum of Polster No. 6 may be a result of the fact that the Ericaceae in the vicinity of Plot 13 appeared to produce relatively few inflorescences, and the bog surface was dry, conditions operating against the production and preservation of pollen.

Some bog plants are definitely present nearby when their pollen appears in the spectrum, for example, *Scheuchzeria, Drosera, Sarracenia*, and *Sphagnum*. These same plants, however, may be present in large numbers without appearing in the corresponding pollen spectrum. Another group of bog and aquatic plants, *Typha, Potamogeton, Sagittaria, Myrica, Myriophyllum, Dryopteris thelypteris*, and *Polygonum*, appear to have their pollen and spores distributed widely over the bog area, but probably not as widely as the regional component of the pollen rain.

Some grains that appear in the spectra are difficult to appraise, for example, *Plantago*, the species of *Lycopodium*, and *Osmunda*; but most likely they belong to the regional component of the pollen rain. The erratic occurrence of the miscellaneous bryophyte spores is also difficult to evaluate, although this could be a result of intermittent fruiting periods.

It is interesting to compare the spectra from these bryophytic polsters with the uppermost level in Sears' (1942) pollen diagram from the peats in Inverness Mud Lake Bog. Only the tree components of the fossil pollen were presented in his profile, and the precise percentages were not shown; but from the profile diagram one can see that the pollen types were ranked as follows (in order of decreasing abundance): *Pinus, Tsuga,* Betulaceae, *Picea, Fagus,* and *Quercus*. *Abies* was not present above the two-foot depth, and *Salix* was not present above the three-foot depth. In contrast, the bryophytic polster spectra show *Quercus* equaling or exceeding *Pinus, Tsuga* much less abundant, and *Abies* present in all but one polster. This difference between the top of the profile of fossil pollen and the modern spectra may have significance with respect to vegetational change over the past 100 years. Pine (*Pinus*) and hemlock (*Tsuga*) have been logged out of the forests and have been destroyed by fires. Oak (*Quercus*) and balsam fir (*Abies*) are prominent in second-growth forest and old field successions, respectively.

In the pollen diagram from Douglas Lake, Cheboygan County, published by Wilson and Potzger (1943), the following approximate percentages for the pollen can be read from the diagram: *Pinus*, 45%; *Betula*, 18%; *Tsuga*, 8%; *Abies*, 8%; *Quercus*, 5%; *Fagus*, 8%; *Larix*, 3%; *Acer*, 2%; *Ulmus*, 1%; *Salix*, 1%. This spectrum is similar to that of Sears for the uppermost level in the Inverness Mud Lake profile, except for the presence of *Abies, Larix, Acer, Ulmus,* and *Salix*, and the absence of *Picea*. But again the predominance of *Pinus* and *Tsuga* over *Quercus* and the larger proportion of *Fagus* is in contrast with the modern spectra from the bryophytic polsters examined in this study.

Conclusions

This study of pollen spectra from seven bryophytic polsters collected in Inverness Mud Lake Bog provides some guidance in the interpretation of pollen profiles from the region, especially with respect to NAP, when such more detailed pollen profiles are available. The regional vegetation indicators are the pollen of *Pinus, Quercus, Fagus, Tsuga, Acer,* Gramineae, and Compositae. Regional indicators which are influenced by local bog or swamp stands of the same entities are *Picea, Betula,* and *Abies*.

Reliable indicators of local vegetation are Cyperaceae, Ericaceae, and *Thuja*; if one of these is present in the pollen spectrum, the contributing plants were growing essentially on that site. Other indicators of local vegetation are less reliable; if the pollen occurs in the spectrum, the contributing plants were probably within a few hundred yards of the depositing area. These include *Larix, Thuja, Scheuchzeria, Epilobium, Potamogeton, Typha, Sparganium, Sarracenia, Drosera, Myriophyllum, Viburnum, Dryopteris thelypteris,* and *Sphagnum.*

Many other entities in the pollen spectra of these bryophytic polsters remain to be explained with respect to their representativeness.

LITERATURE CITED

BENNINGHOFF, W. S. 1947. Use of Trisodium Phosphate with Herbarium Material and Microfossils in Peat. Science, **106**: 325.

BEVIS, FREDERICK B. (in press). Phytosociological Study of Inverness Mud Lake Bog, Cheboygan County, Michigan. Pap. Mich. Acad. Sci., Arts, and Letters, **45** (1959).

BRAUN-BLANQUET, J. 1932. Plant Sociology (trans. by G. D. Fuller, and H. S. Conard). New York: McGraw-Hill Book Co., Inc. 439 pp.

CAIN, STANLEY A. 1939. Pollen Analysis as a Paleoecological Research Method. Bot. Rev., **5**(12): 627–654.

CARROLL, GLADYS. 1943. The Use of Bryophytic Polsters and Mats in the Study of Recent Pollen Deposition. Amer. Jour. Bot., **30**: 361–366.

ELLENBERG, H. 1956. Aufgaben und Methoden der Vegetationskunde. *In* Grundlagen der Vegetationsgliederung, Bd IV of Einführung in die Phytologie, hgb. von Heinrich Walter. Stuttgart: Eugen Ulmer.

ERDTMAN, G. 1943. Introduction to Pollen Analysis. Waltham, Mass.: Chron. Bot.

HANSEN, H. P. 1949. Pollen Content of Moss Polsters in Relation to Forest Composition. Amer. Midl. Nat., **42**: 473–479.

POTZGER, J. E., BR. M. SYLVIO, AND FRANCIS HUEBER. 1956. Pollen from Moss Polsters on the Mat of Lac Shaw Bog, Quebec, correlated with a forest survey. Canadian Jour. Bot., **13**: 24–35.

SEARS, PAUL B. 1942. Forest Sequences in the North Central States. Bot. Gaz., **103**(4): 751–761.

VOSS, EDWARD G. 1954. Preliminary Revised Checklist of the Higher Plants of Emmet and Cheboygan Counties, Michigan. Manuscript report on file at the University of Michigan Herbarium, Ann Arbor.

WILSON, IRA T., AND J. E. POTZGER. 1943. Pollen Study of Sediments from Douglas Lake, Cheboygan County and Middle Fish Lake, Montmorency County, Michigan. Indiana Acad. Sci. Proc., **52**: 87–92.

PAPERS OF THE MICHIGAN ACADEMY OF SCIENCE, ARTS, AND LETTERS
VOL. XLV, 1960 (1959 MEETING)

PHYTOSOCIOLOGICAL STUDY OF INVERNESS MUD LAKE BOG, CHEBOYGAN COUNTY, MICHIGAN[1]

FREDERICK B. BEVIS
The University of Michigan

INVERNESS Mud Lake, a relic bog formation occurring in the transitional region between the coniferous forest to the north and the deciduous forests on the south, has been the subject of several floral and ecological investigations in the past. These studies of Coe, Erickson, and Woollett (1924) and of Gates (1942) contain only species lists of the vascular plants occurring in the area and mention brief descriptions of the vegetation features. Therefore, the major objectives of this investigation were to evaluate quantitatively the vegetation composition and structural relationships of the various plant communities comprising the Mud Lake area; and to utilize the methods of the Zurich-Montpellier School of Phytosociology for this evaluation. Three ecologically distinct communities occurring along the southeast margin of the lake were studied: the *Thuja* seepage, the *Sphagnum* bog extending east from an ice-push ridge near the lake margin, and the quaking *Carex* mat bordering the open water of the lake (Fig. 1).

DESCRIPTION OF THE AREA

Inverness Mud Lake (herein spoken of as Mud Lake) lies within the upper extension of the Carp River drainage basin. This basin projects an arm into the northwest corner of Inverness Township, Cheboygan County, Michigan, and contains Mud Lake. The geo-

[1] Contribution from the University of Michigan Biological Station, Summer Session, 1958; supported in part by a National Science Foundation Grant-in-Aid. I am indebted to the following: Dr. William S. Benninghoff, Department of Botany, University of Michigan, for constructive criticism leading to the final presentation of the paper; to Dr. Howard A. Crum, National Museum of Canada, for the verification of bryophyte collections; and to Harold E. Robinson, Department of Botany, Duke University, for aid in identification of lichen collections.

graphic placement of the area is: T37N, R2W, in parts of Sections 7 and 18.

Cheboygan County, situated at the apex of the northern lower peninsula of Michigan, lies between Lake Michigan to the west and Lake Huron to the east. The area is characterized by late Pleistocene glacial features; namely, glacial lake basins, meltwater valleys, drumlins, ridges, and channeled ground moraines (Spurr and Zumberg, 1956). The principal topographic features are orientated from northwest to southeast corresponding with the direction of ice movement.

The open water of Mud Lake is one fifth of a mile wide and slightly more than twice as long (Fig. 1). An additional 300-foot area of sedge mat is situated at the southeast corner of the lake. This represents a threefold increase from the 1925 mat boundary (Coe, et al.). These same authors report the lake as distinctly alkaline (from 7.0 to 8.0 in pH), "...While the surrounding mat is decidedly

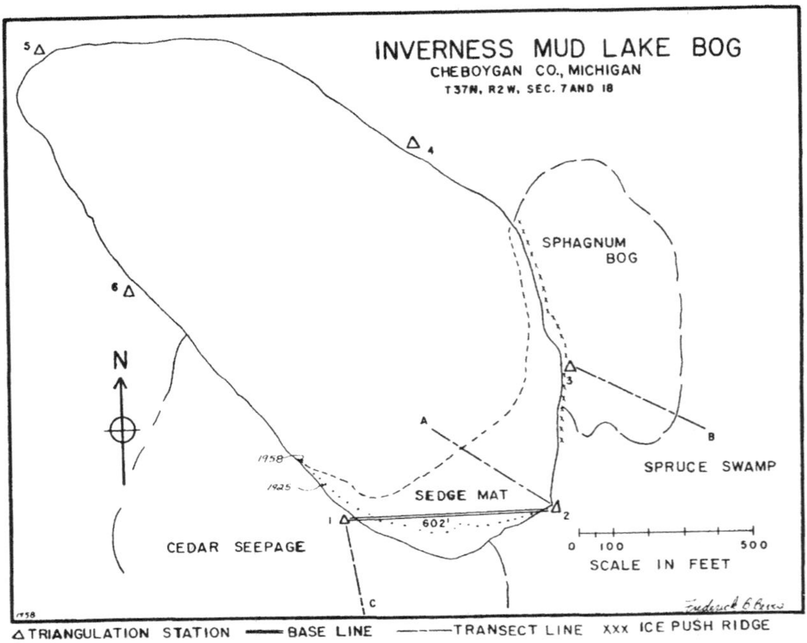

FIG. 1. Map showing the three ecologically distinct communities along the southeast margin of Inverness Mud Lake Bog.

acid." Field readings taken in August of this study show the pH at various stations was: 6.4 in the flocculent mud extending into the open water from the sedge mat; varying from 5.1 to 6.2 on the outer edge of the mat; dropping to 4.0 on the mat proper; varying from 3.8 to 4.4 on the *Sphagnum* bog; and from 6.0 to 6.5 under the heavy growth of *Thuja*.

The only perceptible flow of water into the lake is through the *Thuja* seepage. Here the water table is perched above a sand subsoil lying only two feet below the ground surface. In a few locations the water flowage can be seen through the entangled mass of cedar roots.

In concluding this brief description of the area a few comments on the prevailing climatic conditions are necessary. The climate of the northern lower peninsula of Michigan is largely controlled by three factors: proximity to both Lake Michigan and Lake Huron; the local topography; and the existing vegetation patterns. Local extremes of climate, resulting from the interaction of these features, are characteristic of the area.

Picea mariana (Black Spruce) and other northern coniferous species find suitable habitat primarily in wet lowland areas associated with poor to moderate drainage systems and having thick accumulations of organic materials. The occurrence of these relic species and communities, notably the conifer swamps and *Sphagnum* bogs of the Lake States region, are definite indicators of cooler climates during the past. Relic communities persist in local areas not characterized by the present regional climatic regime. Cain (1944) has remarked, ". . . relic colonies [occur] in local situations where frequently edaphic or microclimatic conditions provide at least temporary compensation for a general climatic unfavorableness."

Methods of Study

Transect lines were placed through the three communities and a series of milacre plots (approximately 2 m. on a side) were marked out 2 m. apart on alternate sides of the transect line. In this way a total of 90 plots was established: 30 in the *Thuja*; 40 in the *Sphagnum*; and 20 covering the *Carex* mat. A total area of 120 square meters in the *Thuja*, 160 in the *Sphagnum*, and 80 in the *Carex* community was studied. Each series of plots was tied into a triangulation station located on the base map of the area (Fig. 1).

TABLE I

SOCIABILITY SCALE (AFTER BRAUN-BLANQUET, 1932)

5	Plants occurring in great crowds (pure populations)
4	Plants in small colonies, in extensive patches, or forming carpets
3	Plants in groups, in small patches, or in cushions
2	Plants in small groups or tufted
1	Plants growing in one place; singly, solitarily

TABLE II

COMBINED COVER-ABUNDANCE SCALE (BRAUN-BLANQUET, 1932)

5	76–100% cover of the plot regardless of the number of plants
4	51–75% cover of the plot regardless of the number of plants
3	26–50% cover of the plot regardless of the number of plants
2	5–25% cover; individual plants common, but low in cover value
1	Less than 5% cover; plants abundant but of insignificant value
+	Less than 1% cover; plants occasional
r	Erratic occurrence; usually one plant on a plot

A plane table survey was conducted around the lake, and the various vegetation patterns immediately surrounding the open water were plotted. Detailed presentation of these vegetational features will follow in the discussion of the phytosociological survey.

The methods of vegetation analysis developed by the Zurich-Montpellier School were utilized in the recording of plot data (see Becking, 1957). Three basic concepts were used; Sociability (Table I), Cover Abundance (Table II), and Carles (1948) system of importance values.

Relative importance (RI) is based on the values obtained in the cover-abundance determinations and reflects the plant's dominance in the community in terms of its number of individuals, their degree of distribution, and their total coverage. The importance scale is as follows: a cover-abundance value of 5 equals 80; one of 4 is 60; 3 is 40; 2 is 20; 1 is 5; + equals 0.1; and r has no value (is 0). The RI values for each species were obtained by totaling the importance scale figure of each cover-abundance number and dividing by the total number of plots in the community. These values were used to facilitate the comparison of the synusiae, to determine the dominant,

TABLE III

LIFE-FORM CLASSIFICATION (AFTER RAUNKIER, 1934)

E	Epiphytes
MM	Mega and meso-phanerophytes (more than 24 feet tall and having exposed overwintering buds)
M	Micro-phanerophytes (6 to 24 feet tall with exposed buds)
N	Nano-phanerophytes (less than 6 feet tall)
Ch	Chamaephytes (less than 25 cm. tall; buds protected just above the ground surface)
H	Hemicryptophytes (buds at the ground level)
G	Geophytes (buds overwintering in the soil)
HH	Hydrophytes (aquatic plants)
Th	Therophytes (annual seed producers)

co-dominant, and character species of a community, and to compare the vegetation structure and composition between the communities.

The frequency concept (per cent occurrence of a species from all the sample plots) and Raunkier's (1934) system of plant life-form expressions (Table III) were utilized to depict the position of the individual species within the community.

PHYTOSOCIOLOGICAL SURVEY

The value of any vegetation investigation depends to a great extent on the accuracy of the method employed and on the degree of unbiased judgment achieved by the observer. The Zurich-Montpellier method largely eliminates the errors of judgment that often result when a high degree of precision is attempted by ocular estimation. Numerical classes are large enough to include the limits of error involved.

Phytosociology does not attempt to analyze the relationships of the individual plant to its environment, nor to study one particular environmental factor, but, more specifically, it confines itself to the study of the plant community. Becking (1957) states that phytosociology studies include "all characteristics of plant communities . . . , their physiognomy, floristic composition, community morphology and structure, community development and change, the multilateral relations of plants to one another and to environment, and the classification of communities." In contrast to general "plant ecology," phytosociology is the science of community relations (i.e., the study of the phytocenose).

Cajander (1903) was the first to introduce the tabulation of plot data by numerical means. At present, practically all aspects of the community are approached by some numerical evaluation; such as, its fidelity, cover abundance, sociability, frequency, constancy, relative importance, and presence. Each species in the communities studied is listed on the Vegetation Plot Data Sheet (Tables IV, V,

TABLE IV

VEGETATION PLOT DATA SHEET—INVERNESS MUD LAKE BOG, CHEBOYGAN COUNTY, MICHIGAN

Community: THUJA SEEPAGE Transect: C Plots: 1 to 30 Total Area: 120 sq. m.

Life-Form	Species	F	C-A.S.	RI
TREE STRATUM Height: 20 to 90 ft., Crown Closure: 80%				
MM	Thuja occidentalis	57	4.1	20.0
MM	Abies balsamea	22	1.1	6.2
MM	Acer rubrum	17	1.1	6.0
MM	Betula allegheniensis	3	r.1	0.7
MM	Picea mariana	3	r.1	1.3
SHRUB STRATUM Height: to 9 ft., Total Cover: 15%				
Ch	Taxus canadensis	30	2.3	19.3
M	Acer spicatum	33	1.1	8.8
N	Amelanchier spp.	10	r.1	4.0
HERB STRATUM Height: 2 in. to 18 in., Total Cover: 35%				
H	Coptis groenlandica	65	2.3	12.5
H	Rubus pubescens	23	1.2	6.7
G	Aralia nudicaulis	27	1.1	5.3
G	Maianthemum canadense	17	1.1	4.2
H	Galium aparine	20	1.2	4.0
Ch	Linnaea borealis	13	+.1	1.5
H	Trientalis borealis	17	+.1	3.3
H	Clintonia borealis	13	+.1	4.5
G	Athyrium filix-femina	13	+.3	2.6
G	Goodyera repens	10	r.1	+
Ch	Pyrola rotundifolia	7	r.1	0.8
Ch	Mitchella repens	3	r.1	+
G	Habenaria hyberborea	3	r.1	+
HERB STRATUM (*Continued*)				
G	Dryopteris spinulosa	3	r.1	+
G	Erythronium americanum	3	r.1	+
G	Dryopteris marginalis	3	r.1	+
MOSS (GROUND) STRATUM Height: to 2 in., Total Cover: 15%				
Ch	Mnium punctatum var. elatum	17	1.5	8.6
Ch	Mnium punctatum	7	+.4	4.0
Ch	Thuidium recognitum	14	+.5	2.9
Ch	Thuidium delicatulum	14	+.5	4.0
Ch	Sphagnum magellanicum	10	+.5	4.0
Ch	Sphagnum wulfianum	10	+.5	4.7
E	Bazzania trilobata	17	+.5	4.0
E	Radula complanata	20	+.4	3.2
E	Cladonia cristatella	3	r.1	+
E	Parmelia sulcata	3	r.1	+
E	Parmelia rudecta	3	r.1	+
E	Parmelia physodes	3	r.1	+
E	Bacidia muscorum	3	r.1	+
E	Usnea florida	3	r.1	+
E	Cetraria saepincola	3	r.1	+
Ch	Heterophyllum haldanianum	3	r.1	+
**	Dicranum rugosum, Hypnum pratense, Calliergon cordifolium, C. stramineum, Pellia epiphylla, Phlyctis agelaea, Cephalozia connivens, Riccardia latifrons, and Lophocolea heterophylla.			

KEY: F = Frequency of Occurrence on Plots (%).
C-A.S. = Cover-Abundance, Sociability Values.
RI = Relative Importance Value.
+ (RI) = RI values less than 0.1.
** Collections made of these species in the moss stratum, no values assigned.

TABLE V

VEGETATION PLOT DATA SHEET—INVERNESS MUD LAKE BOG, CHEBOYGAN COUNTY, MICHIGAN

Community: SPHAGNUM BOG Transect: B Plots: 31 to 70
Total Area: 160 sq. meters

Life-Form	Species	F	C-A.S.	RI
TREE STRATUM Height: 4 to 35 ft., Crown Closure: 2%				
MM	Picea mariana	10	1.1	4.5
MM	Larix laricina	8	+.1	3.0
SHRUB STRATUM Height: to 24 ft., Total Cover: 20%				
Ch	Kalmia polifolia	25	2.2	12.0
Ch	Andromeda glaucophylla	20	1.2	5.8
Ch	Ledum groenlandicum	13	+.2	3.1
N	Vaccinium macrocarpum	8	+.4	3.0
N	Picea mariana	5	+.1	2.5
N	Chamaedaphne calyculata	3	r.4	1.5
N	Vaccinium oxycoccos	13	r.3	0.5
HERB STRATUM Height: 2 in. to 1 ft., Total Cover: 20%				
H	Rhynchospora alba	50	+.1	0.4
G	Smilacina trifolia	20	+.1	0.6
HH	Drosera rotundifolia	13	+.1	0.5
G	Habenaria blepariglottis	10	r.1	0.3
HH	Menyanthes trifoliata	8	+.1	0.1
HH	Nymphaea odorata	3	r.1	1.0
H	Carex pauciflora	5	r.1	0.1
G	Arethusa bulbosa	8	+.1	+
HH	Sarracenia purpurea	8	r.1	+
H	Eriophorum viridi-carinatum	8	+.1	+
HERB STRATUM (*Continued*)				
HH	Potentilla palustris	5	r.1	+
H	Carex lasiocarpa	5	r.1	+
G	Pogonia ophioglossoides	3	r.1	+
Ch	Pyrola virens	3	r.1	+
HH	Drosera intermedia	3	r.1	+
HH	Drosera linearis	3	r.1	+
H	Viola pallens	3	r.1	+
G	Cypripedium acaule	3	r.1	+
H	Carex oligosperma	3	r.1	+
E	Arceuthobium pusillum	10	+.1	+
HH	Caltha palustris	5	r.1	0.1
MOSS (GROUND) STRATUM Height: to 2 in., Total Cover: 25%				
Ch	Sphagnum magellanicum	70	2.5	27.5
Ch	Sphagnum warnstorfianum	55	1.5	6.5
Ch	Sphagnum cuspidatum	30	+.5	0.3
Ch	Sphagnum teres and S. subsecundum	28	+.5	0.3
Ch	Sphagnum capillaceum	15	r.5	+
Ch	Dicranum bergeri	5	r.5	+
Ch	Tetraphis pellucida	3	r.5	+
**	Sphagnum papillosum, S. girgensohnii, Microlepidozia setacea, Mylia anomala, Porella platyphylloidea, Lejeunea cavifolia, Alectoria nidulifera, Parmelia olivacea, Cetraria ciliaris, C. saepincola, Cephalozia pleniceps.			

KEY: F = Frequency of Occurrence on Plots (%).
C-A.S. = Cover-Abundance, Sociability Scale.
RI = Relative Importance Values.
+ (RI) = value less than 0.1.
** = Collections made, no assigned values.

and VI) according to its occurrence in the various strata (synusiae) of the community. Life-form values, frequency of occurrence percentages, cover-abundance and sociability figures, and relative importance values are presented. The cover-abundance and sociability, as well as the relative importance figures are based on the average

TABLE VI

VEGETATION PLOT DATA SHEET-INVERNESS MUD LAKE BOG, CHEBOYGAN COUNTY, MICHIGAN

Community: CAREX MAT   Transect: A   Plots: 71 to 90   Total Area: 80 sq. meters

Life-Form, Species	F	C-A.S.	RI
TREE STRATUM—Absent			
SHRUB STRATUM Height: 3 to 20 ft., Total Cover: 1%			
N Alnus (incana-rugosa)	5	+.3	3.0
N Salix spp.	5	+.4	2.0
N Nemopanthus mucronata	5	r.2	0.2
N Chamaedaphne calyculata	5	r.3	+
HERB STRATUM Height: 2 in. to 3 ft., Total Cover: 60%			
H Carex lasiocarpa	75	3.5	40.0
H Eleocharis olivacea	15	+.2	2.2
H Scirpus cyperinus	15	+.2	1.5
H Carex comosa	15	+.3	1.0
H Dulichium arundinaceum	15	r.2	0.5
G Hypericum virginicum	15	r.1	0.3
H Lycopus virginicus	15	r.2	0.3
HERB STRATUM *(Continued)*			
HH Nymphaea odorata	5	r.2	+
H Carex chordorhiza	10	r.2	+
H Eriophorum virginicum	10	r.2	+
and E. gracile			+
H Epilobium strictum	5	r.1	+
HH Calla palustris	5	r.4	+
H Carex canescens	5	r.2	+
HH Nuphar advena	5	r.2	+
H Galium palustre	5	r.1	+
HH Hypericum boreale	5	r.1	+
H Scheuchzeria palustris		**	
H Carex disperma		**	
H Carex trisperma		**	
MOSS (GROUND) STRATUM Height: to 2 in., Total Cover: 5%			
HH Drepanocladus exannulatus	15	+.5	0.5
HH Calliergonella cuspidata	5	r.3	+
H Cladopodiella fluitans		**	

KEY: F = Frequency of Occurrence on Plots (%).
C-A.S. = Cover-Abundance, Sociability Value.
RI = Relative Importance Values.
+ (RI) = value less than 0.1.
** = Collections made, no assigned values.

value obtained in the summation of the respective values from each plot. The figures represent the value of a species over the total sample area of the community.

DISCUSSION

Synusial formation is a distinctive feature of the *Thuja* seepage, whereas in the *Carex* and *Sphagnum* communities, the development of characteristic belts of vegetation is most pronounced. By accepting the assumption that a forest cover is the ultimate vegetation growth that will occupy wild land areas of the Lake States region, the lack of synusial development serves to illustrate the young succes-

TABLE VII

TOTAL RELATIVE IMPORTANCE

Stratum	*Thuja*	Communities *Sphagnum*	*Carex*
Tree	34.1	7.5	—
Shrub	32.1	28.4	5.2
Herb	46.4	3.1	45.8
Moss	35.4	34.6	0.5

sional character of the latter communities. In contrast, the *Thuja* seepage stands at a more advanced successional position, showing definite synusial structure.

Total relative importance values (TRI) are extremely useful in the evaluation of community structure (Table VII). These values are obtained by summing the RI values for each species as they occur in their respective strata.

The *Carex* community is dominated by the herb stratum with *Carex lasiocarpa* having an RI value of 40.0 in relation to the stratum's TRI of 45.8. The *Sphagnum* community has a dominant moss stratum and also a distinctive heath-shrub layer. *Sphagnum magellanicum* with a RI value of 27.5 from a TRI of 34.6 is an important feature of the community. *Kalmia polifolia* (RI of 12.0) occupies a dominant place in the low shrub synusiae.

The belts of vegetation in the *Carex* and *Sphagnum* areas show considerable stability in their space-age relationships. Each belt (including the Black Spruce islands) was found to have persisted for a relatively long time and had maintained similar vegetative composition for its duration. Cuttings made through the submerged stems of *Chamaedaphne calyculata*, growing between the water-moat and ice-push ridge separating the *Carex* quaking mat from the *Sphagnum* bog, showed 50 or more yearly rings. This *Chamaedaphne* belt extended along the inner edge of the *Carex* mat, being replaced in a few places by patches of *Alnus incana*.

Several increment core measurements of 115 years and 124 years were taken from Black Spruce growing on the ice-push ridge. These ages show that this spruce belt has persisted for a long time and further substantiates the idea of stability and duration of the vegetative belts. *Tetraphis pellucida,* a moss usually found growing on advanced stages of decaying wood, occurs on spruce and *Larix laricina* logs along the ridge.

A wide area of open *Sphagnum* bog stretches from the spruce belt on the ridge to the Black Spruce swamp east of Mud Lake. This flora has close affinities to the boreal bog elements on Isle Royale, Michigan (Adams, 1909; Brown, 1933), and to those of eastern Canada (Dansereau and Segadas-Vianna, 1952). Species lists vary little in these regions; the only change in vegetation readily observable are those differences in cover abundance and, therefore, the RI values that would be determined for the species. These changes may be the result of differences in micro-habitat conditions or from the successional stage of the bog.

In several points on the Sphagnum bog, the ground water level intercepts the vegetative cover, forming small moist depressions. The occurrence of *Sphagnum teres, S. capillaceum, S. papillosum,* and *S. girgensohnii,* as well as *Drosera rotundifolia, Smilacina trifolia,* and *Kalmia polifolia* are closely associated and often restricted to these water-filled depressions. At the other extreme, *Sphagnum warnstorfianum* and *Andromeda glaucophylla* were usually found forming raised hummocks over the bog. *Sphagnum magellanicum,* the dominant species in the community, forms a continuous level carpet over much of the area. Interspersed on this moss carpet were *Rhynchospora alba* and *Habenaria blephariglottis,* and trailing over it were *Vaccinium macrocarpon* and *V. oxycoccos.* The Rhynchospora has a frequency of 50 per cent, but a cover-abundance value of only +.1 and an RI of 0.4. It is a character species of the community with only relatively minor importance.

In the transitional area between the exposed *Sphagnum* bog and the spruce swamp occur islands of spruce and larch surrounded by typical bog vegetation structure. Ages of two Black Spruce from an island were 150 and 200 years; the age of a larch was approximately 200 years. The diameter of these trees showed their abnormally slow growth, ranging from four inches for the larch to six inches for the largest of the spruce. The heights of these trees were 12, 22, and 35 feet, respectively.

The structure of these islands differed considerably from the adjacent bog, both in species composition and synusial development. A representative bog and spruce-larch island plot are compared in Ta-

TABLE VIII

Comparison of a Sphagnum Bog and Spruce Island Plot

Stratum / Species	Sphagnum No. 6 C-A.S.	Sphagnum No. 6 RI	Spruce No. 9* C-A.S.	Spruce No. 9* RI
Tree				
Picea mariana			3.2	40
Larix laricina			3.1	40
Shrub				
Kalmia polifolia	r.1	0	2.2	20
Herb				
Smilacina trifolia	1.1	5		
Habenaria blephariglottis	1.1	5		
Rhynchospora alba	+.1	0.1		
Drosera rotundifolia	+.1	0.1		
Pyrola virens			r.1	0
Viola pallens			r.1	0
Moss				
Sphagnum magellanicum	3.5	40	1.5	5
S. warnstorfianum	1.5	5		
S. cuspidatum	+.5	0.1		
S. teres & capillaceum	+.5	0.1		
S. subsecundum			+.5	0.1

* Lichens growing on the Spruce were: *Cetraria saepincola*, *Cetraria ciliaris*, *Alectoria nidulifera*, and *Parmelia olivacea*. Bryophytes present in the spruce plot but not in the *Sphagnum* were: *Microlepidozia setacea*, *Mylia anomala*, *Porella platyphylloidea*, *Lejeunea cavifolia*, *Dicranum bergeri*, and *Cephalozia pleniceps*. Dicranum and Microlepidozia were found only at the base of spruce, protected by the overhanging branches.

ble VIII. The differences in cover abundance and sociability, relative importance, and synusiae are striking.

Before continuing with a discussion of the structure of the *Carex* mat vegetation, a profile through the peat formation of the *Sphagnum* bog is described. From the bog surface to a depth of nine feet, *Sphagnum* peat in various stages of decomposition is present. Lenses of decaying spruce and larch logs are scattered in the same region. A gelatinous deposit occurs at nine feet and continues to about eleven feet. At this position a distinct marl accumulation is present and at thirteen feet sandy lake bottom sediments occur. Many kinds of shells were in these sandy bottom layers (see also Gates, 1942).

On the *Carex* mat two wide belts of vegetation were present; one of *Carex lasiocarpa* on the stable inner mat, and the other of *Eleocharis olivacea* on the outer area of gelatinous mud. *Dulichium arundinaceum*, the three-way-sedge, occurred in patches on the more solid portions of the Eleocharis belt, while *Carex comosa* occurred in small clumps between the two areas. *Drepanocladus exannulatus* and *Calliergonella cuspidata* were found bordering small water-filled depressions and along muskrat runways in the mat. Additional species listed on the community table (Table VI) were found scattered through the thick mat of *Carex lasiocarpa.*

The similarities in TRI values for each stratum in the *Thuja* seepage community (Table VII) are notable. The dominant tree member is *Thuja occidentalis* (RI of 20.0), while *Abies balsamea* (RI of 6.2) and *Acer rubrum* (RI of 6.0) can be considered as co-dominants. The entangled mass of exposed *Thuja* roots, the proximity of the water table, and the rich accumulation of organic materials, all combine to form a diversity of suitable micro-habitats for the herb and moss synusiae.

Plant distribution patterns in the *Thuja* do not show distinctive belts of vegetation. Rather species occurrence in the lower synusial layers are restricted by the location and suitability of the habitat. *Taxus canadensis* and *Acer spicatum*, the dominant and characteristic members of the shrub layer, occur on the drier sites. There was little development of herbaceous or bryophytic cover under the *Taxus*, possibly because these thick clumps excluded almost all sunlight. *Coptis groenlandica*, the most abundant species in the herb layer, *Maianthemum canadense, Pyrola rotundifolia*, and *Linnaea borealis* were found growing in moist areas at the base of large *Thuja*. *Clintonia borealis, Trientalis borealis*, and *Rubus pubescens* were more restricted in distribution to the hummocks of *Sphagnum magellanicum* and *S. wulfianum*, while *Mnium punctatum* var. *elatum* was confined to moist low hollows. Both the liverworts, *Bazzania trilobata* and *Radula complanata* were associated with the moist, protected, lower trunks of *Thuja.*

Because of the nature of previous floral and ecological work at Mud Lake, there can be no quantitative comparison of data. Therefore, only a limited discussion of general changes in the vegetation will be presented.

Coe, Erickson, and Woollett (*ibid.*) report the presence of an Iris-Aspidium and Chamaedaphne "association" in conjunction with the *Carex* mat. The leatherleaf (Chamaedaphne) association at this time covered a more extensive area lying between the ice-push ridge and the sedge mat and was also a prominent feature in the *Sphagnum* bog. Now, leatherleaf is confined to the border of a few spruce islands, along the eastern margin of the *Sphagnum* bog near the spruce swamp, and in the narrow belt along the water moat on the inner edge of the *Carex* mat. The Iris-Aspidium "association" is no longer present.

At the time of their study (Coe, et al.), mat development was limited to the extreme southeast corner of the lake. The present mat occupies a much greater area (Fig. 1). Gates (*ibid.*) has a photograph of this area taken in 1939 which shows the presence of clumps of vegetation over the Eleocharis belt. He states, "Since neither the *Carex lasiocarpa* nor the Chamaedaphne associations appear to invade the *Eleocharis olivacea* association on this soft organic mud, its successional relationships are not yet clear." *Dulichium arundinaceum* and *Carex comosa* (or a similar clump-forming sedge) are readily identified in this illustration; in all probability these species help form a more suitable substrate for the invasion of *Carex lasiocarpa.* This species appears to become established after early successional species have helped to stabilize marginal areas of the existing mat and the gelatinous mud area. The presence of *Menyanthes trifoliata, Potentilla palustris,* and *Nuphar* pushing out into the mud along the edge of the sedge mat could also aid in preparing a firm base for the extension of the *Carex lasiocarpa.* Perhaps the most prominent successional feature of Mud Lake is the speed by which the sedge mat has enlarged.

Only a few obvious changes in the *Sphagnum* bog community will be noted. The absence of leatherleaf has previously been discussed. A marked decrease in the abundance of *Larix* bears further comment. Few seed trees survived the larch sawfly outbreaks in Michigan during the 1920's (Graham, 1956). Gates (*ibid.*) states tamarack was a dominant tree in the Mud Lake area in 1916, and Coe (Coe, et al.) maps a large distribution in 1924. Today, a few old-aged, poorly developed trees exist in the protected islands of spruce and in the transitional region between the *Thuja* seepage and the Black Spruce swamp, and have a low cover-abundance and relative importance rating (Table V). Further phytosociological study of Mud Lake will aid in the analysis of vegetation pattern, composition, and structure and will thereby disclose successional trends not completely understood.

Summary

The validity and therefore importance of any ecological method rests on its ability to provide an intellectually satisfying account of the factors involved in the structure, function, and history of vegetation formations. The phytosociological approach to vegetation measurement allows extreme flexibility in its analyses of the community complex and offers an opportunity to rapidly compare vegetation on a regional basis.

LITERATURE CITED

Adams, Charles A. 1909. An Ecological Survey of Isle Royale, Lake Superior. Lansing: Mich. Biol. Surv., vii + 468 pp.

Becking, Rudy W. 1957. The Zurich-Montpellier School of Phytosociology. Bot. Rev., 23(7): 411–488.

Braun-Blanquet, J. 1932. Plant Sociology (transl. by H. S. Conard and G. D. Fuller). New York: McGraw-Hill Book Co., Inc., xviii + 439 pp.

Brown, Clair A. 1933. Ferns and Flowering Plants of Isle Royale, Michigan. Washington, D.C.: U.S.D.A., National Park Service, 90 pp.

Cain, Stanley A. 1944. Foundations of Plant Geography. New York: Harper & Bros. v + 556 pp.

Cajander, A. K. 1903. Beitrage zur Kenntniss der Vegetation der Alluvionen des nordlichen Eurasiens. I. Die Alluvionen des unteren Lena-Thales. Diss. Acta Soc. Sci. Fenn., 32 1(6): 182.

Coe, Louise, Elsie Erickson, and Edith Woollett. 1924. An Ecological Study of Mud Lake Bog, Cheboygan County, Michigan. Pap. Mich. Acad. Sci., Arts, Letters, 4: 297–310.

Carles, Jules. 1948. Geographie Botanique. Paris: Presses Universitaires de France, 126 pp.

Dansereau, Pierre, and Fernando Segadas-Vianna. 1952. Ecological Study of the Peat Bogs of Eastern North America. I. Structure and Evolution of Vegetation. Can. Jour. of Bot., 30: 490–520.

Gates, Frank C. 1942. The Bogs of Northern Lower Michigan. Ecol. Mono., 12: 213–254.

GRAHAM, SAMUEL A. 1956. The Larch Sawfly in the Lake States. Forest Sci., 2(2): 132–160.

RAUNKIER, C. 1934. The Life-Form of Plants and Statistical Plant Geography. Oxford: Clarendon Press. xvi + 632 pp.

SPURR, STEPHEN, AND JAMES H. ZUMBERGE. 1956. Late Pleistocene Features of Cheboygan and Emmet Counties, Michigan. Amer. Jour. Sci., 254: 96–109.

Papers of the Michigan Academy of Science, Arts, and Letters
Vol. XLV, 1960 (1959 Meeting)

EFFECT OF SETTLEMENT ON THE VEGETATION OF THE UNIVERSITY OF MICHIGAN BIOLOGICAL STATION[1]

Paul D. Kilburn
The Principia College

The University of Michigan Biological Station, situated between Douglas and Burt lakes in Cheboygan County, Michigan, is one of the most intensively studied ecological areas in the state. Although the effect of settlement during the past century, accompanied by severe forest and logging fires, is known in general terms, yet the precise vegetational history of any particular area in the county is not at all known. Without this information, the investigator is often severely handicapped, both in understanding how present conditions came to be and in predicting what course future changes will take. This paper lists and discusses several new facts that contribute to a more thorough and precise knowledge of this recent vegetational history.

No study of the fire and logging history has ever been made at the Station, nor have the references to these treatments ever been compiled. Most of the major ecological papers are cited by Voss (1956), including the important papers by the early ecologists F. C. Gates and H. A. Gleason. Several of these mention logging and fire, but too few of them record their precise dates, locations, and extent. Donahue (1936) claims the pine land was logged in 1871–72 and the hardwoods in 1912 but does not mention the source of his information. Both dates are reasonable although this paper reports evidence which shows that the pine logging occurred almost ten years later.

Voss (1956) discusses much of the flavor and history of the settlement of the county. Table I gives several county statistics and summarizes the progression of settlement. It is within the framework of these events that the logging, burning, and settlement of the Biological Station took place. The county was formed and the

[1] Contribution from the Biological Station of The University of Michigan.

TABLE I

SAWMILL PRODUCTION, FARM AREA, AND POPULATION CHANGE, CHEBOYGAN COUNTY, MICHIGAN, 1845–1950

Date	Million Board Feet[1]	Population[2]	Per Cent of Area in Farm Use[3]
1845		1 family	
1846	0.5		
1849		6 families	
1850			0.0
1855	3.3	300	
1860		393	
1870	16.0	2,196	
1872	35.0		
1880		8,065	
1890	100.0+	11,986	5.0
1896	200.0		
1900		15,516	20.6
1910		17,872	26.0
1920		13,991	27.4
1930		11,502	24.8
1940		13,644	28.3
1950		13,371	27.0

[1] Before 1880, Robinson and Sprague (1873); 1890 and 1896, Gibson (1944).

[2] Before 1860, Foster et al. (1939); 1860 and after, U.S. Census, Population, 1870–1950.

[3] 1850 data, Robinson and Sprague (1873); Land Area in 1890, Gibson (1944); other data from U.S. Census, Agriculture, 1910–1954.

General Land Office Survey begun in 1840, although the first settler did not arrive until 1845. By 1855, the population had reached 300, and logging of the virgin timber was under way. This year marked the completion of the General Land Office Survey of the subdivision of the townships around Douglas and Burt lakes. Kilburn (1957) describes the vegetation of this time and notes that it was essentially undisturbed by Indians, fire, or settlement. Major changes in the original forest took place after this time.

By 1896, lumbering reached its peak in the county and declined thereafter. The major softwood logging was over around the turn of the century, and lumbering operations concentrated on hardwoods for the next decade. In 1955 three early settlers in the county were interviewed about the early logging and fire history of the region. Their accounts coincided remarkably. The three were David Wolf, 82, of Riggsville, who moved to the county in 1882; Jack Spray, 75, of Riggsville, born in the county; and Noel Linebaugh, 82, of Pellston. These men reported that by 1880 all the better pine had been logged in both the Douglas Lake and Pellston areas. According to Mr. Spray, the last of this logging involved the UMBS land and occurred in the winter of 1879–80. He reported that the winter was snowless. To transport the logs, tracks had to be made out of logs, a car built, and the logs carted downhill in railroad fashion to Reese's Bog and Burt Lake and thence by water to Cheboygan. Most of the trees logged were white pine.

Hemlock logging was in progress during the 1870's, and at this time a tannin extraction factory was in operation on Burt Lake, according to Mr. Wolf. In logging, the trees were cut and the bark "ringed" around the trunk at four-foot intervals, then split down the sides and peeled. The wood was then cut up for rough lumber. The usefulness of hemlock, plus its susceptibility to fire, helped convert it from the most abundant tree in 1855 (Kilburn, 1957) to an uncommon one in 1955.

From 1880–1900 small operators were salvaging poor quality trees and dead pine snags on cutover areas, according to Mr. Linebaugh. By 1900 the conifers were gone, and attention was directed to the high-grade hardwoods. In 1902 Tindle & Jackson built a big mill in Pellston and began to cut hardwoods west of Pellston. Early hardwood lumbering concentrated on large sugar maples for veneer logs but gradually turned to other species and smaller trees as the availability of more desirable timber diminished. By 1920 hardwood logging on a large scale had ceased. Location of some of these later operations is described by botanists (Gleason, 1918; Gleason and McFarland, 1914).

The interval between 1880 and 1920 was described by both Mr. Wolf and Mr. Spray as one of constant fires. No timber remained in the pine lands, and no attempt was made to guard against fires or even to put them out. In fact, the natives regarded the fires as favorable, since they stimulated growth of blueberries and huckleberries. Mr. Spray remarked that, at times, the berries were so thick that they were removed with pronged dustpans and the leaves later removed by fanning berries and debris.

Fires kept the aspens and other timber down, so that from the hill overlooking Douglas Lake, one could see down the uncompleted railroad grade to the Douglas Lake Hotel to the west and to Germaines (west of Riggsville) to the east.

Mr. Spray claims that forest fires often started when farmers set small fires to clear the land. Mr. Wolf disagrees, claiming that farmers were usually careful of fire. He attributes most fires to carelessness by transients. Despite their disagreement on the cause of fires, the two men do agree on their frequent occurrence.

Early biologists at the Biological Station attest to the frequent forest fires both on and off the property. Voss (1956), citing Frank C. Gates in a personal communication, notes major fires in the region in 1892, 1901, 1911, and 1923. The latter date was a severe fire year and is described by Gleason (1923, p. 274), who observed that "twelve fires were counted . . . in seven miles" on one summer day. Fire scars on living trees throughout the area support the statements of both the early settlers and the early biologists. These scars attest accurately both the extent and intensiveness of former fires and thus prove more valuable and reliable than the eyewitness accounts. An examination of stumps remaining from a cleared powerline strip through the Biological Station disclosed fires in the area just north of Reese's Bog and east of Carp Creek (Kilburn, 1957) in 1880, 1890, 1896, 1899, and 1907. Only a few of these trees were older than 1880, but none of those that were showed fire scars before this date. These fire scars support the eyewitness account of logging of this area in 1879–80. Furthermore, many of the smaller red pines began growth after this date. It is possible that the 1880 fire could have been a slash fire which occurred immediately after logging.

These eleven fires were recorded during a forty-four year interval, or one every four years. This high fire frequency does not mean that the entire area was burned over this often, for the data suggest numerous small fires rather than extensive ones. It merely shows the high fire frequency at the Biological Station and illustrates that reconstruction of the fire history for this area may prove extremely difficult.

Several conclusions as to the effect of settlement on the original vegetation of the Biological Station can be made. First of all, the softwood logging of pine and hemlock on the upland sandy soils of Cheboygan County took place in the interval between 1850 and 1900.

The softwood logging of the Biological Station uplands took place most certainly between 1870 and 1880. Aside from the logging of white pine in the hardwood stands and bogs, logging operations in these forest types took place at a later date. For forty years after the original logging, fires occurred at least once a decade, eliminating most of the small pines and practically all the hemlocks. In the 1920's, with the area now fire-controlled, the second-growth forests sprang up dominated by bigtooth aspen. Today, most of the older aspen stems are between thirty and fifty years of age, although, as the plant is clonal, the clones may be much older. Each forest stand may well have its own fire history, and care must be taken in transposing fire scar dates from one place to another.

It is now possible in ecological investigations at the Biological Station to proceed with some understanding of the past treatment of the vegetation; yet caution is urged, since it is apparent that any particular locale may have a very complex history.

LITERATURE CITED

DONAHUE, R. L. 1936. A Forest Soil Study of the University of Michigan Biological Tract. Pap. Mich. Acad. Sci., Arts, Letters (1935) 21: 269–278.

FOSTER, Z. C., ET AL. 1939. Soil Survey of Cheboygan County, Michigan. U.S. Dept. Agr. Soil Survey Series. 1934. No. 15. 47 pp.

GIBSON, D. L. 1944. Socio-economic Evolution in a Timbered Area in Northern Michigan. (A case study of Cheboygan, Michigan, 1890–1940). Mich. Agr. Expt. Sta. Tech. Bull., 193. 76 pp.

GLEASON, H. A. 1918. Development of Two Plant Associations in Northern Michigan. Plant World, 21: 151–158.

——. 1923. Botanical Observations in Northern Michigan. Jour. N. Y. Bot. Gard., 24: 273–283.

——, AND F. T. MCFARLAND. 1914. The Introduced Vegetation in the Vicinity of Douglas Lake, Michigan. Torrey Bot. Club Bull., 41: 511–521.

KILBURN, P. D. 1957. Historical Development and Structure of the Aspen, Jack Pine and Oak Vegetation Types on Sandy Soils in Northern Lower Michigan. Unpublished Ph.D. dissertation, University of Michigan. 267 pp.

ROBINSON, G., AND R. A. SPRAGUE. 1873. History of Cheboygan and Mackinac Counties: Business and Manufacturing Statistics, Soil, Timber, Prospects for Settlers and c. Detroit. 45 pp.

VOSS, E. G. 1956. A History of Floristics in the Douglas Lake Region (Emmet and Cheboygan Counties), Michigan, with an Account of Rejected Records. Journ. Scientific Laboratories, Denison University, 44: 16–75.

Papers of the Michigan Academy of Science, Arts, and Letters
Vol. XLV, 1960 (1959 Meeting)

PHYTOSOCIOLOGICAL STUDIES OF SOME BEECH-MAPLE STANDS IN MICHIGAN'S LOWER PENINSULA

William S. Benninghoff and A. I. Gebben

The University of Michigan and Calvin College

Contrasts in the composition and reproduction of beech-maple stands in the northern and southeastern extremities of the Lower Peninsula were demonstrated by routine phytosociological surveys made in the summers of 1957 and 1958. Sixty-two phytosociological lists provide a means for testing the ideas of B. E. Quick (1924) and E. Lucy Braun (1950) about the contrasting composition of beech-maple communities from south to north by comparison of observations in these two areas.

In this paper the emphasis will be placed on the tree species only. The extent of sampling accomplished thus far is too small to permit analysis of shrub and herb strata components. The tree reproduction in the shrub and herb strata, however, provides an indication of the relative successional status of the various stands. In subsequent studies it is contemplated that shrub and herb species will be compared when additional phytosociological surveys have been completed, including those of the vernal phase as well as the aestival phase.

The authors wish to acknowledge assistance from members of the University of Michigan Biological Station ecology class of 1957, whose phytosociological lists were used for an analysis of the northern stands. The 1958 study of stands in the southeastern area is part of the senior author's current phytosociological survey of southeastern Michigan supported by National Science Foundation Grant G-5578.

The taxonomic nomenclature is that of *Gray's Manual of Botany*, 8th ed. (Fernald, 1950).

Localities and Methods

The localities and extent of sampling of the eight stands is shown in Table I. Stand numbers in this table correspond with the num-

TABLE I

BEECH-MAPLE STANDS SURVEYED

Stand No.	Name of Locality	County	Township	Town, Range, and Section	No. of Plots	Plot Size
1	Bois Blanc Island	Mackinac	Bois Blanc	Sec. 9	6	10 × 10 m.
2	Island View	Emmet	Redmon	T.37N, R.6W. Sec. 18	14	10 × 10 m.
3	Colonial Point	Cheboygan	Burt	T.36N, R.3W. Sec. 21	14	20 × 20 m.
4	Haven Hill	Oakland	White Lake	T.3N, R.8E. Sec. 19	8	20 × 20 m.
5	Proud Lake	Oakland	Commerce	T.2N, R.8E. Sec. 17	4	20 × 20 m.
6	Irwin's Woods	Washtenaw	Sharon	T.3S, R.3E. Sec. 4	4	20 × 20 m.
7	Brookville Road	Washtenaw	Salem	T.1S, R.7E. Sec. 28	6	20 × 20 m.
8	Beck's Woods	Wayne	Northville-Plymouth	T.1S, R.8E. Sec. 33	8	10 × 10 m.

bers on the map in Figure 1. Observations were recorded in the form of complete species lists, with cover-abundance and sociability estimates according to the Braun-Blanquet method (Braun-Blanquet, 1932). The scale of cover-abundance ratings used for the surveys is given in Table II.

Phytosociological lists of the plots surveyed were combined into tables for each stand. Space limitations prevent presentation here of the complete tables, which will be published elsewhere. Frequency class for each species in each stand was computed according to the following scale:

Frequency Class	Frequency Percentage
1	Less than 20%
2	20–40%
3	40–60%
4	60–80%
5	80–100%

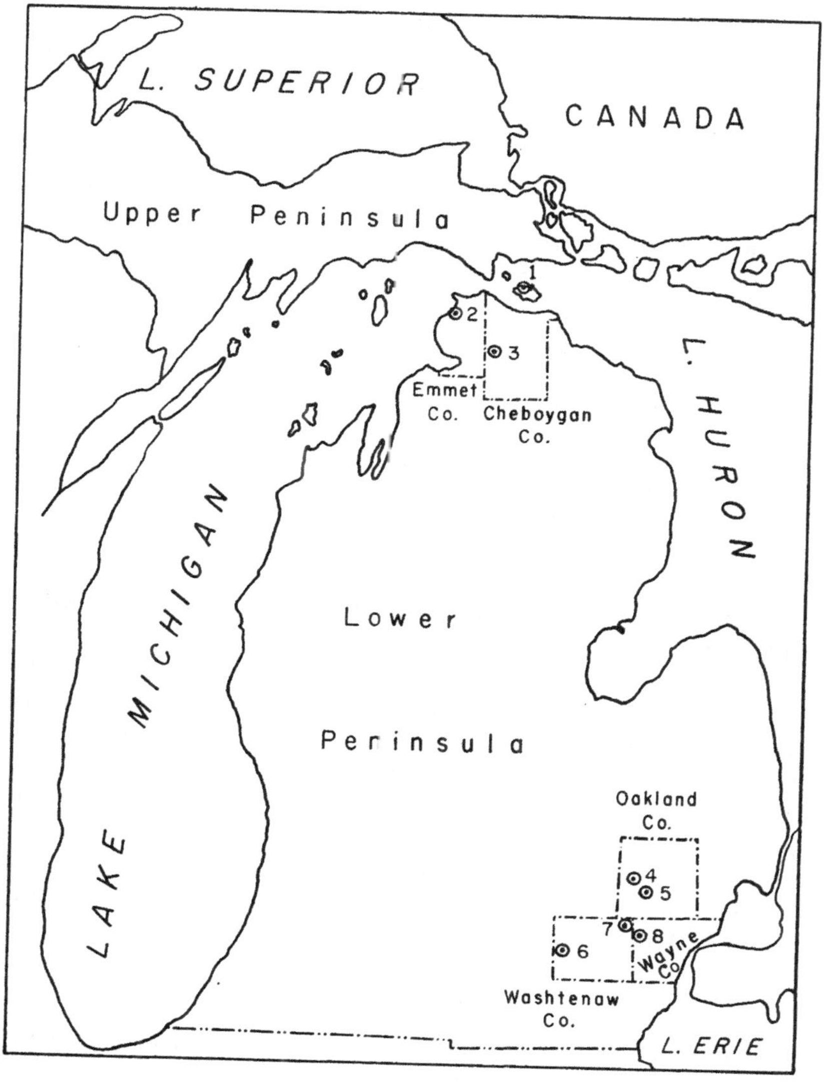

Fig. 1. Index map of beech-maple stands surveyed in the Lower Peninsula of Michigan. 1. Bois Blanc Island; 2. Island View; 3. Colonial Point; 4. Haven Hill; 5. Proud Lake; 6. Irwin's Woods; 7. Brookville Road; 8. Beck's Woods.

TABLE II

SCALE OF COVER ABUNDANCE WITH CONVERSION VALUES FOR CALCULATING IMPORTANCE (MODIFIED FROM BRAUN-BLANQUET, 1932)

Cover-Abundance Rating	Interpretation of Cover-Abundance Ratings	Conversion Value for Importance
5	Covers 75–100% of the plot, irrespective of number of individuals.	80
4	Covers 50–75% of the plot, irrespective of number of individuals.	60
3	Covers 25–50% of the plot, irrespective of number of individuals.	40
2	Numerous individuals, but covers only 5–25% of the plot.	20
1	Plentiful, but covers less than 5% of the plot.	5
+	Sparse, cover very small.	0.1
r	Rare individuals, cover insignificant.	0

Histograms of the frequency classes of tree species are presented in Figure 2.

Figure 3 presents histograms of importance values for tree species based on cover-abundance data. For calculation of importance values, a numerical value was assigned to each cover-abundance rating (*cf.* Carles, 1948) as indicated in Table II. By substituting the importance value for the cover-abundance rating for each plot, it was possible to obtain a mean importance value for each species in each stand. Figure 3 presents the mean importance values of the predominant tree species, plotted on a scale based on geometric progression. Values less than 0.25 are indicated by a dot.

DISCUSSION

Figures 2 and 3 show strong contrasts in the composition of the northern and southeastern groups of stands. Each of the following species occurs in at least two of the three northern stands: *Pinus strobus, Abies balsamea, Tsuga canadensis,* and *Betula papyrifera.* These species do not occur in the southern stands. Conversely, *Prunus serotina, Ulmus rubra, Carpinus,* and *Carya* were not recorded for the northern area stands. *Nyssa sylvatica, Juglans ci-*

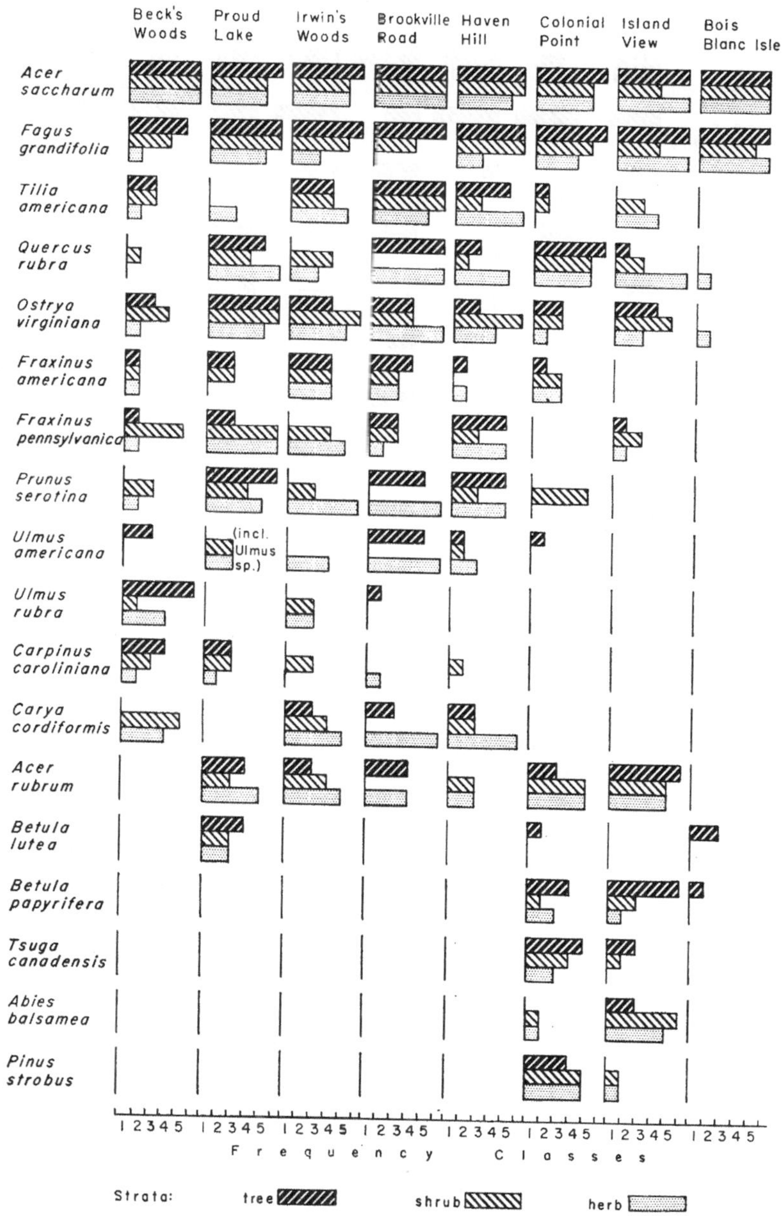

FIG. 2. Analysis of tree species in beech-maple stands by frequency classes.

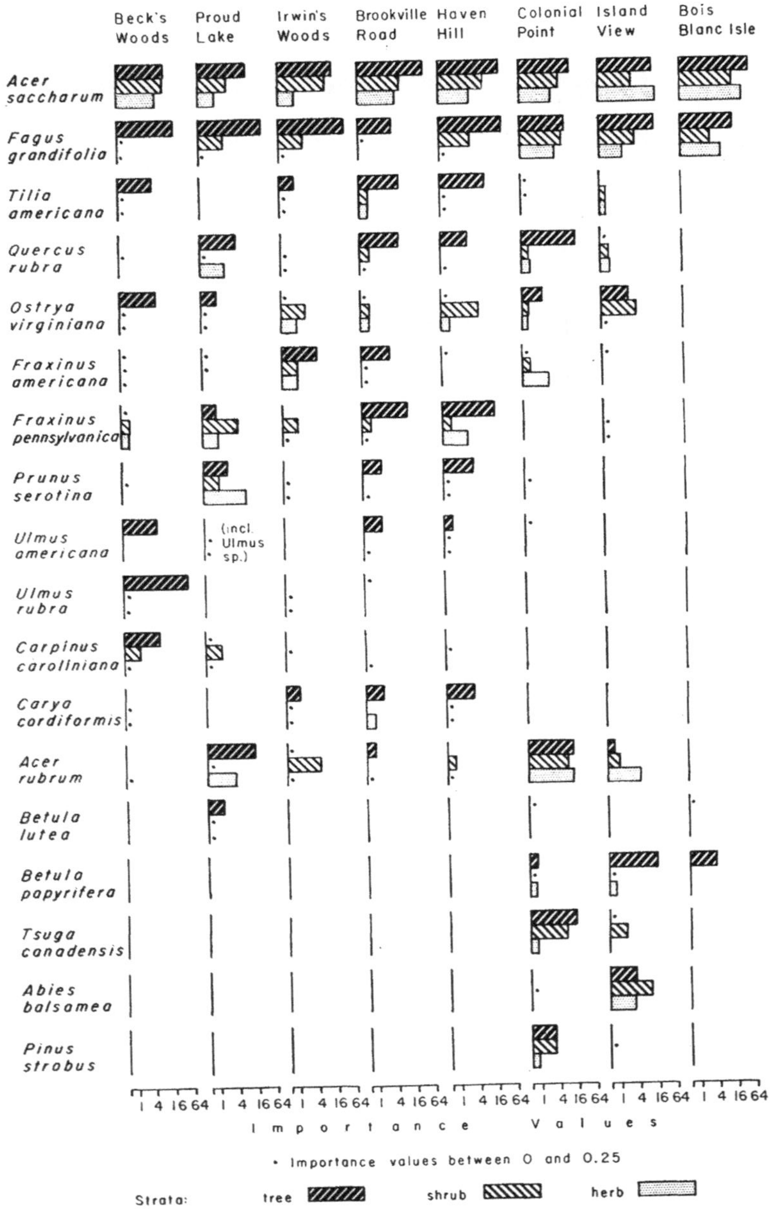

FIG. 3. Analysis of tree species in beech-maple stands by importance values based on abundance-cover estimates.

nerea, Quercus macrocarpa, and *Q. Muehlenbergii,* recorded by Quick (1924), did not occur in the stands observed.

Similar contrasts appear to exist in the shrub stratum of the two sets of stands, but the present sampling is inadequate for sorting out relicts from previous successional stages and transient species in this stratum. *Acer pensylvanicum* and *Vaccinium angustifolium,* for example, occurred in many of the northern plots but were not observed in the southern localities.

The differences in species composition and frequency values in the southern stands reflect some of the site differences in that area. Beck's Woods, which lies on the border between the Lake Plain and the Morainic Upland and is on a surface spotted with small, poorly drained depressions, shows high frequency and importance values for *Ulmus rubra.* Although the evidence is not presented here, *Lindera benzoin,* a predominant shrub in this stand, does not occur in any of the other stands.

The Proud Lake stand is situated on a ridge adjacent to a swamp forest. The soil profile exhibited a heavy accumulation of decaying humus over a thin podzol in the A horizon and a red-brown B horizon. This stand exhibits swamplike characteristics by the presence of *Betula lutea,* the relatively high frequency and importance values for *Acer rubrum,* and poor representation of *Tilia americana.*

The stands at Irwin's Woods and Haven Hill represent the more "typical" expression of beech-maple in southeastern Michigan. They exhibit relatively high frequency values for *Tilia americana* in all strata as well as for the beech and maple. *Fraxinus americana* and *F. pennsylvanica* are important species in these stands. *Acer rubrum,* although present, does not achieve high importance values.

The Brookville Road stand exhibits many features common to the stands at Irwin's Woods and Haven Hill. Observation of the frequency classes in the shrub stratum of this stand reveals that five species—*Acer rubrum, Prunus serotina, Quercus rubra, Ulmus americana,* and *Carya cordiformis*—are not present in the shrub stratum. As this particular feature is exhibited by a group of species, it is inferred that some disturbance (e.g., grazing) interfered with stand reproduction in its recent history.

In the northern area, the Colonial Point stand, near Burt Lake in Cheboygan County, and the Island View stand on a morainic ridge near the shore of Lake Michigan in Emmet County possess very similar tree species composition. Both stands contain *Betula papyrifera, Tsuga canadensis, Abies balsamea,* and *Pinus strobus.* Balsam fir and white birch have higher importance values at Island View, whereas hemlock and white pine have higher importance values at Colonial Point.

The Bois Blanc Island stand is poor in species. The shrub and herb strata also lack the diversity of species characteristic of beech-maple communities. Several factors may contribute to this condition. The soil profile comprises a thin humus layer overlying a pebbly limestone till that is underlain by limestone bedrock at depths of less than ten feet. The climate is surely more severe on the island than it is on the mainland.

Although this study is based on a relatively small sample of beech-maple stands, and the sampled stands are not on sites of uniform character, the analyses of tree species components demonstrate fundamental differences in composition between beech-maple communities in the southeastern Lower Peninsula and those in the northern Lower Peninsula.

Conclusions

From the above observations we conclude that:

1. This analysis of tree species by frequency and importance values bears out the nonuniform composition of the beech-maple communities in the Lower Peninsula of Michigan, as recognized by E. Lucy Braun (1950) and Quick (1924). To determine whether the variants observed are subassociations or different associations in the European sense, or parts of two climaxes in the sense of Clements, will require more extensive surveys.

2. The contrasts in the stratal composition between the northern and southern stands indicate different successional characteristics. In the southeastern area, oaks, hickories, and white ash of the uplands and elms, red maple, and green ash of the lowlands are members of facies of the beech-maple community. In the northern area, balsam fir, white pine, hemlock, white birch, and yellow birch are characteristic of different successional histories of the stands as well as indicative of different site conditions.

3. B. E. Quick's (1924) concept of a degree of unity in the beech-maple communities for the Lower Peninsula is borne out by these observations; but it seems that the climax concept is greatly

strained, if not denied, by the strong contrasts in composition and successional origins.

4. E. Lucy Braun's (1950) division of beech-maple communities into a "beech-maple region" and the "hemlock-white pine-northern hardwoods, Great Lake Section" places the beech-maple communities of the Lower Peninsula in two separate forest regions (Clements' "associations"?). The basis for this division resides in evidence of the kind presented in this paper. However, the unifying character of the dominant roles of beech and maple in the stands examined here would argue against this treatment.

5. The core of the problem is the anomaly of uniformity of the dominance of beech and maple in these communities along with non-uniformity in accompanying tree species between northern and southeastern Lower Peninsula stands. We suggest that, within the Lower Peninsula, beech-maple communities are of the nature of edaphically selected segregates within the oak-hickory region in the south and within the hemlock-white pine-northern hardwoods in the north. Further investigations should seek to determine the extent to which edaphic factors, especially soil moisture regimen and type of humus, are correlated with occurrence of beech-maple communities.

LITERATURE CITED

Braun, E. Lucy. 1950. Deciduous Forests of Eastern North America. Philadelphia: Blakiston Co.

Braun-Blanquet, J. 1932. Plant Sociology—The Study of Plant Communities. (Transl. by G. D. Fuller and H. S. Conard.) New York: McGraw-Hill Book Co., Inc.

Carles, J. 1948. Geographie Botanique (Étude de la Vegetation). Series "Que Sais-je?" No. 313. Paris: Presses Universitaires de France.

Fernald, M. L. 1950. Gray's Manual of Botany, 8th ed., New York: American Book Co.

Quick, B. E. 1924. A Comparative Study of the Distribution of the Climax Associations in Southern Michigan. Pap. Mich. Acad. Sci., Arts, Letters, 3(1923): 211–243.

PAPERS OF THE MICHIGAN ACADEMY OF SCIENCE, ARTS, AND LETTERS
VOL. XLV, 1960 (1959 MEETING)

FORESTRY

PRELIMINARY REPORT ON THE SURVEY OF FOREST PLANTATIONS IN NORTHERN LOWER MICHIGAN

ROBERT N. STONE

Lake States Forest Experiment Station

IN the last few years the need for a comprehensive study of forest plantations and their yields in Michigan has become increasingly apparent. The beginnings of reforestation date back more than fifty years, and many of the plantings, especially on public forests, are now twenty to thirty years old. Mill expansion has increased the demand for conifer pulpwood, and small amounts of wood from plantations are already arriving at the mills. During 1956 Christmas trees cut from plantations numbered 525,000.[1] Furthermore, the money spent for tree planting has been one of the largest investments made in forestry in the state. Many people are wondering how this investment looks.

In 1957 and early 1958 the Lake States Forest Experiment Station, Lower Michigan National Forest, Michigan Department of Conservation, and the Consumers Power Company surveyed the plantations in northern Lower Michigan and Muskegon, Ottawa, and Allegan counties. This survey was designed to yield mainly inventory information, such as area planted during the last fifty years; area now in established forest plantings by forest type, stand size, and stocking level; and present volume and growth. In addition, some miscellaneous information was collected about site, damage, and past treatment.

The project consisted of two parts, the first being to determine the acreage planted. We compiled existing planting records to determine planted area on public and industrial lands. It was not so simple to get acreage planted on private lands. We first had

[1] Lee M. James, "Resurvey of Christmas Tree Marketing in Michigan, 1957," Mich. Agr. Exp. Sta., Special Bull. No. 419, illus.

to find out how many trees were shipped to private planters from State and Clark-McNary nurseries, and how many were distributed through the Soil Conservation districts from other nurseries. Our summaries include all trees shipped to the study area through these organizations in units of 500 trees or more. Trees raised in private nurseries were not picked up unless they were sold through an SCS district. After summing the numbers of trees shipped, we converted these numbers to acres by temporarily assuming 1,000 trees planted an acre.

In the second major phase of the project, over 3,000 $\frac{1}{50}$-acre plots were measured in sample plantations, which were randomly selected from all plantings. The samples were distributed by decade to obtain a good sample in the early plantations where acreages were small. From this sample, area of established plantings was computed by forest type, stand size, and stocking level. To correct for the 1,000-trees-per-acre assumption, we measured the area planted with each traced shipment and found that 1,450 trees were shipped for every acre of forest plantings established on private lands.

Area Planted

The record summaries showed an estimated 660,600 acres planted to trees in northern Lower Michigan from 1906–56.

Areas of forest planting varied widely by counties, although some patterns were apparent. Counties with high public ownership tend to have large plantation areas. For example, Iosco, Oscoda, Wexford, Lake, Manistee, and Newaygo counties (Fig. 1), each with large acreages planted, all have large public ownership. Generally, planting on private lands declines across the survey area from southwest to northeast. The major planting effort is now concentrated in the western part of the survey area, with Wexford County presently having the highest planting rate.

Peak planting rates were reached on the public forests in the thirties (Fig. 2). On private lands the planting rate climbs each year. Since 1950 at least three-quarters of all planting has been on private lands.

Established Forest Plantings

The field sample indicated the present condition of the planted area. By applying the results of the field sample to the total area

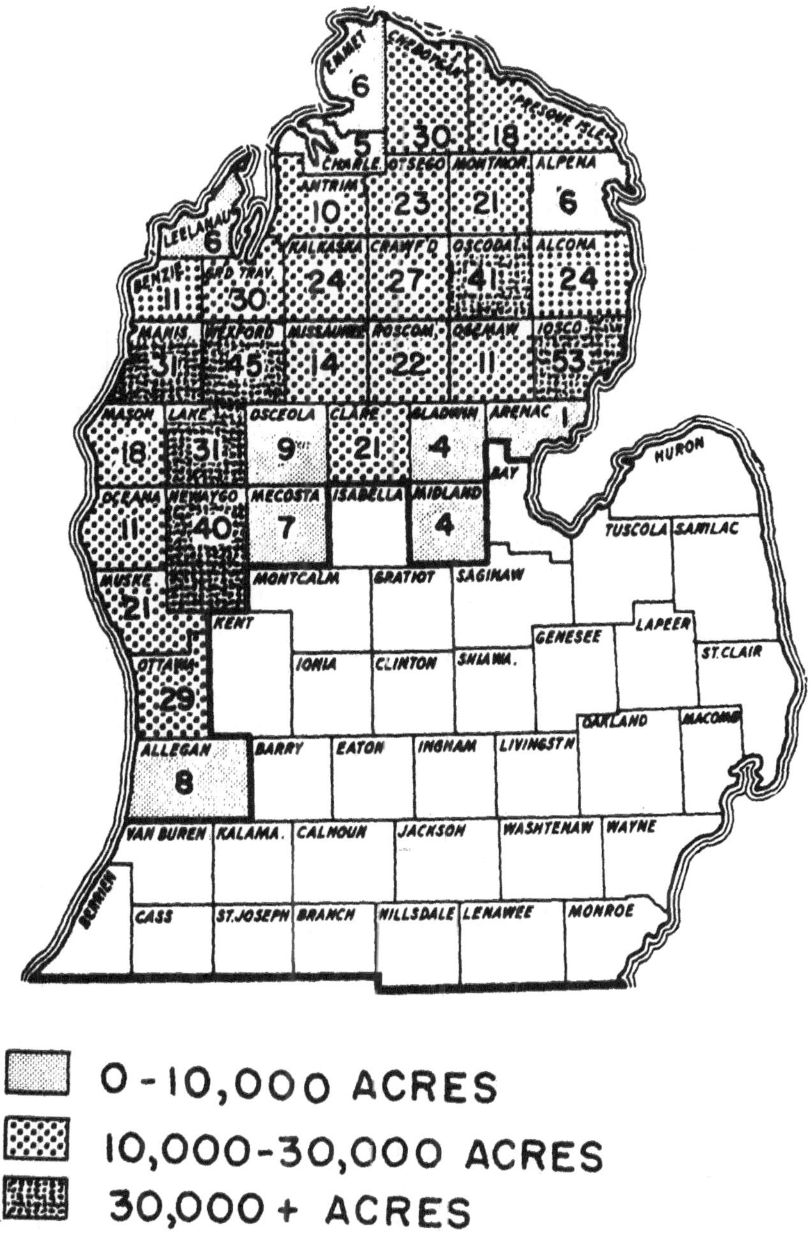

Fig. 1. Forest plantings by county, 1900–1956. The figures within each county indicate total area planted in thousands of acres.

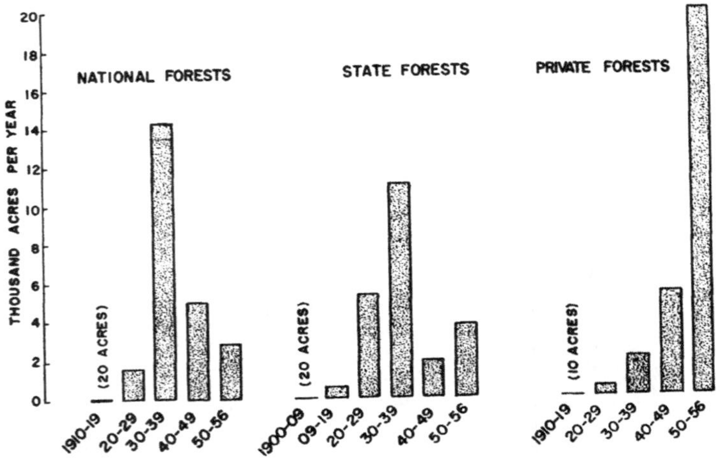

FIG. 2. Planting trends in northern Lower Michigan by owner, 1900–1956.

planted, we arrived at present acreage by forest type, stand size, and stocking level. Established forest plantings totaled 516,900 acres, or 6.6 per cent of the commercial forest land in the study block. The difference of 143,700 acres between estimated planting and the established plantations is not all due to failures; some trees were planted in windbreaks and narrow shelterbelts, while others were used to thicken up natural stands. In addition, some areas were not classed as plantations, since the planted trees were less important than the native conifers. These sweetening plantings, while only a small component of the stands they are in, should improve the stocking and ultimate yields from the area.

Only under extreme circumstances, such as burning, was a planting a complete failure. Less than 3 per cent of all planted area has reverted to grass and brush. Thirty-nine per cent of the established plantations are on the Lower Michigan National Forest, another 31 per cent are on the State forests, and 30 per cent are privately owned (Fig. 3). Distribution of established plantations by type is: red pine 53 per cent, jack pine 35 per cent, white pine 11 per cent, and spruce 1 per cent.

Fully 31 per cent of the plantations have significant native hardwood overstories, and 8 per cent have native conifer overstories (Fig. 3). Many of these plantings need release if they are to produce commercial timber in a reasonable time. Although there is a tendency to think of plantations as solid blocks of a single species where every tree is a conifer and all are nearly the same size, few of the plantations visited were pure. The average plantation is constantly being invaded by aggressive hardwood species in addition to the trees that were there when the planting was made. It contains numerous small openings and often has a brush fringe around its border. As the plantings mature, more native trees will tend to invade the stands.

Plantations on about 62,400 acres have reached pole-timber size or larger; the rest are seedlings and saplings. However, stands on large acreages planted during the thirties are rapidly growing into pole timber; within 10 years another 74,000 acres of plantations will reach this size.

Most of the present pole timber is jack pine, totaling 40,700 acres. Other pole-timber stands include red pine with 17,400 acres, and

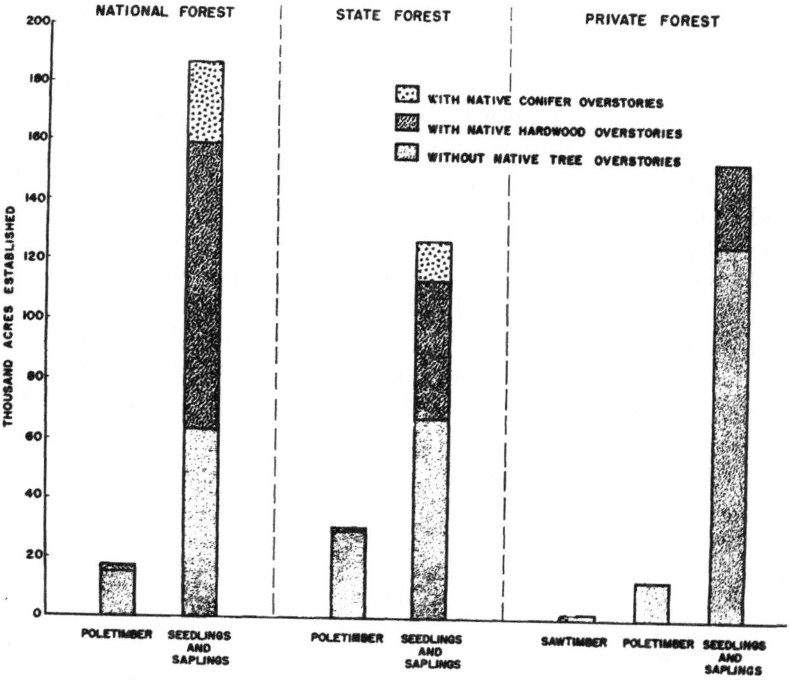

FIG. 3. Established forest plantation acreages in northern Lower Michigan by owner and size.

white pine with 3,300 acres. It is from this pole-timber acreage that the first yields will come.

Total plantation volume is increasing rapidly. Volume in planted trees over 5 inches d.b.h. is now about 585,000 cords. We obtained this estimate by determining mean volume per acre for each 10-year age class and species, then multiplying each volume by the appropriate acreage. An estimate of volume in 10 years was obtained by advancing the average age 10 years and multiplying the new mean volume by the appropriate acreage. By this method total volume in 10 years is estimated at 1,527,000 cords, and the annual growth (plus ingrowth) at 94,000 cords. The growth rate at a specific age is assumed to be the same during the near future as it was in the past 10 years.

The first yields from plantations will be as thinnings in pole-timber stands, with a minimum of about 3 cords per acre of material not needed for growth. Growth seems favorable in the younger stands where basal area is maintained in the neighborhood of 90 to 110 square feet per acre. With basal area growth of several feet per year, stands can be cut back to about 75 to 85 square feet. Present volume, in stands with 3 cords or more above this level, totals 85,000 cords. In addition, the thinnable volume on new acres reaching pole timber each year must be considered. When these are included, the volume that can be thinned from plantations during the next decade is about 15,000 cords annually.

If bolts to a 3-inch top were accepted at the mills this volume would be in the neighborhood of 15 per cent higher. One problem in profitably harvesting thinnings is that much of the volume is in "one-stick" trees where the cost of felling each tree must be charged to a single bolt. If the stands go unthinned until only 2-stick trees need be cut, growth will lag. This dilemma might be partially side-stepped if the markets would accept half sticks and sticks to 3 inches on the small end, although this would create stacking and debarking problems at some mills.

Stand Development

Little is known about the ultimate development of plantations in Michigan. Even the earlier plantings are still relatively young. For example, Beal Pinetum just east of the Michigan State University campus is believed to be one of the oldest white pine plantings in the State, and it is only 62 years old.

A number of factors impede the favorable development of a plantation as it progresses through the seedling and sapling stages into pole-timber size and ultimately to saw timber. Heavy ground cover of grass, low shrubs, and brush immediately slow the growth of a seedling. Browsing, insect or disease damage, and overstory competition delay it further. As a plantation closes and begins making greater demands on the soil, site quality becomes a critical factor. Development may be slowed by overstocking when needed thinnings are not made.

We found that native overstories seemed to be the most serious impediment to the transition of plantings from the seedling and sapling size to pole-timber size (Fig. 4). Where native overstories were present, growth was slower, volumes per acre were lower, and height and diameter growth was less than in plantations without

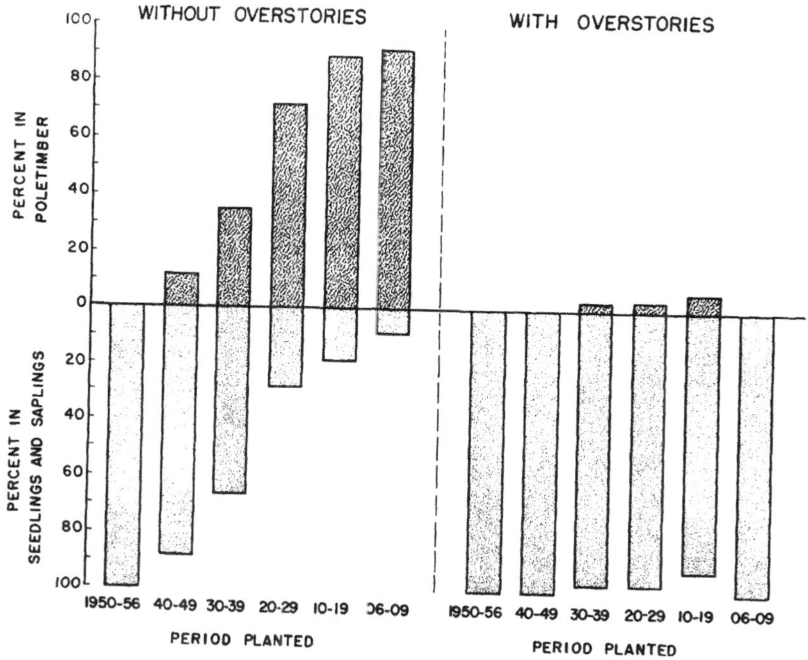

FIG. 4. Development of plantations without native overstories, and with an overstory of 3 cords or more per acre. Northern Lower Michigan, 1906–56.

native overstories (Figs. 5, 6, and 7). Over half of the plantations without overstories were of pole-timber size in 30 years, but less than 3 per cent of those with overstories of 3 cords per acre were pole timber in this time. Furthermore, planted trees under overstories

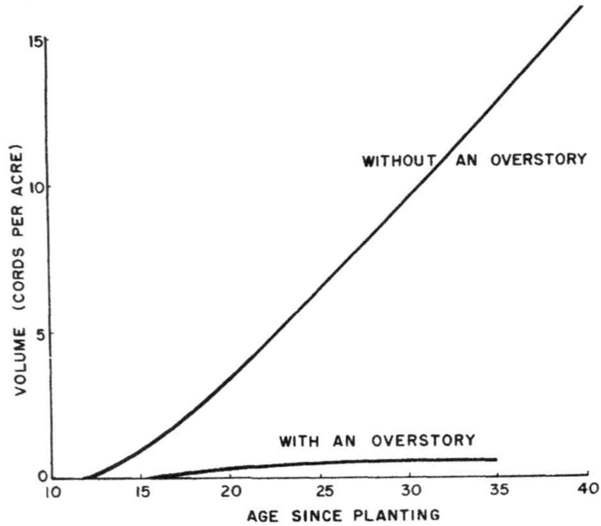

FIG. 5. Volume of red pine in plantations without a native overstory and with a native overstory of 3 cords or more per acre.

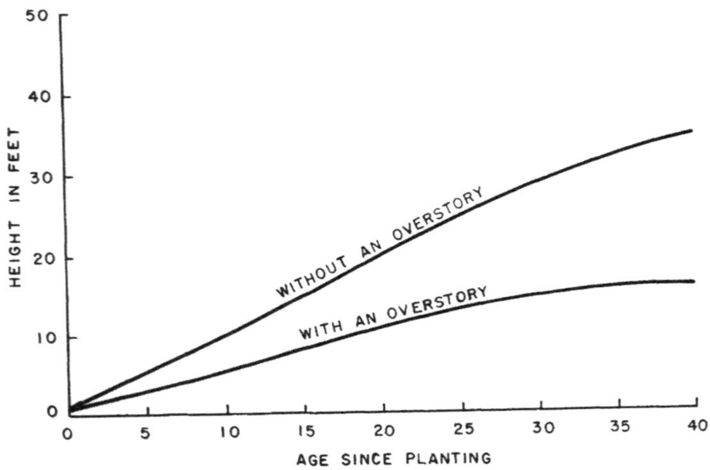

FIG. 6. Height growth of red pine in plantations without a native overstory and with a native overstory of 3 cords or more.

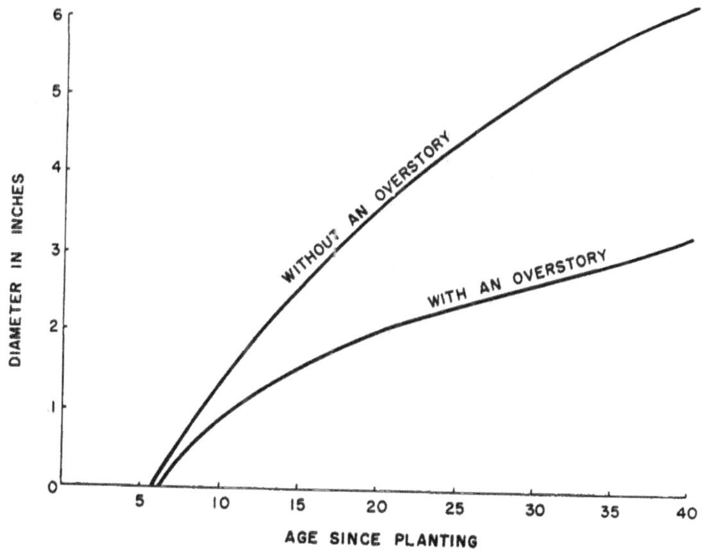

Fig. 7. Diameter growth of red pine in plantations without a native overstory and with a native overstory of 3 cords or more per acre.

are generally slim, with frail branches and thin crowns. They are easily damaged by sleet and heavy snow.

Factors other than overstories, such as insect damage, are important problems in certain localities and particular stands.

Conclusion

Our survey revealed that 660,600 acres had been planted in the study area during 1906–56. The planting rate on private land is rising; on public forests the rate seems stable. About 516,900 acres are in established forest plantings, with 70 per cent on public land and 30 per cent privately owned. Nearly 40 per cent of the reforested land also contains native overstories, seriously reducing the development of the planted trees. Red pine has been most widely planted, occupying 53 per cent of the established plantation acres.

Volume of trees 5 inches and larger in 1957 was about 585,000 cords. In 10 years it will be 1,527,000 cords. Growth plus ingrowth amounts to about 94,000 cords per year. Plantations of pole-timber size or larger total 62,400 acres. This will more than double in 10 years. About 15,000 cords can be thinned annually from plantations during the next decade.

The results of the plantation survey of northern Lower Michigan will be published by the Lake States Forest Experiment Station in four Technical Notes and a Station Paper. Three of the Technical Notes, presenting information on acreage planted and acreage established, are available now. The last Technical Note will indicate volume and age relationships. The Station Paper, which will be a complete report of the study and its findings, is now in preparation.

Papers of the Michigan Academy of Science, Arts, and Letters
Vol. XLV, 1960 (1959 Meeting)

PALEONTOLOGY

A RADIOCARBON DATE FOR A WOODLAND MUSK OX IN MICHIGAN

Claude W. Hibbard and Frank J. Hinds

The University of Michigan and Western Michigan University

On October 29, 1956, Ray Coville of Climax, Michigan,[1] took three large vertebrae that he had found in a marl spreader to Frank Hinds of the Department of Biology, Western Michigan University, for identification. The marl in which the vertebrae occurred had been taken from a marl pit on the Coville farm in the southeast corner of the SE. ¼ Sec. 18, T. 3 S., R. 9 W., Kalamazoo County, Michigan.

Frank Hinds, the junior author, made two trips to the Coville marl pit prior to November 14, 1956, but failed to find additional musk-ox bones. The senior author was invited to aid in a study of the vertebrae and of the locality where they were found. On November 14, 1956, we visited the site but further work produced no other fossil bone.

The marl pit is situated in a small depression previously occupied by a lake of post-Cary age. The pit was dug in a marshy area. A small stream flowed from the pit into a nearby slough. A layer of muck, ranging from one foot to eighteen inches thick, covered the marl deposit. The depth of the marl varied from six to eight feet. Ray Coville reported that at the time of the removal of the marl which contained the vertebrae, they were operating the power shovel from four to six feet below the water surface. Due to the spring the

[1] We wish to express our gratitude to Mr. Coville for donating the musk ox vertebrae to the University of Michigan and to Carlton W. Angell for making the drawing of the woodland musk ox and the American mastodon. We wish also to express our appreciation for permission to study specimens under the care of William H. Burt and Emmet T. Hooper, of the Division of Mammals of the University of Michigan Museum of Zoology, and to William R. Farrand for making a comparison of the vertebrae with specimens in the collection at Ohio State University.

pit was full of water so that examination of the walls or floor was not possible.

The three vertebrae, No. 34124, University of Michigan Museum of Paleontology, are the atlas, axis (Pl. II, Fig. 3), and the third cervical of a large musk ox, *Symbos cavifrons* (Leidy). The vertebrae were slightly damaged by the marl spreader. It is evident that all three had been in close contact because the articular faces were clean and free from marl, and that the atlas was in contact with the skull at the time of the removal of the marl from the pit, since the proximal articular faces were clean and showed clearly the outline of the condyles of the skull. But further excavation has never revealed other bones of the skeleton.

Without doubt numerous skeletal elements of musk oxen have been encountered in excavations in Michigan. But, because of their small size when compared with the bones of the mammoth or mastodon, they have been considered merely bones of deer, elk, or domestic animals, all of which are common in muck areas. The remains of caribou went unrecognized for many years in Michigan owing to their small size.

The dorsal surface and the wings of the atlas have been broken. The anteroposterior width of the atlas from the anterior articular surface to the posterior articular surface is 77.0 mm. The atlas is more massive than the atlas of a large male *Bison bison* (No. 102162, University of Michigan Museum of Zoology). The vertebral foramen for the spinal cord is larger, and more nearly round in *Bison*, than the foramen in the atlas of *Symbos*. The atlas of *Symbos* is also larger than the atlas (No. 102448, UMMZ) of a large male *Ovibos*, the recent musk ox. The depth of the ventral wall of the atlas of *Ovibos* measured from the articular surface of the ventral arch for the dens of the axis to the tip of the ventral spine is 45.0 mm. In *Symbos* the depth is 51.0 mm.

The atlas assigned to *Symbos cavifrons* is too large and massive for *Ovibos* and *Boötherium*. It fits the skull (UMMP 3450) of *Symbos* reported by Case (1915 and 1921) from the farm of William J. Schlecht, Washtenaw County, Michigan.

The axis was the least damaged of the three vertebrae. The dorsal spinous process and the two transverse processes are missing (Pl. II, Fig. 3). The shape of the axis is shown in Plate II in comparison with the axis of *Ovibos* and *Bison*.

MEASUREMENTS OF AXES OF SYMBOS, OVIBOS, AND BISON

	Symbos No. 34124	*Ovibos* No. 102448	*Bison* No. 102162
Greatest width across anterior face..	123.5 mm.	111.5 mm.	115.0 mm.
Depth from neural canal to anterior ventral spinous process............	67.0	58.5	51.0
Anterior width of neural canal.......	28.0	24.0	35.5
Anterior height of neural canal......	25.5	21.0	30.5
Greatest width across transverse processes.........................	99.0	82.0	74.9
Distance from midline of ventral surface of centrum to dorsal surface of a posterior articular process.	130.5	105.5	107.0

The third cervical vertebra was used to obtain a radiocarbon date. The vertebra was submitted to the University of Michigan Memorial-Phoenix Project Laboratory and given No. M-639. Professor R. H. Crane gave us the date as 13,200 ± 600 B.P. in C^{14} years for the radiocarbon sample.

Very few Pleistocene mammals available for radiocarbon dates have been found in Michigan since 1950. The specimens previously collected had been treated with shellac, which made them useless for such dates. The remains of one other late Pleistocene mammal from Michigan have been dated by radiocarbon: an American mastodon (*Mammut americanus*) palate and tusk (No. UMMP 29276) recovered from beneath 2 ft. 4 in. of muck on the J. M. Bruggeman farm in Lenawee County. This adds to the evidence that some of the large mammals of Michigan became extinct only a few thousand years ago. Part of the tusk was submitted for radiocarbon dates. The outer portion of the mastodon tusk (No. M-281) gave a date of 7,820 ± 450 B.P. C^{14} years. The inner part of the tusk (No. M-280) dated 7,070 ± 240 B.P. C^{14} years. The beaver-cut limb of aspen (No. M-282) upon which the mastodon bones rested, gave a date of 9,568 ± 1000 B.P. C^{14} years (Crane, 1956; and Crane and Griffin, 1958). It may be noted that the teeth marks of the beaver on the aspen limb were the size and shape of those of *Castor canadensis* Kuhl now living in Michigan.

The above date indicates that this specimen of the woodland musk ox lived in Michigan prior to the development of the Port Huron

Moraine (Flint, 1956, p. 272). *Symbos cavifrons* was considered by Hibbard (1951) to be a woodland form, and he concluded that they foraged along the edge of the lakes and forests, with the woodland caribou in Michigan. Kitts (1953) from his study of the musk oxen arrived independently at the conclusion that *Symbos* was a woodland or steppe form.

Since Case (1915) reported the first skull of *Symbos cavifrons*

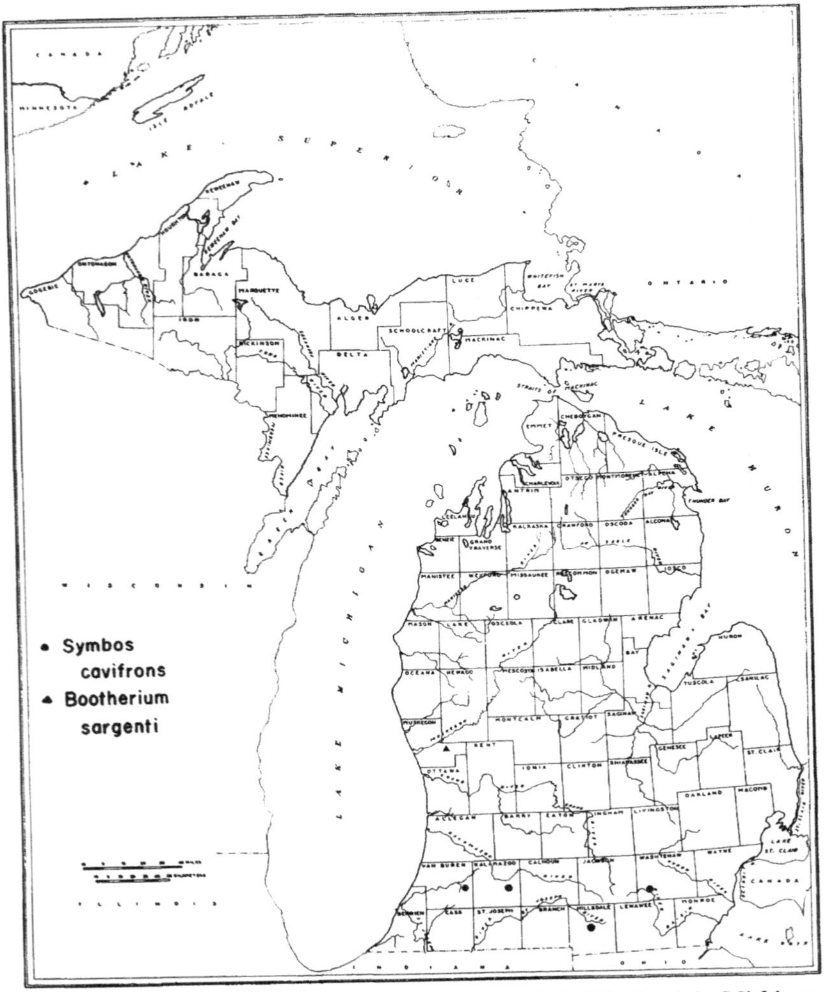

FIG. 1. Locality records of musk oxen (*Symbos* and *Boötherium*) in Michigan.

found in Michigan, parts of two other skulls have been recovered. The skull (No. UMMP 3450) reported by Case was taken about 0.5 mile northwest of Manchester, near the center of the NW. ¼, NW. ¼ of Sec. 1, T. 4 S., R. 3 E., Washtenaw County (see O. P. Hay, 1923, p. 250). The skull was found beneath four feet of muck and lay on clay. The frontal part of a skull with horn cores (No. UMMP 10566) was found in 1919 about two miles south of Hillsdale, Michigan, while a drainage ditch was being dug in a muck area. It was recovered from four feet below the surface. The posterior part of a skull with imperfect horn cores (No. UMMP 23113) was recovered from a marl pit in the farm of Max Schwanke in Antwerp Township, Van Buren County, Michigan, in 1942. For the distribution of the woodland musk oxen in Michigan, see Figure 1.

Gidley (1908) described a small and different skull type of musk ox (*Boötherium sargenti*) from the farm of Charles McKay, near Moorland, Muskegon County, Michigan. The skull was taken from two or three feet below the surface in a marsh area. It was found beneath the pelvis of an American mastodon (see Plate I). The specimen is in the Museum of Natural History at Grand Rapids, Michigan (see O. P. Hay, 1923, p. 250).

It is very likely that *Boötherium* is the female woodland musk ox since all specimens of *Symbos* based on skulls are considered as those of bulls. Kitts (1953) gives a good distributional map of all records of fossil *Ovibos, Symbos,* and *Boötherium* reported from North America. We do not consider *Euceratherium* reported by Kitts to be a musk ox but to belong to the tribe *Euceratherini* instead of to the tribe *Ovibovini* to which the musk oxen belong (see Hibbard, 1955, p. 80, Pls. 7–9).

LITERATURE CITED

Case, E. C. 1915. On a Nearly Complete Skull of *Symbos cavifrons* Leidy from Michigan. Univ. Mich. Mus. Zool., Occ. Pap. No. 13: 1–3, 2 pls.

——. 1921. Something about the Paleontological Collections in the University. Mich. Alumnus, 27: 292–300, 8 figs.

Crane, H. R. 1956. University of Michigan Radiocarbon Dates I. Science, 124: 664–672.

——, and James B. Griffin. 1958. University of Michigan Radiocarbon Dates II. Science, 127: 1098–1105.

Flint, Richard F. 1956. New Radiocarbon Dates and Late Pleistocene Stratigraphy. Am. Jour. Sci., 254: 265–287, 1 pl., 1 fig.

Gidley, J. W. 1908. Description of Two New Species of Pleistocene Ruminants of the Genera *Ovibos* and *Boötherium*, with Notes on the Latter Genus. Proc. U. S. Nat. Mus., 34: 681–684, 3 pls.

HAY, O. P. 1923. The Pleistocene of North America and Its Vertebrated Animals from the States East of the Mississippi River and from the Canadian Provinces East of Longitude 95°. Carnegie Inst. Wash. Publ., No. 322: 1–499, 41 pls., 25 figs.

HIBBARD, CLAUDE W. 1951. Animal Life in Michigan During the Ice Age. Mich. Alumnus Quart. Rev., 57: 200–208, 4 figs.

——. 1955. Pleistocene Vertebrates from the Upper Becerra (Becerra Superior) Formation, Valley of Tequixquiac, Mexico, with Notes on Other Pleistocene Forms. Univ. Mich. Mus. Pal., Contrib., 7 (5): 47–96, 9 pls., 5 figs.

KITTS, DAVID B. 1953. A Pleistocene Musk-ox from New York and the Distribution of the Musk-Oxen. Am. Mus. Nat. Hist. Novitates, No. 1607: 1–8, 2 figs.

PLATE I

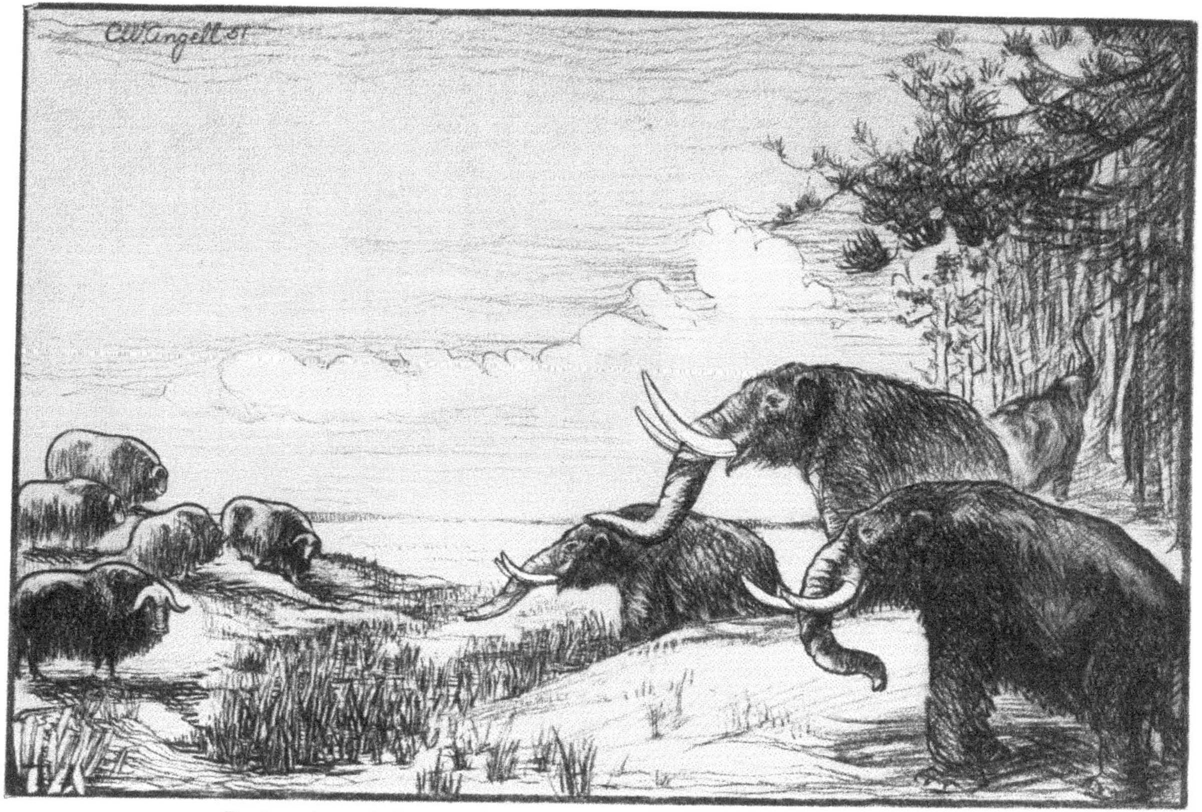

Restoration of the woodland musk ox and the American mastodon.

PLATE II

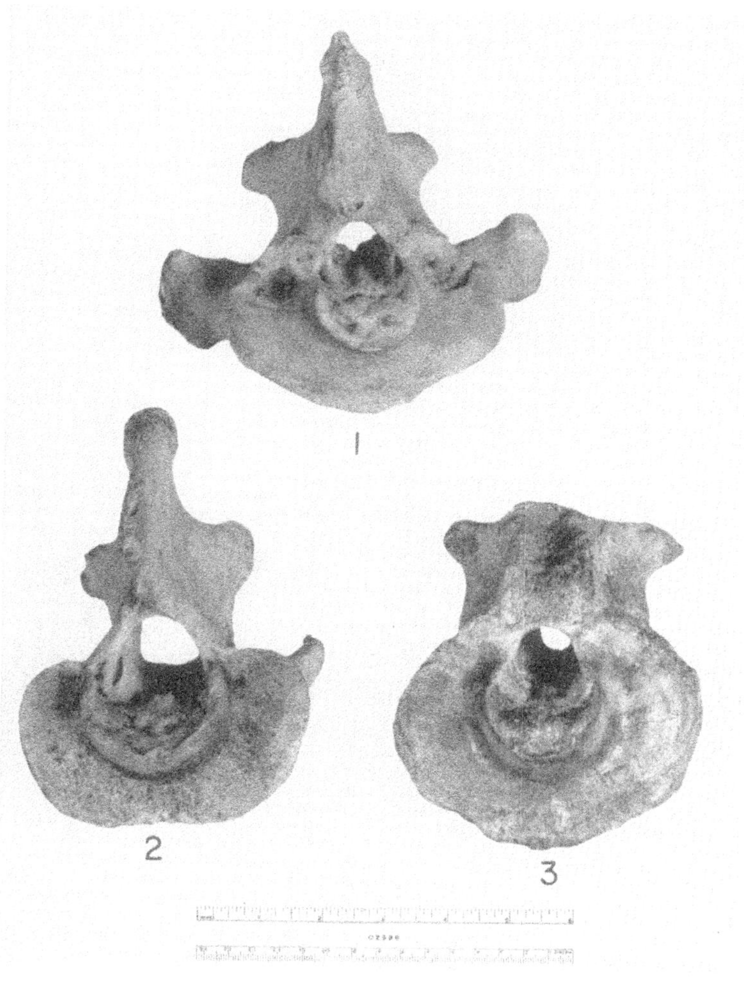

FIG. 1. *Ovibos moschatus*, axis, UMMZ 102448, anterior view. Reduced.
FIG. 2. *Bison bison*, axis, UMMZ 102162, anterior view. Reduced.
FIG. 3. *Symbos cavifrons*, UMMP 34124, anterior view. Reduced.

PAPERS OF THE MICHIGAN ACADEMY OF SCIENCE, ARTS, AND LETTERS
VOL. XLV, 1960 (1959 MEETING)

ADDITIONAL NOTES ON THE LIZARDS OF THE CRAGIN QUARRY FAUNA

RICHARD ETHERIDGE

The University of Michigan

NOT long ago I published an account of some lizard fossils from the Cragin Quarry fauna of Meade County, Kansas (Etheridge, 1958). Hibbard (1955) places the fauna in the upper part of the Kingsdown formation, referred to the Sangamon interglacial of the Pleistocene. Six species of lizards, all of which are extant, were identified: *Crotaphytus collaris, Holbrookia texana, Phrynosoma modestum, P. cornutum, Eumeces obsoletus,* and *Cnemidophorus sexlineatus.* Two of these forms, *H. texana* and *P. modestum,* do not occur in the area today but range to the southwest. Two other forms, *Sceloporus undulatus* and *H. maculata,* are now common in the area but have not appeared in the Cragin Quarry fauna.

Additional lizard fossils that have become available from this locality seem to warrant some further comments on the subject. The original sample studied consisted of about 100 fossils, of which 54 could be identified. The additional material includes 140 fossils of which 71 are identifiable. All of the newly acquired fossils represent species already reported from the fauna but a number of them are elements not present in the original sample. These permit a more certain identification of one species, *Holbrookia texana,* give a clearer notion of the order of abundance of the various species, and have led to the discovery of a curious condition of sexual dimorphism in the skull of *Crotaphytus collaris.*

The fossil *Holbrookia* of the Cragin Quarry fauna was tentatively referred to *Holbrookia texana.* I was unable to find any structural differences, in the elements available, between *H. texana* and *H. maculata.* Although the latter species occurs in southwestern Kansas today and the former does not, the large size and associated lizards indicated a more probable identification of the fossils as *H. texana.* In addition to the original sample, I now have two scapulocoracoids, two maxillaries, seven frontals, one articular, one sacral vertebra, and twenty pelves. The new cranial elements permit a more certain identification of the species. The fossil articular is like that of *H. texana* in that the retroarticular and angular processes of the element are better developed than in *H. maculata*. In addition, the new material confirms a maximum size of the fossil lizards as great as that found in modern *H. texana* and greater than that found in *H. maculata*.

A number of new cranial elements of *Phrynosoma modestum* are now available. These consist of two jugals, four squamosals, one quadrate, one parietal, three supra-angulars, one frontal, and one scapulocoracoid. Nearly all of the skull roofing bones and elements of the lower jaw are easily identified because of their ornamentation. The newly acquired elements appear to be identical with those of recent *P. modestum*.

No additional fossils of *Phrynosoma cornutum* have been recovered. The locality of the only known fossil of this species from the Cragin Quarry fauna was reported, in error, as the Cragin Quarry. The correct locality is the Cragin Quarry horizon of Mount Scott, a short distance from the Cragin Quarry itself.

Additional fossils of *Cnemidophorus sexlineatus* consist of a part of an ilium and two articulars. They differ in no way from those of recent lizards of this species.

Eumeces obsoletus is represented by an additional two maxillaries, one dentary, twelve pelves, and one caudal vertebra. The caudal vertebra is from the autonomous section of the caudal series but was recovered unfragmented, apparently because of the presence of consolidated matrix in the neural canal.

Fragments of two dentaries of *Crotaphytus collaris* are of especial interest in that they have led to the discovery of a curious type of sexual dimorphism in the lower jaw of this form. The fossil dentaries are so unusual that I originally thought the elements represented an undescribed genus of lizards. Their identity became obvious after I had compared them with skeletons of large adult males of *Crotaphytus collaris*. Eastern populations of this species exhibit considerable sexual dimorphism. The males are larger; the skull and especially the lower jaws are more robust than in the female. As in most iguanid lizards, the anterior teeth of the dentary are simple and pointed. In *Crotaphytus* this series generally contains from ten to twelve teeth, ending about at the position of the last mental foramen. In adult males, the first few teeth of the anterior simple series are very long and slightly recurved. Following these, there is a gradual reduction in size. The posterior marginal teeth are moderately short, robust, and weakly trilobate. They become smaller toward the beginning of the series. The last few teeth of the anterior series overlap the first few teeth of the posterior series, the former being lingual to the latter. The overlap occurs about one third of the way back from the symphysis to the end of the entire tooth series. The teeth of the two series may alternate or be opposite one another. The lateral alveolar border in this region bulges out over the superior lateral face of the dentary. I know of no other lizard in which such overlap occurs. Females and subadult males of the species differ from adult males in that all of the marginal teeth are of about the same height, and there is no overlap between the simple and trilobate series.

The two fossil fragments (Fig. 1) are parts of a left and right dentary in the region of overlap. The teeth are badly worn; otherwise the positions of the mental foramina, the open Meckel's groove, and the general configuration of the elements are identical with those of recent adult male *Crotaphytus collaris.*

A total of 125 lizard elements have now been identified from the Cragin Quarry fauna. These are divided among the species as follows: *Holbrookia texana* 51, *Eumeces obsoletus* 28, *Phrynosoma modestum* 24, *Crotaphytus collaris* 20, *Cnemidophorus sexlineatus* 4, and *Phrynosoma cornutum* 1. The order of abundance based upon the minimum number of individuals which must be present in the total sample (that is, the number of the most abundant single element for each species) is as follows: *Holbrookia texana* 12 right pelves, *Eumeces obsoletus* 8 right pelves, *Crotaphytus collaris* 5 right dentaries, *Phrynosoma modestum* 4 left squamosals, *Cnemidophorus sexlineatus* 2 right dentaries, and *Phrynosoma cornutum* 1 parietal. This order of abundance is in agreement with that derived from the total number of elements of each species, except that the positions of *Phrynosoma modestum* and *Crotaphytus collaris* are reversed. These figures were not taken into account in my consideration of the paleoecological implications of the fauna, yet they are in general agreement with the conclusions drawn from other data. Of especial interest is the rarity of *Cnemidophorus sexlineatus.* The order of abundance is about what one would expect in a modern lizard fauna composed of the same species, except that *Cnemidophorus sexlineatus* would probably be one of the more common forms. I had previously

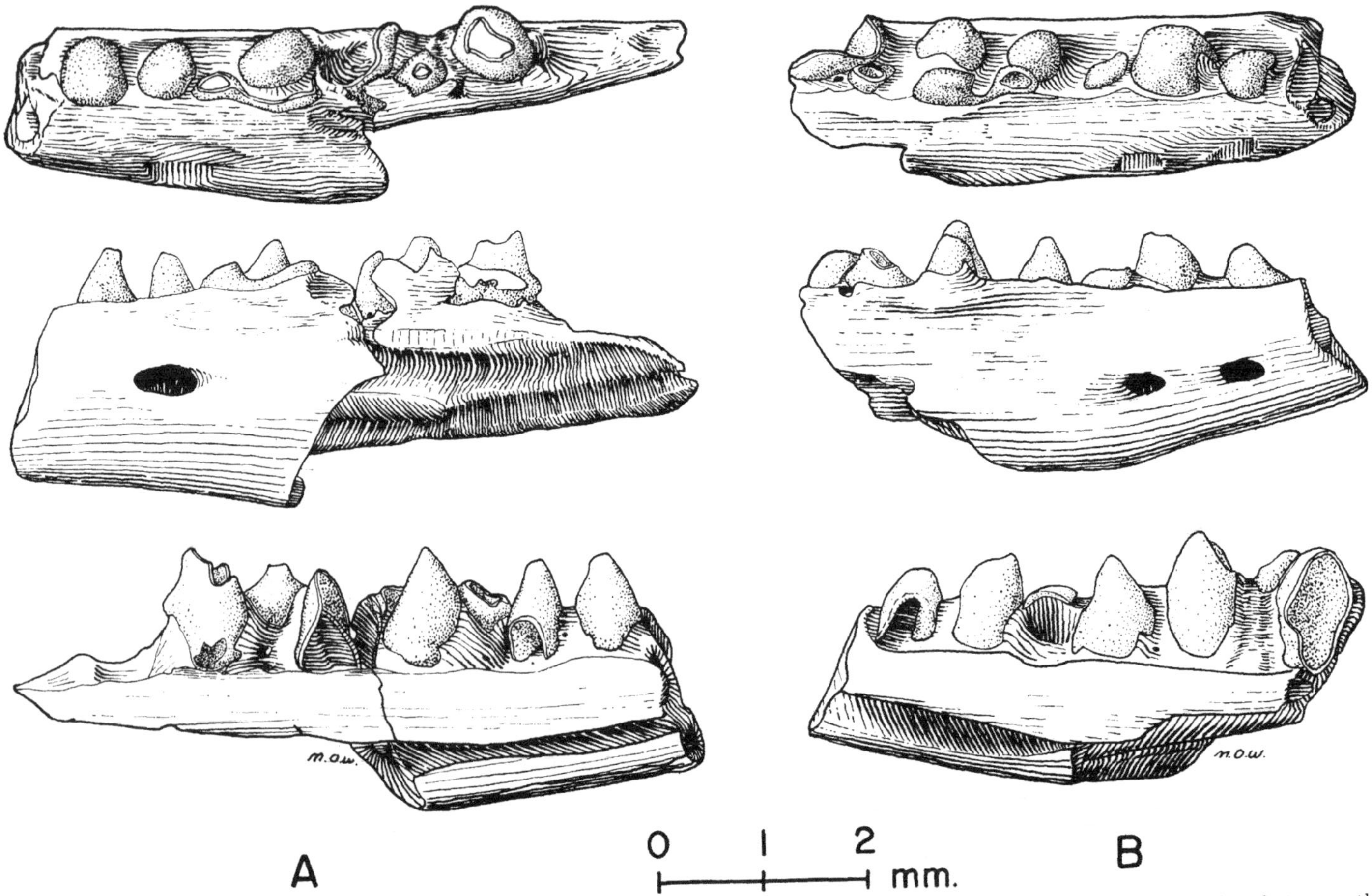

FIG. 1. A. *Crotaphytus collaris*, UMMP No. 39694, adult male, fragment of a left dentary at the region of overlap between the anterior and posterior tooth series. B. UMMP No. 33830, fragment of a right dentary from the same region.

suggested that a habitat suitable for this species was not present in the immediate vicinity of the Cragin Quarry locality during Sangamon times, a conclusion supported by the above figures for the order of abundance.

All specimens are in the University of Michigan Museum of Paleontology. The line drawings by Michael O. Woodburne were supported by the National Science Foundation (Project G-5635).

LITERATURE CITED

ETHERIDGE, RICHARD. 1958. Pleistocene Lizards from the Cragin Quarry Fauna of Meade County, Kansas. Copeia, 2: 94–101, 3 figs.

HIBBARD, CLAUDE W. 1955. The Jinglebob Interglacial (Sangamon ?) Fauna from Kansas and Its Climatic Significance. Univ. Mich. Mus. Pal., Contrib., 12(10): 179–228.

Papers of the Michigan Academy of Science, Arts, and Letters
Vol. XLV, 1960 (1959 Meeting)

GEOLOGY

AN EXAMINATION OF PARTS OF THE SOUTHERN COMPLEX NEAR PALMER, MARQUETTE COUNTY, MICHIGAN[1]

R. A. Long, R. A. Vehrs, and A. Sahakian

Michigan State University

The Southern Complex is composed of pre-Cambrian igneous and metamorphic rocks, predominantly granites, gneisses, schists, and basic intrusives. It embraces an area of approximately 1,400 square miles south of the Marquette synclinorium, bounded by folded Huronian rocks to the west and south and covered to the east by flat-lying Paleozoic sediments. For a century the nature and age of the Southern Complex have been subjects of controversy.

Van Hise and Leith, summarizing work to 1909, concluded that the granites were Archean and post-Huronian. Lamey (1931–1935) considered the granites mainly post-Huronian and pre-Keweenawan, and introduced the term "Republic Granite," which has since been more or less abandoned. Dickey (1936–1938) recognized Laurentian granite intruding Keewatin-type schists to form an injection gneiss; a later Laurentian granite porphyry; a post-Lower, pre-Middle Huronian granite (since proven to be pre-Huronian); and a post-Huronian, pre-Keweenawan granite, which he called Killarney. Tyler, Marsden, Grout, and Thiel (1940) recognized two pre-Huronian granites, a Huronian granite, and a post-Huronian granite. They determined these ages by the type of zircon present, the hyacinth variety indicating pre-Huronian and the malacon variety post-Huronian. Ayers (1943) considered the Republic granite porphyritic and post-Huronian.

Along the northern boundary of the Southern Complex, immediately south of the Palmer area is located the Palmer Gneiss. This narrow belt of highly metamorphosed rocks was considered by early

[1] The writers wish to express their appreciation to Dr. Justin Zinn for his help and encouragement throughout their study.

workers to be pre-Huronian. Van Hise and Leith (1911) relegated parts of the Palmer Gneiss to the Huronian period. Lamey (1935) concluded that essentially all of the Palmer Gneiss represented metamorphosed Lower and Middle Huronian.

This brief introduction perhaps illustrates the complexities with which one is faced in unraveling the problems of the Southern Complex. In reality, little detailed work has been done, considering the large area involved. The present writers have selected small areas near the northern border of the Complex (Fig. 1).

A part of Sec. 34, T. 47N., R. 26W. was mapped by Richard A. Long; Sec. 36, T. 47N., R. 26W. was mapped by Armen Sahakian; and Sec. 1, T. 46N., R. 26W. was mapped by Robert A. Vehrs.[1] It is hoped that the information thus obtained will be amplified by future work until a clearer and more detailed picture of the Southern Complex emerges.

Geology

The topography in Sec. 34 is rugged with good bedrock exposure. Only infrequent and scattered outcrops occur in the level sand plain of Sec. 1 and Sec. 36.

Granite.—The predominant rock in the area is an orange-colored albite granite. This granite frequently becomes gray at contacts. The granite is medium-grained and often exhibits a gneissoid appearance. In thin section the rock has a xenomorphic texture. The average composition is 35 per cent quartz, 25 per cent microcline and microperthite, 35 per cent albite, and 5 per cent biotite. Accessory minerals are zircon (malacon), pyrite, sphene, apatite, and rutile.

Quartz in the main part of the granite mass is relatively undeformed; but toward contacts it becomes quite strained and in places mylonitic. Frequently the quartz contains lines of highly birefringent elliptical inclusions which appear to show a preferred orientation.

Albite has an anorthite content of only about 3 per cent. It exhibits extensive alteration to paragonite in the entire area and locally to minor amounts of calcite. Combination twins of Albite, Carlsbad, and Pericline types are abundant.

Microcline is fresh and unaltered. Perthitic blades of albite often occur in the cleavage traces. In some parts of the granite a great

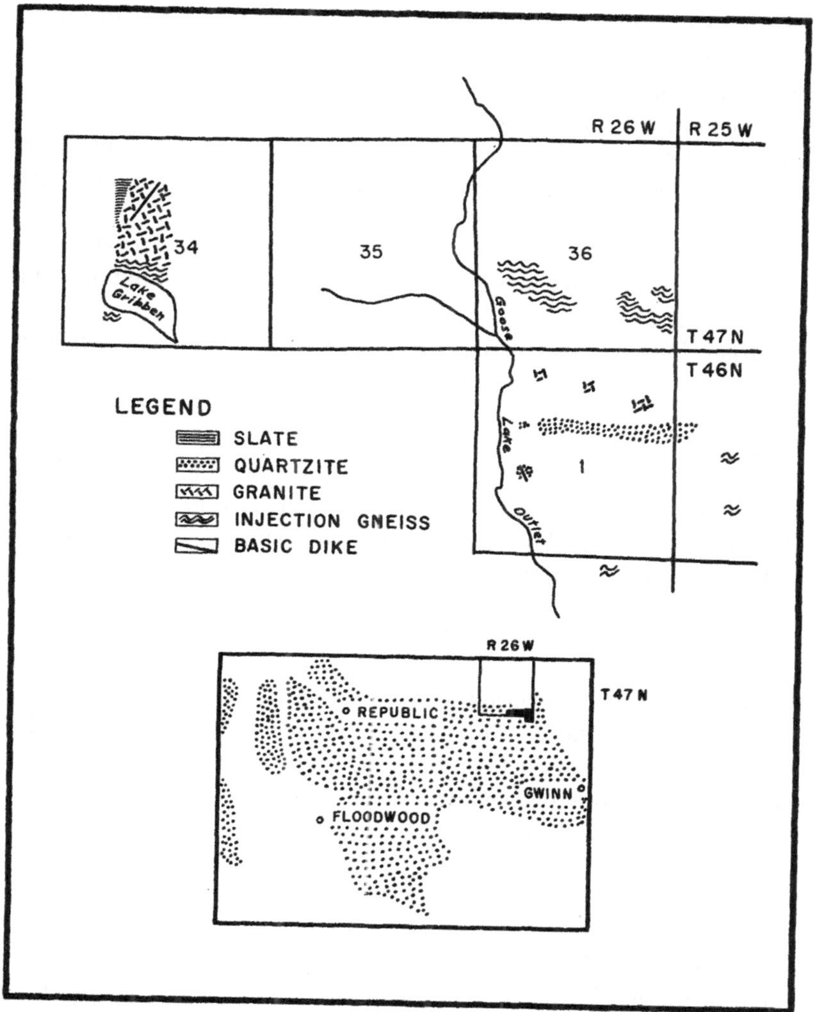

FIG. 1. Map of the area and its location in the Southern Complex of Michigan.

deal of the feldspars exhibit more complex perthites, some even showing rims of albite. Very little orthoclase was observed in any of the samples of the granite.

Biotite comprises about 5 per cent of the rock. It occurs in long, slightly oriented blades. It is always observed in some stage of alteration to penninite and magnetite, with elliptical masses of quartz developed along the cleavage. Locally it shows alteration to calcite and muscovite.

Zircon of the malacon type occurs in rounded to euhedral grains, almost always surrounded by a dark brown reaction rim. Rutile was observed locally as a sagenitic intergrowth in quartz. Sphene and apatite are locally abundant.

The granite is considered magmatic because of the occurrence of complex twin combinations in the plagioclase and the abundant complex perthites.

Other intrusives.—Cutting the granite are a brown-weathering, olivine diabase dike and numerous veins of aplite, pegmatite, and quartz.

Age of granite.—It is difficult to postulate an age for the granite without a more detailed knowledge of the surrounding area. We do know that the granite is cut by an olivine diabase dike which is probably Keweenawan. And that the granite is similar in some respects to Lamey's post-Huronian Republic granite.

Metamorphic Rocks

Contacts of the granite with several rock types were observed in Sec. 1 and Sec. 34. Toward the north in Sec. 34 the granite intrudes a green, quartzose slate. The slaty cleavage strikes E-W. and dips 70° N. Just north of Gribben Lake, again in Sec. 34, occur quartzose and graywacke-like rocks which exhibit a great deal of injection phenomena and are gneissose.

Quartzite.—A linear ridge of quartzite trends in an easterly direction across Sec. 1 and also outcrops in small hills slightly to the east and west. As shown in Figure 1, the granite also outcrops in several places in the northern part of Sec. 1. Stringers of this granite are found intruding the quartzite in the eastern portions. This quartzite throughout most of its exposure is a white, fine-grained, dense, homogeneous rock. Quartz is the only apparent mineral under the hand lens. Broken surfaces exhibit a sugary texture. The rock has been extremely deformed and mashed and is cut extensively by quartz veins. No evidence of sedimentary features was observed either in the field or in thin section. Red hematite staining is frequently seen.

Under the microscope the rock is seen to be composed entirely of small quartz fragments in a chlorite-sericite matrix. Cataclasis of the quartz is prominent in all of the sections examined. Undulatory extinction and ruptural strain shadows in the quartz grains are very prevalent. Chlorite content appears to increase with increased cataclasis, reaching 20 per cent in some samples.

A quartzite outcrop west of the ridge is extremely sheared and brecciated and in places resembles a sheared conglomerate. Under the microscope this rock is seen to be large cataclastic quartz grains and fragments in a mylonitic groundmass of quartz (0.0125 mm.) and stringy bands of chlorite.

Approaching a granite contact the quartzite becomes greenish in color and scattered feldspar grains occur. Under the microscope it is seen that chlorite becomes prevalent. The quartz grains become more deformed, and small grains of albite appear. Granite intrusion and quartz veins appear to be interrelated.

A dull-green weathering dike was found cutting the quartzite. It strikes N. 45 W. and dips 70° NE. It is highly altered and is composed of chlorite, sericite, quartz, and prevalent grains of limonite.

Age and origin of quartzite.—It would be difficult at present to give an age to this quartzite or to make a definite statement regarding its origin. It resembles a quartzite described by Lamey which occurs in the Palmer Gneiss to the northwest of Sec. 1. He considers this to be Mesnard. There is no evidence, however, of a sedimentary origin for this quartzite. Jointing relationships strongly suggest at least two major deformations. This quartzite may possibly be volcanic in origin. The age problem would be simplified if the age of the intrusive granite could be determined more conclusively.

Injection gneiss.—An injection gneiss outcrops south of Gribben Lake, and in Sec. 36, and in areas adjacent to Sec. 1. It is characterized by alternating dark bands of ferromagnesians and light bands of a granitic material.

Hornblende comprises 35 to 65 per cent of the rock. The hornblende often appears in prismatic sections, mostly subhedral to occasionally euhedral. It is often accompanied by chlorite altering from biotite. Chlorite content averages 10 to 25 per cent. Some unaltered biotite is observed. There is a profusion of sphene and epidote at the contacts of the dark and light bands. Large euhedral crystals of biotite are present.

The light bands are comprised essentially of feldspar and quartz. The dominant feldspar is albite-oligoclase; it usually constitutes 50 per cent of the light bands. Orthoclase generally amounts to 15 per cent and quartz accounts for the remainder. The feldspars generally are heavily sericitized and not distinct in the dark bands; while in the light bands they appear to be relatively fresh, with original boundaries preserved, and with Albite, Carlsbad, and Pericline twinning more distinct. Quartz appears to be mylonitized at contacts. Subhedral to rounded grains of malacon zircon are present in the light bands.

The above data seem to indicate that this rock is a lit-par-lit injection gneiss. It appears that a highly liquid intrusive material invaded the cleavage or bedding planes of a rock rich in ferromagnesians. This rock is similar to the Laurentian injection gneiss described by Dickey.

LITERATURE CITED

AYERS, V. L. 1943. Republic Area: Strategic Minerals Investigations in Marquette and Baraga Counties. Mich. Dept. Cons., Geol. Surv. Div., Prog. Rpt. 10, pp. 31–46.

DICKEY, R. M. 1936. The Granitic Sequence in the Southern Complex of Upper Michigan. Journ. Geol., 44: 317–340.

——. 1938. The Ford River Granite of the Southern Complex of Upper Michigan. Journ. Geol., 46: 321–335.

LAMEY, C. A. 1931. Granite Intrusions in the Huronian Formations of Northern Michigan. Journ. Geol., 39: 288–295.

——. 1933. The Intrusive Relations of the Republic Granite. Journ. Geol., 41: 487–500.

——. 1934. Some Metamorphic Effects of the Republic Granite. Journ. Geol., 42: 248–263.

TURNER, F. J. 1951. Observations on Twinning of Plagioclase in Metamorphic Rocks. Am. Mineralogist, 36: 581–589.

TUTTLE, O. F. 1952. Origin of the Contrasting Mineralogy of Extrusive and Plutonic Sialic Rocks. Journ. Geol., 60: 107–124.

TYLER, S. A., MARSDEN, R. W., GROUT, F. F., AND THIEL, G. A. 1940. Studies of the Lake Superior Pre-Cambrian by Accessory-Mineral Methods. Geol. Soc. America Bull., 51: 1429–1538.

VAN HISE, C. R., AND LEITH, C. K. 1909. Pre-Cambrian Geology of North America. U.S. Geol. Surv. Bull., 360: 140.

——. 1911. The Geology of the Lake Superior Region. U.S. Geol. Surv. Mono., 52.

PART II

SOCIAL SCIENCE

PAPERS OF THE MICHIGAN ACADEMY OF SCIENCE, ARTS, AND LETTERS
VOL. XLV, 1960 (1959 MEETING)

FOLKLORE

FOLKLORE AND PLANT DRUGS[1]

E. H. LUCAS

Michigan State University

WHEN thinking of folklore, most people instinctively correlate this term with something that springs from knowledge of, or usage by, ignorant people. This is only partly true. The origin of so many things that fall under the classification of folklore is obscure, and for this reason the personality of the originator has become unessential whereas the mere fact that the substance of the lore has become the property of the people is the prevalent and overshadowing essence. The possibility remains—and in the case of plant drugs and their use as remedies for diseases it seems to me to be a certainty—that individuals who stood out because of their imagination, curiosity, and inventiveness were mostly responsible for the discoveries.

It is easy to understand that word of mouth remained one of the most important means of handing down information if we consider the destructibility of written information and the great odds against retaining any records unless they were compiled in books. It is most likely, however, that whenever someone undertook the writing of a herbal, such information as had been passed on either in writing or by word of mouth became recorded therein. Very few of the old herbalists claimed to have originated the remedies. They simply state their existence and use, without mentioning the source of information. Thus it is more than likely that what we commonly call folk remedies actually were prescribed by the equivalent of the modern doctor, although in practically all cases their names have been forgotten. Yet, in many instances the remedies listed go back to the ancient Greeks and Arabs, as far as the Western World is concerned, while in the East it is the ancient Chinese and Indian civilizations that originated them. Thus, many of our folk remedies can be traced back to outstanding individuals of the civilizations of Baby-

[1] Dr. Eugene Hans Lucas died on January 28, 1959; his paper was delivered at the Annual Meeting by Dr. Joseph A. Stevens, March 28, 1959.

lon, at least twenty centuries prior to the birth of Christ, and of Assyria, of Egypt, of Greece, and of Rome. How far remedies of the Far East have entered western folklore is questionable. Much of the present-day plant lore in China is based on the classical Chinese herbal, *Pen-tsao,* compiled by Li Shih-chen in 1597. But this comparatively recent work takes much of its information from sources almost 4,000 years older (Sen-nung, 2738–2698 B.C.). It is quite likely that the great migration of Asiatic tribes into Europe may have transferred some of the medicinal plant lore into the Western World. And the fact that the Russian Empire covered large parts of Europe as well as of Asia has always created a bridge between the West and the East. In the United States the considerable amount of Indian folklore must be added. Whether this lore has developed independently or has been influenced by elements originally from the ancient East has not been investigated sufficiently, as far as I know. Bridges may have existed here too, as well as from South America, where another independent center of plant lore appears to lie. In addition, local plant lore must have developed in practically every inhabited part of the world, since we know of the use of drug plants in many parts of Africa as well as on the islands of the Pacific and the Australian continent. Further studies dealing with a comparison of these plant lores may bring out similarities which could be due to independent observation or to unknown connections.

For all these plant-lore centers it can be assumed that alert individuals have originated the use of the various plants in connection with different diseases. That names can be forgotten while facts are handed down can be easily seen by looking at the history of modern medicine. How many physicians know, for instance, the name of the originator of aspirin in spite of the fact that in this case it is not a source of a drug but a manufactured drug we are dealing with? And aspirin, even now, only half a century after its discovery, has already acquired the earmarks of a folk remedy.

If we, then, agree that many of the remedies of the past centuries were discovered by persons who were far ahead of their times and their contemporaries and who primarily were keen observers, able to discern causes and effects, we will be inclined to judge much less harshly the potential merits of these medicines. After all, they must have helped some people, and, though we have no data for comparison, the percentage of effectiveness of a plant drug may not have been far different from that which modern drugs show. We must consider one very important factor also; namely, the great amount of physiological variability among the members of the species Homo sapiens, which must have greatly increased by continuous hybridization and mutations over a period of several thousand years. We may then assume that plant remedies which had a predictable effect in a more uniform population may have much less chance when used on members of a population of greater diversity. One may even assume that some remedies which were found useful in one area of the world may have no effect whatever when administered to the population of another area. An example for the differential response of races is for instance the difference in sensitivity to poison ivy; it is known that people from the Far East, especially from China, are hardly susceptible to the poison of this plant.

If one further considers that the science of medicine, as we know it today or have known it for the past, say, seventy to eighty years did not exist, that there did not stand an army of physiologists and biochemists behind the healers as is now the case, one must admire human curiosity, intuition, and ability to observe, which uncovered so many of nature's products that could be used for human relief.

With the rapid development of chemistry in the past century, it was only natural that the original plant constituents were often replaced by other compounds, when man was able to "improve on nature"; but the original "inspiration," as it were, came from the plant. It is well known that even now we are tremendously dependent on these natural sources, and on the often scant information we may have about their effect on certain human ailments, in order to fill the gaps in our pharmacological knowledge.

The well-known instances in which a remedy was brought into use by medicine and has continued to remain are far outnumbered by those which are not commonly recognized. Remedies like the digitalis glycosides, quinine, colchicine, morphine, and the more recent Rauwolfia alkaloids, have captured the imagination of the professional man as well as the layman because theirs was, or is, an application in cases where the physician had been more or less helpless without the drug. There are innumerable other instances, however, where folklore is the basis of a commonly used drug—and again aspirin is a handy example: salicylates in plant decoctions have been used for thousands of years in cases for which aspirin is indicated.

This discussion is not intended for the purpose of bringing up the merits of accepted drugs. Its primary purpose is to mention that there exists a large number of drug plants recorded in folklore but consistently counted useless, apparently because casual administration has not shown the effects claimed for them. There are others which have never been tried out in modern times, either because the lore of them was not commonly known or because they have simply been overlooked. And finally, modern society has more or less decided to draw a line between the time when it was proper and necessary to recognize and record lore about drug plants and the enlightened time in which we live when seemingly we can dispense with such a primitive activity.

It is therefore a rewarding task to study the following three areas: (1) the drugs which have been tried in modern times but were found to be ineffective and therefore discarded, (2) the drugs which apparently have been overlooked, and (3) the folklore referring to plants which has never been recorded. I shall give an example representative of each group.

In the group comprising drug plants which have been tried off and on but obviously had not been found worthy of continued study, there is almost too wide a variety of material that might be investigated. The example I will mention shows very clearly how readily the effect of a drug may be misjudged. The case in question is quite clear-cut, because the drug is not one of those which are recommended for a dozen ailments but simply for one. A piece of old Russian folklore which is supposed to have originated in the province of Olonyets in northwestern Russia mentions a fungus growing on birch trees (popularly called birch fungus) being effective in the treatment of internal cancers. One always approaches a folk reference to cancer with utmost caution, since in most cases reference is made to external growth of any type and in many cases to nonmalignant ones. In this particular case, however, references pointed to internal growth and the descriptions allowed the conclusions that at least some of them were neoplasms. The folklore in question is found in Siberia, the Baltic area, and Finland. It seems to coincide with the areas in which the fungus occurs. The folklore records that decoctions of abortive sporophores of this fungus were given to patients in the form of a "tea" until the condition improved. Only the abortive fruiting bodies are thought to be effective, and it is claimed that these develop on birch trees only under certain environmental conditions, particularly when the trees grow on low ground.

The lore came to our attention several years ago, and we first tried to identify the fungus. There is no clear-cut evidence as yet that there may not be several species whose occasionally abortive sporophores may serve as sources of the drug. The main source, however, appears to be a fungus which was designated as *Polyporus annosus* Fr. by Dragendorff. A recent book on Russian fungi by A. C. Bondartsev lists this particular fungus as *Inonotus obliquus.*

We were able to obtain some of this particular fungus from Finland and proceeded to make preparations which were to be injected into mice in which small pieces of Crocker mouse sarcoma 180 had been previously implanted. (The test was performed by the Sloan-Kettering Institute for Cancer Research, New York City.) The preparations were made using a variety of solvents and conditions. When tested, none of these showed any indication of retarding the tumor growth. Preparations were then made by subjecting the fungus material to prolonged boiling. These preparations retarded the test tumor. Work along these lines is being continued in our laboratory at the present time. It might be said that for methodological reasons we have not tried yet to administer these preparations by mouth. But even without results from this mode of administration, it appears to be clear that boiling the plant material is a decisive factor in mobilizing the active principle. It is worth mentioning that standard methods of plant drug testing, as they may have been developed in laboratories engaged in this type of work, often dispense (as it happened in our case) with the use of heat-treated preparations—perhaps out of an old biochemical prejudice that one might destroy labile substances; it should be considered that such treatment can also activate therapeutic materials. It is quite remarkable that folklore has discovered this, in fact, seems to apply the principle quite frequently; many of the remedies listed in herbals are administered as "teas."

A search for overlooked drugs, which constitute the second group previously mentioned, would primarily single out those bypassed because too many uses for them are mentioned in folklore. Many a modern pharmacologist, biochemist, or physician immediately becomes suspicious when he reads that a certain plant drug is recommended for a wide variety of ailments. Drugs thus introduced by folklore are being set aside in favor of others that are supposed to be useful in just one respect. The reaction which guides in such a case is perfectly natural and in line with modern scientific concepts. But it often misleads. In spite of the seeming improbability of a plant being so versatile that it may contain agents effective against a variety of pathological conditions, there are examples of such plants. And I personally am convinced that there are many more of them than we realize. Rediscovering them might lead to the rehabilitation of a good number of overlooked plant drugs.

As an example of a plant in this group I should like to mention *Hypericum perforatum*, the common weed called St.-John's-wort. We had started working with this plant about seven or eight years ago, but because of insufficient supply of the plant material we needed at that time, namely the flowers, we turned to an ornamental shrub of the same genus, *H. prolificum*. We had ascertained that the flowers of this species served our purpose as well as those of the weed. I might add that this is not true of all species of the genus *Hypericum*, wild or cultivated. As mentioned, we were primarily interested in the chemistry of the flowers. The ornamental shrub produces an abundance of flowers in July and August, and we were able to collect an ample supply which was frozen and stored at 0° F. This plant had attracted our attention because of its very old reputation as a drug plant, and as early as 1948 we had established that the flowers contain a very potent antibacterial substance.

A word about the plant and its history: The origin of the name Hypericum is obscure, but one of several interpretations traces it to the Greek words "hyper" (meaning over or beyond) and "eikon" (meaning image); thus the term is supposed to express that the value of the plant is beyond imagination. However, I find considerably older designations, which make this interpretation quite doubtful. Arabic-Persian authors call a plant, which might be the same, Hinfarikun; other Arabic sources record it as Hujufarigum. That the word Hypericum (or in Greek Hyperikon) might be derived from one of these two words is possible. Be this as it may, there is no doubt that the plant has been known for thousands of years and ever since has had a reputation as a medicinal herb.

Folklore recommends St.-John's-wort against liver and kidney ailments; gout; diarrheas; hemorrhages; bronchitis; bed-wetting; and nervous headache. It is to be taken in the form of a decoction (tea) —about 1 ounce to a quart of water; 1–2 cups per day, taken slowly. The decoction is prepared from the entire top of the plant. An essential oil is extracted from the flowers (folklore prescribes that the petals be used only and the calyx and pedicel be removed) by immersion of 1 pt. of the petals in 1 qt. of olive oil, letting it stand exposed to sunlight for two weeks, and finally filtering. It is recommended externally against lumbago, sciatica, gout, and burns. Internally (6–8 drops on sugar) it is supposed to help intestinal pains and circulatory disturbances. In addition, the decoction as well as the oil are used in combination with other plant drugs against a variety of complaints.

Offhand one would be inclined to dismiss a plant recommended in this manner, as a panacea whose uselessness would be implied by the variety of applications. A study of the plant material has taught us to reconsider.

We were primarily interested in the antibacterial substance mentioned before. I will not go into cumbersome details. I merely want to mention a few points: First, we found that the antibacterial substance was present primarily in the petals of the flowers; the folklore recommendation that calyx and pedicels be discarded appears entirely justified if the antibacterial substance was the one that was sought. But this substance is by no means the only one extracted in this manner. Further tests with a very active preparation indicate that we are still dealing with an impure product. Further purification may yield two compounds, both highly antibacterial. The most active preparations were inhibitory to *Micrococcus pyogenes* at about 0.25 γ per ml. In addition, the substance is inhibitory to *Mycobacterium tuberculosis* and exceedingly active against *Xanthomonas phaseoli.* Some preparations tested were inhibitory at a level of 0.01 γ per ml.

The presence of such highly antibacterial principles may explain a number of applications which folklore recommends. Several of these seemingly unrelated conditions may be due to bacterial infections, and a highly antibacterial substance may go to the source of the trouble. Since some of the recommendations date back at least several centuries, but more likely thousands of years, this would constitute a very early utilization of antibiotic action.

There may be far more to the esteem that this plant enjoys in folklore, however. Some of the names that the plant has been given in European countries indicate the purposes for which it was thought useful. One of them is the German word "Hexenkraut," which means "witch wort," and another "Jagdenteufel," which means "chase the devil"; still another one is "Teufelsflucht," which means "devil's flight." As a matter of fact, references can be found here and there stating that the plant drives out witches, which led people in some areas of Europe to hanging up dried plants in front of their houses as a protection against evil spirits. The herb was considered so obnoxious to them that a whiff of it would cause them to fly off. This type of lore, although generally applicable to disease (as caused by evil spirits), often points specifically to mental disease. And it is interesting in this respect that there have been reports that a substance derived from *Hypericum* has been found useful in the management of nervous conditions and depressive states.

More recently, we have obtained evidence that another factor is present in the wild as well as in the cultivated plant, again distinctly different from the ones mentioned before. This factor appears to have an effect against certain viruses. An antiviral action might explain curative effects in those cases which are not due to bacterial infection. They could also interpret the action of the plant extracts in cases where mixed infections are present, cases which are often known to be aggravated by antibiotic therapy which may upset a biological balance between bacteria and viruses, thus activating viral action.

Although we have no chemical information about this substance, its nonidentity with the other compounds is established by distinct differences in solubility; the antiviral agent is water-soluble, while the other substances are insoluble in water.

The third area to be studied, potentially perhaps the largest, will require the attention of collaborators outside the field of plant science and medical sciences. Folklorists, ethnologists, anthropologists, and historians could greatly aid in this endeavor. It deals with the discovery of folklore which so far has not been recorded. It may amount to unearthing lost information, thus be dealing with old folklore; collecting information that has been passed on by word of mouth in geographically localized areas and therefore has not transpired. And, quite possibly, it may deal with the acquisition of entirely new experiences.

That new folklore can be discovered even in the twentieth century is shown by the following case: About 1934 I visited a tiny mountain village in the Bohemian Forest, just across from the Bavarian border, in an area where there is an abundance of mushrooms in the spruce-fir forests. The lumberjack inhabitants of this isolated village believed that eating one kind of mushroom (which occurs there) prevents cancer. Actually, there are very few cases of cancer found in this area, and hardly any cases among the poorer people. The people in question are rather uneducated; none of them have more than a few years of elementary schooling. They are, however, closely linked to nature and probably very good observers. They collect and dry several species of *Boleti*, but they are very outspoken in stating that only one, *Boletus edulis*, prevents cancer.

At the time when I came across the information, not being actively engaged in research in drug plants, I paid scant attention to it. Later, when working with drug plants I remembered the lore and decided to put it to a test in a way that was accessible to us.

We obtained dried material of *B. edulis* in 1948 and, in collaboration with the Sloan-Kettering Institute for Cancer Research, tested first crude water extracts, later purified preparations, against Crocker mouse sarcoma 180. The effects were quite striking, and it was soon obvious that *B. edulis* was the only one of the various *Boleti* of that region which exhibited this activity. All other *Boleti* obtained from the same area failed to show tumor inhibition. I should mention that later we did find some American boletes which had tumor-inhibitory properties, but for practical reasons (unlimited supply of *B. edulis*, but only occasional availability of the American boletes) we continued our work dealing with the identification of the active principle using the European material. As far as we know at the present time, the active material is a polypeptide containing a sequence of nine amino acids, and possibly three more.

The positive correlation between the lore and the scientific findings led us to investigate other fungi to which folklore had attributed tumor-inhibiting properties. Our most notable achievement in this investigation was correlating that folklore which attributed skin-cancer-retarding properties to the spores of puffballs. We did not find any tumor-retarding activity in the mature puffballs which contained spores, but we did find a very strong tumor-retarding agent in the immature (white) sporophores of the giant puffball, *Calvatia gigantea*.

It must be kept in mind, however, that these preliminary tests are on Crocker mouse sarcoma 180, and that many months of testing are necessary before the material will be tested against human neoplasms.

It is hardly necessary to say that the field we are dealing with is of enormous breadth and depth. Its possibilities are indicated by one interesting fact: In one of the most comprehensive compilations of medicinal plants in modern time, a book entitled *The Medicinal Plants of the Various Peoples and Ages* by Dragendorff (c. 1898), about 12,700 species are listed. Even if we assume that some species were confused and overlapped, we can consider a figure of 10,000 as reliable. All these plants were listed on the basis of folklore. If we make the very conservative assumption that only 1 per cent of these plants may have any value at all as sources of drugs, we arrive at the figure of 100. And even without going as far as Osler went when he said that he used one medicine for a dozen diseases, we could cover a fairly wide range of human ailments. But the very depth and breadth of the field to be searched also means that merely to scratch its surface will take a lot of pretty sharp plowshares.

Papers of the Michigan Academy of Science, Arts, and Letters
Vol. XLV, 1960 (1959 Meeting)

GEOGRAPHY

CHANGES IN SUGAR BEET PRODUCTION IN MICHIGAN, 1899–1958

JOHN M. BALL

Central Michigan College

Change, which has characterized the Michigan sugar beet industry during its first sixty years, has never been greater than in the last decade. Many of its aspects warrant intensive study; for example, the shifts in governmental policy toward beet sugar production,[1] the lessening reliance of the beet farmer on migrant labor,[2] the improving of beet varieties and farming techniques. But the purpose of this paper is to trace only the changes in beet production, in producing areas, and in the location of processing plants in Michigan from 1899 to 1958.

If the comments of people in some parts of the state are valid, the sugar beet industry of Michigan is dying.[3] Actually, although annual production of sugar beets has fluctuated greatly in the past, 1958 with a harvest of 1,112,000 tons was one of the better years for the industry (Fig. 1).[4] Any year that a million or more tons of beets are produced is a good one. There have been only fourteen of these since 1909. The acreage devoted to sugar beets has also fluctuated but

[1] John E. Dalton, *Sugar: A Case Study of Government Control* (New York: Macmillan Co., 1937), p. 25. I am indebted to Professor Paul C. Morrison, Department of Geography, Michigan State University, for his helpful criticism of this paper.

[2] James Rowen, "Mechanization of the Sugar Beet Industry of Scottsbluff, Nebraska," Economic Geography, XXIV (July, 1948), 180.

[3] In 1955 Michigan ranked fourth in the United States behind California, Colorado, and Idaho in tons of sugar beets harvested. U.S. Department of Agriculture, *Agricultural Statistics, 1956* (Washington: U.S. Government Printing Office, 1957), p. 71.

[4] Source: years 1899 and 1909, U.S. Bureau of the Census, *Thirteenth Census of the United States: 1910. Agriculture* (Washington: U.S. Government Printing Office, 1911). For other years, from the files of Michigan Cooperative Crop Reporting Service, 210 Post Office Building, Lansing, Michigan, C. J. Borum, in charge.

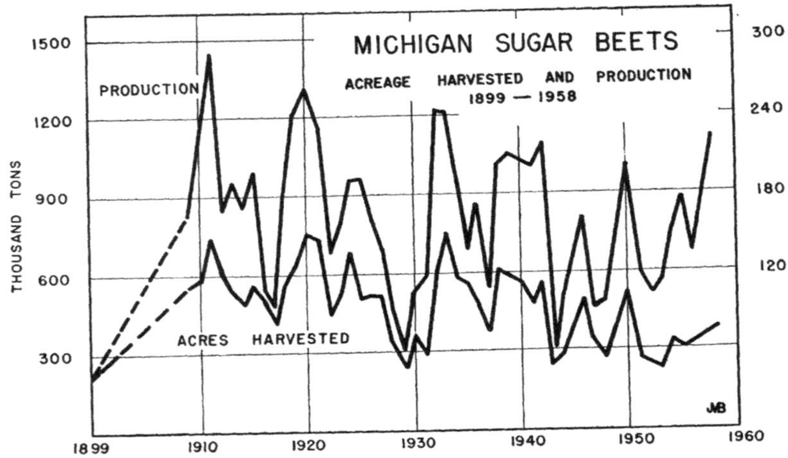

FIG. 1.

TABLE I

SUGAR BEET PRODUCTION DATA FOR MICHIGAN

	1899	1909	1919	1929	1939	1949	1958
Counties producing beets	52	57	47	35	44	37	26
Less than 1,000 t.	31	25	18	12	13	10	4
1,000 to 10,000 t.	17	17	12	15	12	13	10
10,000 to 50,000 t.	3	10	10	8	12	9	8
50,000 to 100,000 t.	1	4	3	0	5	4	0
Over 100,000 tons	0	1	4	0	2	1	4*
Ton yield per harvested acre	8.9	7.3	9.8	5.8	8.6	9.6	15.6†
Average ton yield per acre previous decade	—	—	8.2	7.7	8.2	8.7	12.3
Sugar plants available for operation	9	17	16	16	16	13	5

* Three of four produced over 200,000 tons and the four together 81 per cent of Michigan's total.

† An all-time high yield per acre for Michigan.

seems to have become more nearly stable the last few years. During these same years, yield per acre has increased rapidly, producing more beets from fewer acres (Table I).[5]

[5] Source: *Ibid.*, plus data on sugar plants from, U.S. Department of Agriculture, Sugar Branch, *Beet Sugar Factories of the United States* (Washington: Mimeographed, 1950), pp. 17–18.

Producing Areas and Processing Centers

In 1899 the sugar beet business was new to Michigan.[6] The first successful sugar refinery in the state had started operating at Bay City the previous year.[7] During 1899 eight more plants were opened —at Benton Harbor, Kalamazoo, Rochester, Holland, Alma, Caro, Bay City, and West Bay City (Fig. 2). Cultivation of beets was widely distributed over the southern peninsula, although eighteen counties produced nothing and twenty-nine others provided less than one thousand tons each. The counties in which processing plants were located, plus adjoining counties, ranked highest. Bay was the only county that produced more than 50,000 tons of beets, with the near-by counties of Tuscola, Saginaw, and Gratiot the next largest producers.

Thus, at the turn of the century there was a widespread distribution of beet-producing areas. These supplied nine scattered beet refineries that had just begun operation and were struggling for a supply of beets, the technical knowledge, and the capital necessary to keep in business. The Saginaw Bay region had the greatest concentration of beet production and sugar plants, with southwestern Michigan an area of secondary importance.

The years from 1899 to 1909 were ones of rapid expansion of the sugar industry in Michigan. Twelve new plants were constructed—at Blissfield, Mount Clemens, Marine City, Croswell, Sebewaing, Carrollton (Saginaw), Lansing, Owosso, St. Louis, Bay City, Charlevoix, and Menominee (Fig. 2). During this same time plants were closed at Benton Harbor, Kalamazoo, Rochester, and one at Bay City, leaving a total of seventeen at the end of the period.[8] Sugar beet produc-

[6] For a good early description of the sugar beet industry in Michigan see: F. A. Stilgenbauer, "The Michigan Sugar Beet Industry," *Economic Geography*, III (October, 1927), 486–506.

[7] Actually the first sugar beet plant in Michigan was constructed and put into operation at White Pigeon, St. Joseph County, in 1839, but it ceased operation in the following year. This was a community-instigated and -financed project which, like many other sugar plants in the early days of the industry, failed for lack of knowledge of the extraction process and because of the inability to get a supply of beets. Farmers and Manufacturers Beet Sugar Association, *The Story of Beet Sugar* (Saginaw: by the author, 1953), p. 5.

[8] Two other plants, one at Saginaw and one at East Tawas, were built, put into operation, failed, and were removed during this decade, and thus do not appear for the year 1909.

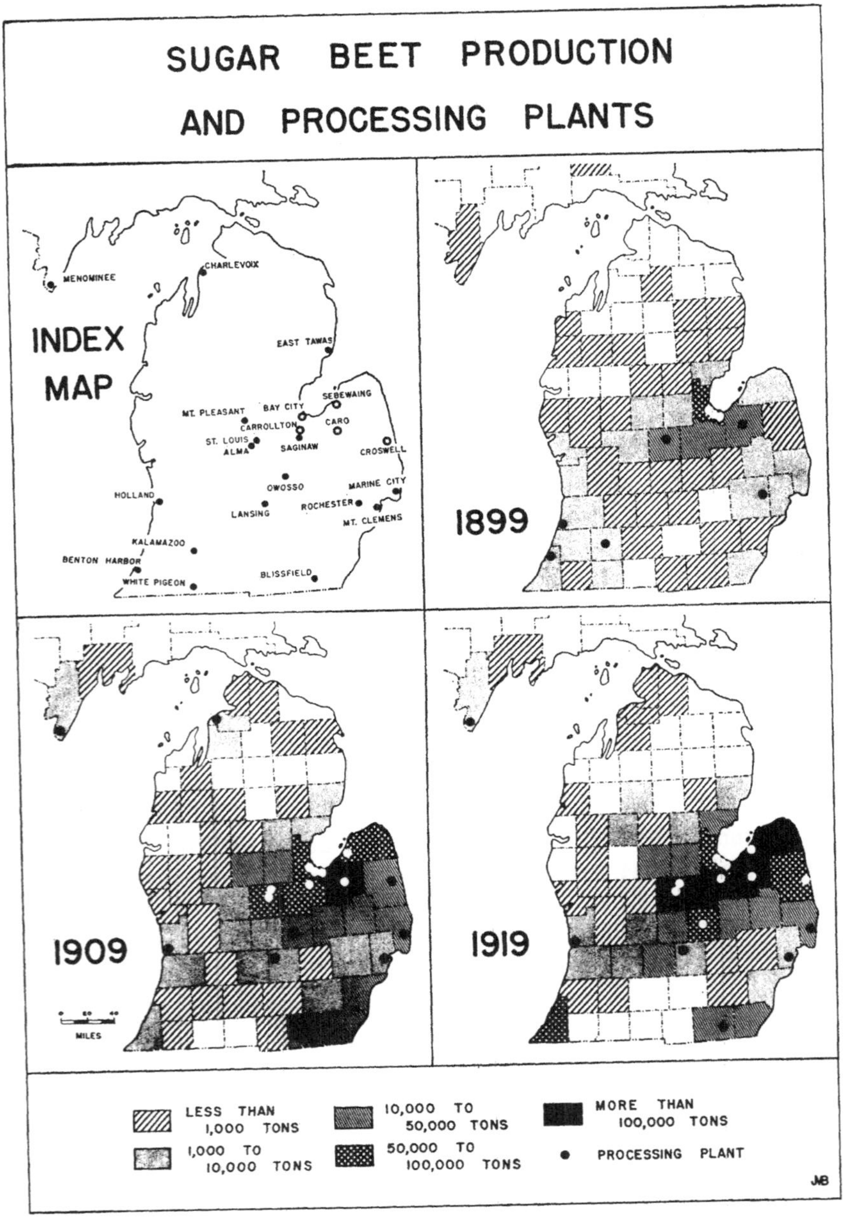
SUGAR BEET PRODUCTION
AND PROCESSING PLANTS
INDEX
MAP
MENOMINEE
CHARLEVOIX
EAST TAWAS
SEBEWAING
MT. PLEASANT
BAY CITY
CARROLLTON
CARO
ST. LOUIS
ALMA
SAGINAW
CROSWELL
OWOSSO
MARINE CITY
HOLLAND
LANSING
ROCHESTER
MT. CLEMENS
KALAMAZOO
BENTON HARBOR
WHITE PIGEON
BLISSFIELD
1899
1909
1919
MILES
LESS THAN 1,000 TONS
1,000 TO 10,000 TONS
10,000 TO 50,000 TONS
50,000 TO 100,000 TONS
MORE THAN 100,000 TONS
PROCESSING PLANT
JMB

FIG. 2.

tion more than doubled during the decade, with fifteen counties producing more than 10,000 tons of beets each, as compared to four ten years earlier. Much of this increase was in the Saginaw Bay and Thumb area. This section was the center of greatest concentration for the industry in 1909, although there was also significant production in southeastern Michigan. There was still a general wide distribution, with twenty-five counties producing less than one thousand tons each. Charlevoix and Antrim counties in the northern part of the southern peninsula and Menominee County in the northern peninsula were producers of some local importance. Generally, whereever a sugar beet plant was established, the local beet production increased. And, not infrequently, the inability of the area to provide raw material adequately or quickly enough for the new plant has been an important cause of its failure.

Although there was great fluctuation in production of beets during the decade 1910–19, there was little change in the areal distribution of production or in the location and number of processing centers (Figs. 1 and 2). No new refineries came into operation, and only the one at Charlevoix was removed. The Saginaw Bay and Thumb area continued to gain in significance relative to other parts of the state. In 1919 seventeen counties produced 10,000 or more tons of sugar beets, a growth of two over 1909, while the number of counties producing less than 1,000 tons declined from twenty-five to eighteen during the same period.

The next ten-year period, 1920–29, ended with the lowest production of sugar beets since 1899 (Fig. 1). Cultivation of beets had declined considerably in all sections of the state and had disappeared from the northern part of the southern peninsula (Fig. 3). Sixteen plants were still available for operation, although several had been idle for varying periods of time. The one at Holland, for example, did not operate from 1926 to 1933, reflecting the decline in beet production in the western part of the state. The only changes in plant location were the completion in 1920 of the new plant in Mount Pleasant, after thirteen years of intermittent work, and the closing of the one at Marine City.

Between 1929 and 1939 there were no changes in the number or location of sugar refineries (although several were not operating at the beginning of the decade), but beet production recovered from the 1929 slump and remained strong for most of this period (Figs. 1 and

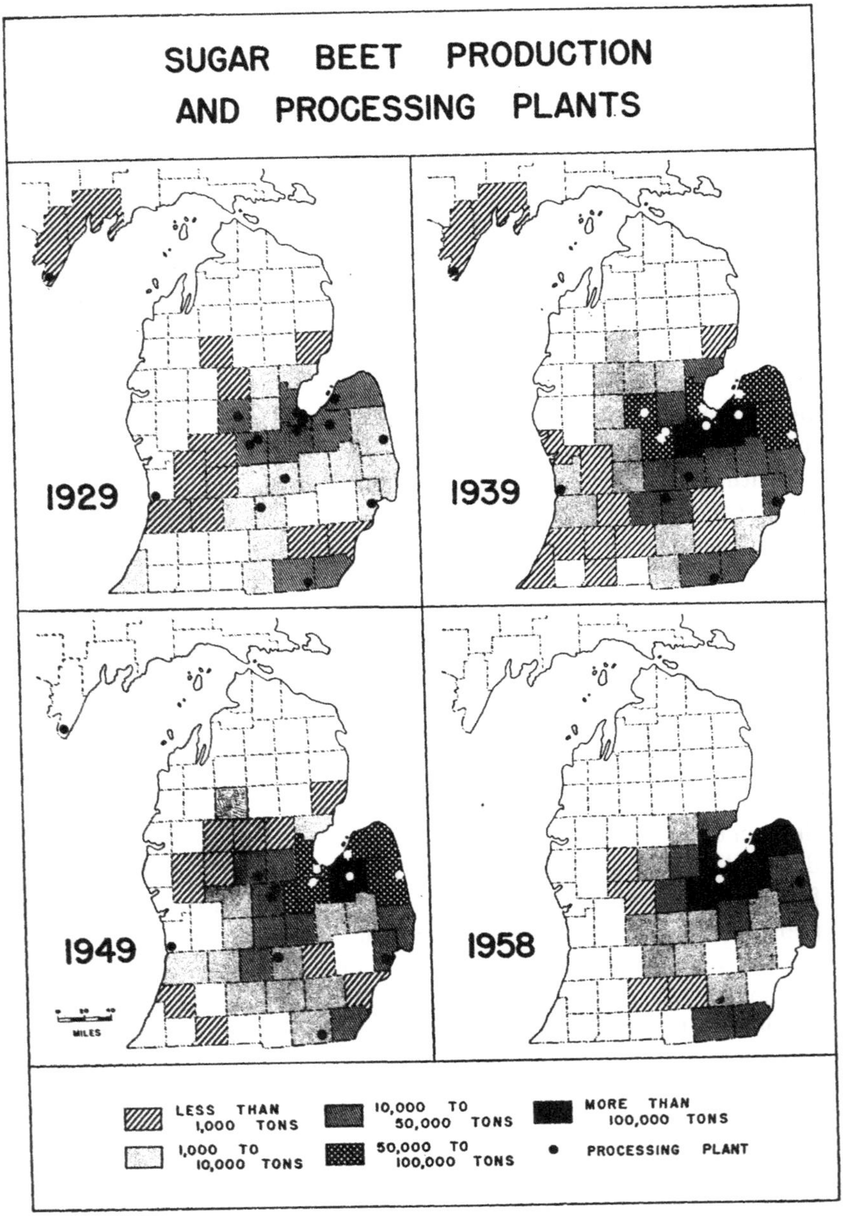
SUGAR BEET PRODUCTION
AND PROCESSING PLANTS
1929
1939
1949
1958
MILES
LESS THAN 1,000 TONS
1,000 TO 10,000 TONS
10,000 TO 50,000 TONS
50,000 TO 100,000 TONS
MORE THAN 100,000 TONS
PROCESSING PLANT

FIG. 3.

3). The depression days of the 1930's, with the surplus of cheap farm labor, and the new, special contract made in 1931 between grower and processor, whereby the two parties shared in the sugar, pulp, and molasses, helped to revive the industry.[9] More counties produced beets in 1939 than in 1929, but these were fewer than during the earlier decades of the century. Nineteen counties now produced 10,000 tons or more. Production and processing of beets tended to concentrate in the Thumb and Saginaw Bay area more than ever before.

From 1938 to 1942 inclusive, beet production exceeded a million tons each year. But the 1943 production was as low as it had been in 1929. By the end of the decade, production had partially recovered from this wartime slump. In 1949 the concentration of sugar beet production and the location of processing plants remained essentially unchanged from 1939, although there were seven fewer counties producing beets and the number of counties producing less than 1,000 tons had declined from thirteen to ten (Fig. 3). Three refineries closed during this decade—at Bay City, West Bay City, and Owosso where the plant had not operated since 1933. Several others that had closed during World War II were not working in 1949, although still available for operation. Owners of these, in most cases, were hoping for a post-war recovery of sugar beet production in their respective areas, but this did not materialize. For the first time no beets were produced in the northern peninsula where the refinery at Menominee had been idle since 1942, and the fact that the plant at Holland last operated in 1942 is reflected in fewer counties producing beets in western Michigan.

There is little question that the greatest changes in the sugar beet industry in Michigan have come during the 1950's. Gone are the processing plants at Blissfield, Mount Clemens, Lansing, Holland, Alma, St. Louis, Mount Pleasant, and Menominee, leaving only five operating in 1958 (Fig. 3). Located at Caro, Sebewaing, Carrollton (Saginaw), Bay City, and Croswell, each of the refineries except the one at Croswell is in a leading beet-producing county and all are in the Saginaw Bay and Thumb area.

This region was most important for beet production in Michigan from the earliest days of the industry and has become more and more concentrated with the passage of time. The number of counties pro-

[9] George N. Fuller, *Michigan Centennial History*, Vol. I (Chicago: Lewis Publishing Co., 1939), p. 481.

ducing sugar beets decreased from thirty-seven in 1949 to twenty-six in 1958, whereas those yielding less than 1,000 tons of beets declined from ten to four (Table I). Not only are sugar beets now grown in fewer counties, but the relative importance of the counties with the large production is greater than at any time this century. Beets harvested in these four counties amounted to 901,275 tons, or 81 per cent of the total for the state. These counties and the tons of beets each produced were Tuscola (269,703), Saginaw (218,954), Huron (218,-408), and Bay (194,210). It is significant that three counties actually produced more than 200,000 tons each. Counties in southeastern Michigan, which formerly supplied the now defunct Blissfield sugar plant, shipped their beets to plants in Ohio, so they still formed a relatively important producing area.

Summary and Conclusions

I have attempted to demonstrate that although sugar beet production in Michigan has fluctuated widely in the last sixty years, it is still of as great, or nearly as great, importance today as at almost any time in the past. Meanwhile, as the industry has become more highly specialized, requiring larger capital investments and increased production by both grower and processor, marginal producing areas and processing plants have ceased to function. Both the average yield per acre of sugar beets and the average beet acreage cultivated per grower have increased significantly in recent years. Many farmers are now cultivating over fifty acres and some more than 200 acres of beets. This is quite a different scale of activity than was common prior to World War II, with the result that the number of producers has declined without lowering production.

The decline in number of growers, in counties producing beets, and in processing plants, without lower output, are all aspects of the long trend toward increased concentration of the industry in the Saginaw Bay and Thumb area. In the future this concentration promises to become even more pronounced, with the areal extent of beet cultivation shrinking further toward that part of the state. Moreover, it is likely that the number of growers will continue to decline and their average beet acreage to increase, as their operations become even more mechanized. Judging from the past, wide fluctuations in sugar beet production can be expected in the years to come, but the industry is hardly a dying one, nor will it probably decrease much in importance in the foreseeable future.

Papers of the Michigan Academy of Science, Arts, and Letters
Vol. XLV, 1960 (1959 Meeting)

THE CHICORY INDUSTRY OF MICHIGAN

EARL J. SENNINGER, JR.

Flint Community Junior College

The ground root of the chicory plant, *Chichorium intybus L.*, a tall herbaceous perennial, is used by the beverage, confection, condiment, and baking industries of the United States. The general public appears to know very little about this agricultural product except that it can be used as an additive to coffee, of which it is said to improve the flavor. But chicory has also been used by manufacturers of chocolate, soybean, and Worcestershire sauces for flavor, body, and color.[1] Since chicory is hygroscopic (absorbs and retains moisture from the atmosphere) it is used by home and commercial bakers to keep baked goods fresh. At one time the tobacco industry used chicory to help keep its products moist.[2] Although the chicory coffee market is most highly concentrated in the Gulf and Atlantic states, it extends to all parts of the United States.

Chicory was a significant cash crop in Michigan between the 1890's and 1954.[3] During this period the Saginaw Valley and Thumb areas of the state accounted for between 90 and 100 per cent of the chicory production in the United States (Fig. 1). Since 1954, there has been no commercial production of chicory in Michigan or any other state in the United States, and it also appears that chicory will not be grown again in the United States unless European competition can be minimized.

The purpose of this paper is to describe the development and importance of the chicory industry of Michigan from the 1890's to 1954 and the causes and effects of its collapse. This study is necessary because several current publications of wide circulation dealing with Michigan geography and American agriculture continue to describe this industry in the present tense.

[1] Lloyd Stouffer, "Production, Processing and Uses of Chicory," (*Food Industries* (October, 1943), n.p.

[2] *Ibid.*

[3] Floyd A. Stilgenbauer, "Chicory: Michigan's Infant Monopoly Crop." *Economic Geography,* VII (January 1931), 84–100.

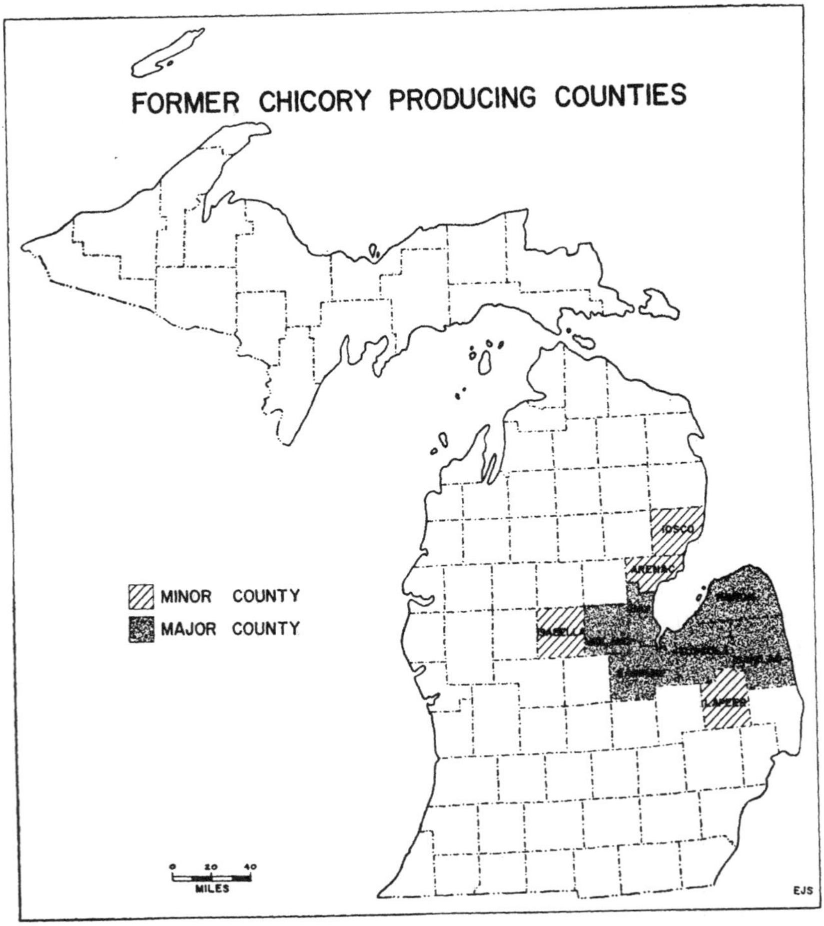

FIG. 1. Former chicory producing counties.

HISTORICAL DEVELOPMENT

Before the 1890's there was no known commercial production of chicory in the United States. The demand for this product increased as large numbers of Europeans immigrated to this country; the initial demand for chicory was the result of the food and beverage habits of the country's immigrant population. All chicory consumed in the United States prior to the 1890's had been imported from Europe.

Sometime during the early 1890's chicory was first grown in the

Saginaw Valley and Thumb areas of Michigan. Production of this crop in these areas was influenced by three factors: (1) nationality of the immigrant farmers, (2) climate, and (3) soil.

The national backgrounds of the immigrant farmers who settled these areas of Michigan were the most important single factor in the development of the chicory industry. These people primarily came from Poland, Germany, and Belgium—countries well known for production and consumption of chicory. Climate and soil were of secondary importance because chicory and sugar beets require approximately identical growing conditions, and since this was so, it was possible to produce chicory in the existing sugar beet areas of the United States. Chicory has also been grown in New Jersey, Wisconsin, Illinois, Nebraska, and California. The total chicory production in these states, however, was insignificant when compared to that of Michigan. The best chicory yields in Michigan were obtained from fertile, well-drained, clay loam soils.[4] The moderate summer temperatures and the well-distributed rainfall of Michigan's Saginaw Valley and Thumb areas were ideal for growing chicory.[5] In addition to the similarity of growing conditions, there is also a close resemblance in the physical appearance of chicory and sugar beets.

The first chicory production in Michigan also coincides with increased federal protection of small domestic industries by protective tariffs. A tariff of 2 cents per pound was enacted in 1890 on burned, roasted, ground, or granulated chicory imports.[6] In 1897, the tariff was increased to 2½ cents per pound on processed chicory,[7] and 1 cent per pound on the dry roots. These tariff rates obviously had a positive effect on Michigan's infant chicory industry.

As early as 1899, there were 1,143 farms reporting chicory production in the United States.[8] Although it is impossible to determine how many of these farms were in Michigan, a conservative estimate would place at least 90 per cent in the state. This estimate also applies for the data in Table I.

[4] C. E. Cormany, *Chicory Growing in Michigan*, Special Bulletin No. 167 (East Lansing: Agricultural Experiment Station, Michigan State College, December, 1927), p. 4.

[5] *Ibid.*

[6] Stilgenbauer, p. 84.

[7] *Ibid.*

[8] U.S. Department of Commerce, *Census of Agriculture*, Vol. II, Chapter 7 (Washington: Government Printing Office, 1954), p. 577.

TABLE I

UNITED STATES CHICORY PRODUCTION*

1919, 1929, and 1939

Years	Growers	Acreages	Green Root Tonnages
1919	480	1,784	9,010
1929	774	3,909	16,461
1939	784	3,329	22,053

* U. S. Tariff Commission, *Chicory, Ground, Or Otherwise Prepared;* Report on Escape-Clause Investigation No. 33 under Section 7 of the Trade Agreements Extension Act of 1951 (Washington: September, 1954), p. 24.

TABLE II

MICHIGAN CHICORY PRODUCTION*

1948–1958

Years	Growers	Acreages	Green Root Tonnages	Dollar Value
1948	202	652	3,549	70,980
1949	120	596	3,124	62,480
1950	176	965	5,792	127,424
1951	117	904	5,495	120,890
1952	131	892	4,560	91,200
1953	121	734	3,147	62,940
1954	111	876	4,569	91,380
1955–58	0	0	0	0

* H. G. McMorran, E. B. Muller & Company, Port Huron, Michigan, personal letter, May 21, 1958.

Tables I and II indicate that chicory could never have been considered a major crop. Farms usually devoted from one to five acres to chicory; since cultivation requires a great deal of hand labor, the proportion of acreage was relatively small. A comparison of Tables I and II also shows production increases from 1919 to 1939, and a decline after World War II. Prior to 1941 there were, according to McMorran, normally 3,500 to 5,000 acres of chicory grown in Michigan by 700 to 1,000 farmers, returning $300,000 to $400,000 annually.[9] Chicory was clearly a specialty crop of great significance to the farmers producing it.

The ten counties prominent in Michigan's chicory production are shown in Figure 1. It will be noted that Midland, Bay, Saginaw, Tuscola, Huron, and Sanilac were the major producing counties, and Iosco, Arenac, Isabella, and Lapeer were of minor importance.

Processing

An elaborate system was developed for gathering chicory roots after harvest. Before truck transportation was common, the chicory areas had many weighing stations ten to twelve miles apart along the railroads. These stations were so located that farmers could quickly transport their chicory crop by wagon with a minimum of shrinkage, since green chicory roots contain approximately 75 per cent moisture.

After being weighed, the chicory roots were then loaded into railroad cars and shipped to the drying plants, which reduced their bulk and eliminated spoilage. A total of nine drying plants are known to have existed in Michigan. These plants were well distributed throughout the growing area (Fig. 2), though most of the nine plants were small and operated only during the harvest season.[10] After truck transportation became common, the weighing stations were abandoned and the green roots were trucked directly from field to drying plant. Several processing steps followed drying—roasting, grinding, and packaging. Most of the Michigan crop was processed in a factory at Port Huron. A small amount, however, was sent to a factory in Flushing, New York.

Foreign Competition

Today 100 percent of all chicory used in the United States is imported from Europe, apparently the result of changes in tariff regulations during 1951. Crude chicory enters the United States under Tariff Paragraph 776 at 1 cent per pound, the duty having been reduced in trade agreement negotiations from 2 cents per pound, the rate established in the Tariff Act of 1930. Chicory, ground, or otherwise

[9] H. G. McMorran, E. B. Muller & Company, Port Huron, Michigan, personal letter, February 6, 1958.

[10] *Ibid.*, personal interview, July 14, 1958.

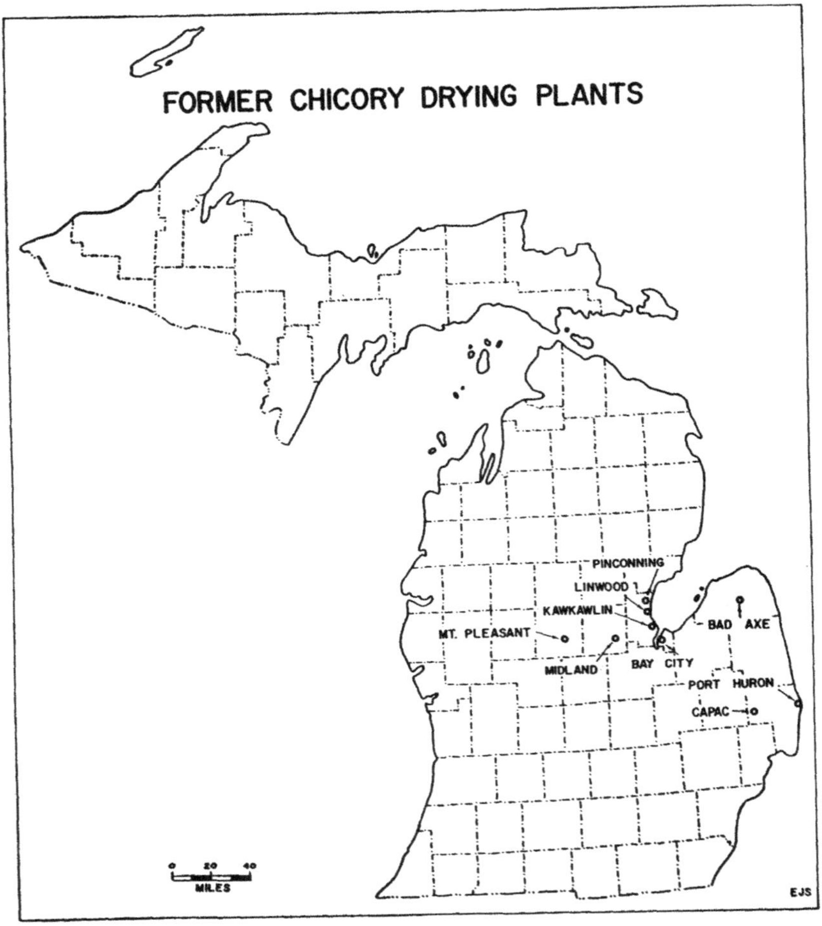

FIG. 2. Former chicory drying plants.

prepared, now enters under the same paragraph at 2½ cents per pound, the duty having been similarly reduced from the 1930 level of 4 cents per pound.[11]

It is noteworthy that European exports of chicory to the United States especially from Belgium, France, and the Netherlands, the

[11] H. P. Macgowan, Advisor on Trade Agreement Policy, Office of Economic Affairs, U.S. Department of Commerce, Washington, D.C., personal letter, January 3, 1958.

TABLE III

United States Imports of Ground and/or Prepared Chicory*
1937–39 and 1946–56

Year	Belgium	Netherlands	France	All Others	Total
		Quantity (pounds)			
1937	3,448	100	551	1,529	5,628
1938	104,877	—	551	1,005	106,433
1939	105,390	48,973	—	422	154,785
1946	—	201	—	—	201
1947	883,152	29,335	—	—	912,487
1948	1,132,523	2,112	—	—	1,134,635
1949	1,449,858	127,530	—	—	1,577,388
1950	1,997,183	697,469	—	—	2,694,652
1951	3,293,599	1,137,916	—	98	4,431,613
1952	4,080,944	1,009,224	500	—	5,090,668
1953	5,406,546	808,515	295,075	98	6,510,234
1954	5,343,613	724,507	3,521,625	—	9,589,745
1955	6,009,039	338,094	4,007,925	—	10,355,058
1956	6,893,941	411,094	6,860,175	—	14,166,004
		Value (dollars)			
1937	189	33	35	271	528
1938	2,833	—	32	104	2,969
1939	2,596	1,169	—	45	3,810
1946	—	22	—	—	22
1947	102,883	3,464	—	—	106,347
1948	87,567	198	—	—	87,765
1949	95,020	7,709	—	—	102,729
1950	124,263	42,503	—	—	166,766
1951	218,155	79,363	—	30	297,584
1952	226,907	66,857	38	—	293,802
1953	320,887	49,759	19,565	27	390,238
1954	555,873	63,981	380,283	—	1,000,137
1955	383,565	25,181	296,433	—	705,179
1956	412,902	27,620	534,700	—	975,222

* U. S. Tariff Commission, *Chicory, Ground, Or Otherwise Prepared*; Report on Escape-Clause Investigation No. 33 under Section 7 of the Trade Agreements Extension Act of 1951 (Washington: September, 1954), p. 38.

TABLE IV

PRICES PAID TO FARMERS FOR GREEN CHICORY ROOTS*

1951–1953

(In dollars per short ton)

Year	Belgium	France	Netherlands	United States
1951	$9.98†	$15.67	$9.29	$20.00
1952	9.98†	15.67	9.55	20.00
1953	9.98†	15.67	10.75	20.00

* U. S. Tariff Commission, *Chicory, Ground, or Otherwise Prepared*; Report on Escape-Clause Investigation No. 33 under Section 7 of the Trade Agreements Extension Act of 1951 (Washington: September, 1954), p. 61.

† Average price for the period 1947–53.

largest suppliers, increased markedly after reduction of the tariff rates (Table III). Reduced tariff rates as yet do not directly apply to Communist or Communist-dominated nations. According to Mr. H. G. McMorran, however, reduced tariffs have helped kill Michigan's chicory industry, since chicory grown in the U.S.S.R. and Poland comes through Belgium and Holland and is marked as the product of those countries; hence taking the lower tariff rates. In any event, Mr. McMorran believes, it is impossible for American growers and labor to compete with the lower costs of those countries.[12]

The ability of these European nations to compete favorably with Michigan's chicory production is well evidenced in Table IV, which indicates the large differentials between the cost of producing a ton of chicory root in Europe and that in the United States. A comparison of wage rates in the chicory-processing industry of Europe and the United States also reveals the same thing. In 1953, the average hourly wage in American chicory-processing plants was $1.40 whereas the European averages varied from 32 cents to 52 cents an hour for the same year.[13]

At present, 85 percent of the chicory entering the United States is ground by European processors because of the existing differential in

[12] H. G. McMorran, personal letter, *op. cit.*

[13] U.S. Tariff Commission, *Chicory, Ground, or Otherwise Prepared*; Report on Escape-Clause Investigation No. 33 under Section 7 of the Trade Agreements Extension Act of 1951 (Washington: September, 1954), p. 62.

wages.[14] The remaining 15 per cent arrives in unprocessed form and is then ground in this country.[15]

Effects

The end of the domestic chicory industry has meant a loss of 50 full-time, 200 seasonal, and 300 to 500 migrant jobs in Michigan.[16] During an average year, sales of processed chicory normally ran between ten and twelve million pounds, which sold at an average price of 10 cents per pound. Therefore, the monetary loss of Michigan chicory sales has been at least $1,000,000 annually,[17] and the former chicory root producers have had to make crop adjustments on their farms to compensate for the loss of this specialty crop.

Conclusions

The growing and processing of chicory was, for more than fifty years, important to the economy of the Saginaw Valley and Thumb areas of Michigan. The origin, development, and collapse of the chicory industry was influenced by a combination of social, geographic, economic, and political factors. This industry completed its life cycle in 1954, and the evidence presented in this paper indicates that a rebirth of chicory production should not be anticipated.[18] In addition, the imports of Russian and Polish chicory which pass through Western Europe show how Communist nations can indirectly defeat tariff regulations of the United States.

[14] McMorran, personal interview, *op. cit.*

[15] *Ibid.*

[16] *Ibid.*

[17] *Ibid.*

[18] Most of the information for this study was obtained from Mr. H. G. McMorran of E. B. Muller & Company, and Professor H. L. Kohls of the Department of Farm Crops, Michigan State University. I am indebted to the Faculty Research Fund of the Flint Community Junior College for support in this research.

Papers of the Michigan Academy of Science, Arts, and Letters
Vol. XLV, 1960 (1959 Meeting)

THE SEQUENT OCCUPANCE OF SKAGWAY, ALASKA

DONALD EUGENE BRUYERE

Modesto Junior College

Skagway is a small town situated in the southeastern section or "Panhandle" of Alaska. Several factors have been influential in varying the human occupance of Skagway. The town's economies, population, distribution of population, and pattern of buildings, roads, and railroads have undergone considerable changes and modifications since the town was founded. Yet these changes in Skagway have been neither more numerous nor severe than the economic changes that have occurred there.

The effects of different economic eras upon man in determining how and where he is to live provide a basis for a study of sequent occupance. It is worth while to understand each pattern of human settlement in a series of economic eras because each era in turn has determined not only the "how" and the "where," but also the "why there."

Skagway lies at the head of the Lynn Canal, the long fiord of the northern end of the Inland Passage, and at the lowest pass to the Yukon Valley (Fig. 1). Strategic location gave the town a greater advantage and greater prominence during the gold-mining days and again during World War II than it enjoys at present. The prominence of Skagway during the two periods is the result of many forces, some of which are still operative. Others, though inactive at present, have left traces in Skagway and may possibly come alive again in another boom.

The population curve shows clearly the eras of prosperity and decline in Skagway (Fig. 2). The curve also indicates a possible new upward trend of the population. Two years, 1898 and 1943, were population peaks. The years 1929 and 1950 were low points. The rise of 1943 did not nearly reach the level of 1898, and the decline of 1950 was not so great as that of 1929. Though the second era of population growth was not so great as the initial boom of 1898, the second period of decline has tended to be less drastic than the previous one, and it is possible that the extremes of population growth and de-

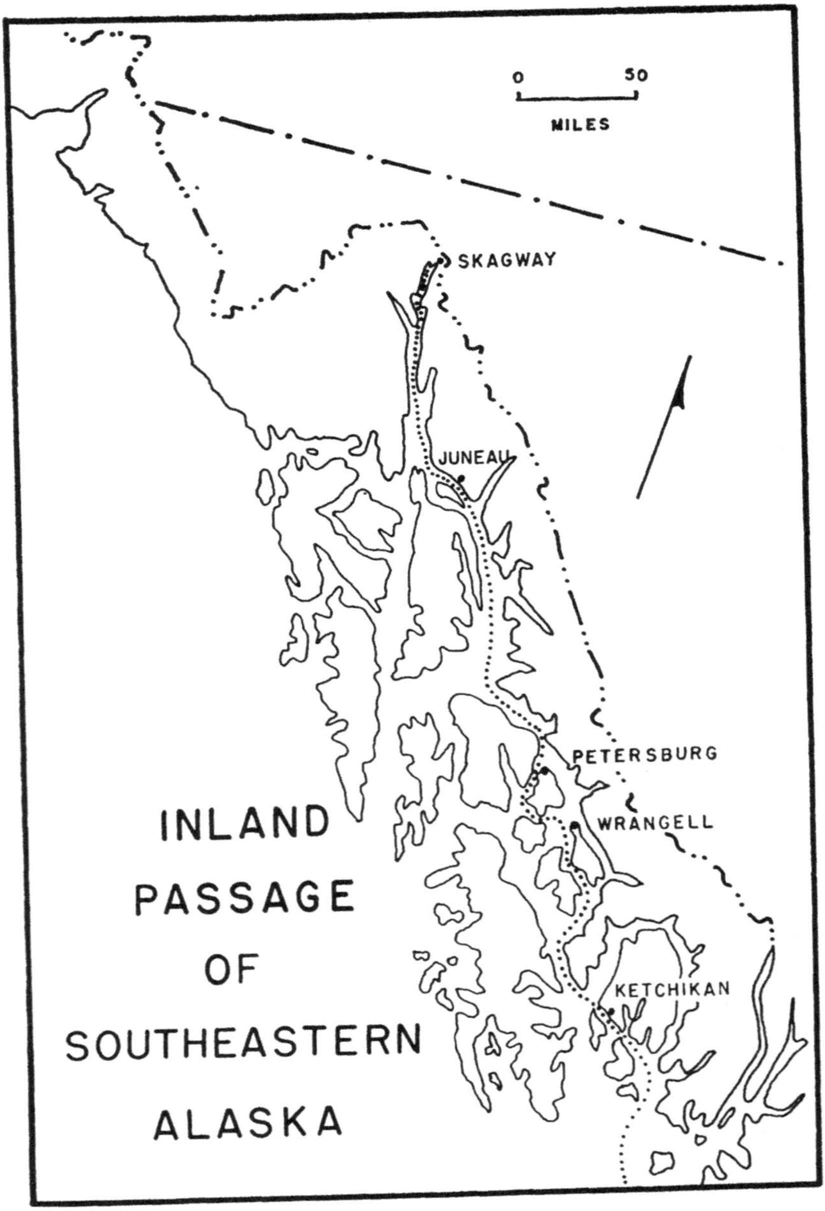

FIG. 1. Inland passage of Southeastern Alaska.

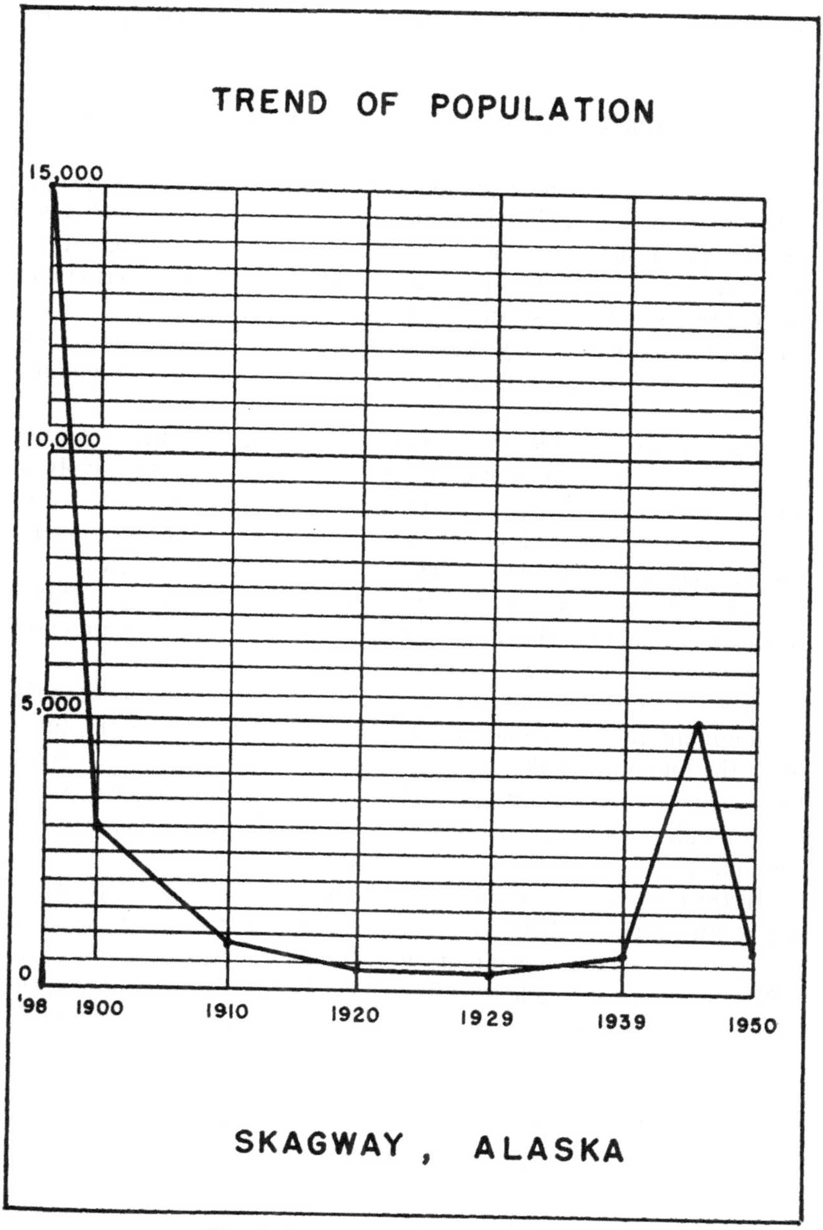

FIG. 2. Population curve for Skagway.

cline will become less and less divergent. The convergence of the extremes may mean an era of relative population stability for Skagway.

It was during the Klondike gold rush of 1898 that Skagway handled from fifteen to thirty thousand people traveling to the Upper Yukon.[1] A large permanent population served these transient people. After the mines were closed the population of Skagway quickly declined. Forty-five years later, during World War II, Skagway accommodated from five to ten thousand soldiers and construction men for building the Alaska-Canada Military Highway (Alcan).[2] After completion of the highway, the significance of Skagway as a transportation center diminished and there was a second drop in population.

It is noteworthy that the importance of Skagway during the two eras of prosperity and population growth was owing exclusively to the location of Skagway with respect to points in Canada. The events of the Upper Yukon during a particular era had a direct influence upon the town. Dependence on Canada was to Skagway's advantage during times of prosperity in that part of Canada. But it is similarly true that the diminishing importance of the area created an era of decline in Skagway.

It is, therefore, the physical setting of Skagway that has been the predominant factor in shaping the sequent occupance of the town. Since the founding of Skagway, the physical setting has not changed. But the fluctuating advantages of that setting has sent Skagway whirling through a sequence of events unmatched by any other Alaskan community.

In 1884, a post was established at the site of Skagway for the trading of furs with the local Indians. But the fur trade here was not prosperous so that the community of Skagway was very small. In 1897, however, when gold was first found in the Klondike River, the seed of prosperity was sown for the outpost community of Skagway for soon it was a major port of entry to the gold fields.

At the time of the gold discovery, the major route from the United States to the Yukon Territory was by ship to the mouth of the Yukon

[1] Edwin J. Foscue, "The Development and Decline of Skagway, Alaska," *Economic Geography,* X (1934), p. 425.

[2] An estimated figure told to the writer by a civilian employee of the government who worked in Skagway at the time.

River and thence up the river by boat. The all-water route was convenient and easy. It was also very long and slow. The discovery of gold meant an emphasis on speed in getting to the Klondike. Only the first few would get the choice claims. The route up the Inland Passage to the head of Lynn Canal and then through the Coast Range to Whitehorse was hundreds of miles and several weeks shorter than the all-water route by the Yukon River. Lynn Canal now attracted ships crowded with men from all parts of the world and funneled them to the site of Skagway (Fig. 3).

The Skagway Valley resembles an elongated triangle wedged into the mountains. The river, which carved the valley and leveled it with alluvial material, flows for the most part along the west side. Scarcely a half mile wide at its mouth and reaching only three miles into the mountains, the Skagway Valley sets limits on man's occupance. Nature constructed a deep-water highway to the door of the Skagway River and up the river placed the lowest pass through the Coast Range to the interior of Yukon Territory. At the precise point where valley and ocean met was the level mouth of the Skagway Valley. If man was to utilize the convenient situation of the Lynn Canal-White Pass Route he must choose the valley mouth. He did just that.

For a short while the town of Dyea at the mouth of the Taiya Valley, which also enters the head of Lynn Canal, surpassed Skagway as port for the interior gold fields (Fig. 3).[3] To be sure, the mouth of the Skagway Valley had housed a small homesteading community when the Taiya Valley was uninhabited. But local Indians told the first arrivals about Chilkoot Pass up the Taiya Valley, and the route, already traversed by an Indian trail, was immediately put to use by white men. White Pass at the head of the Skagway Valley was little known and thought to be too precipitous for the passage of men and goods. Though crossed by an Indian trail, the pass was seldom used and its exact location was known to but very few people. As a result, in the early months of 1897 the town of Dyea was founded at the mouth of the Taiya Valley just beyond the tidal limits of the Lynn Canal. Dyea quickly grew to considerable size. During this time the community of Skagway remained as nearly unaffected by civilization as it had always been.

[3] Merle Colby, *A Guide to Alaska, Last American Frontier* (New York: Macmillan Company, 1939), p. 183.

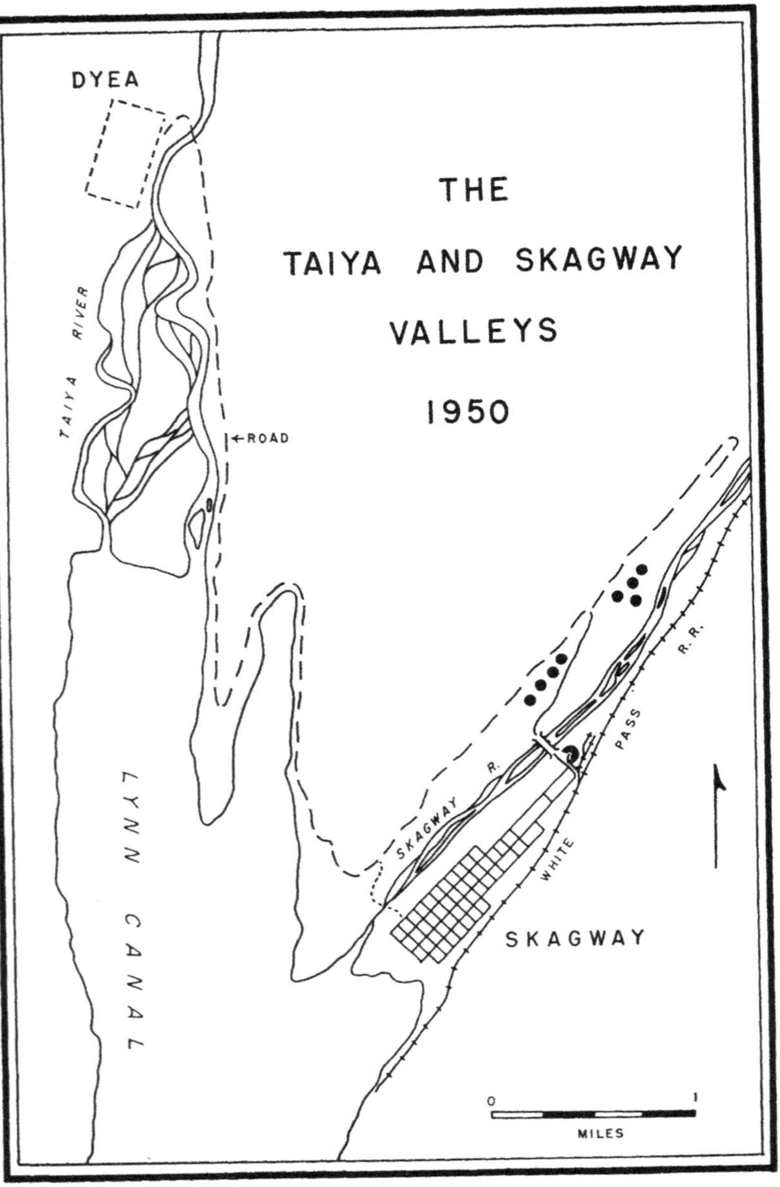

FIG. 3. The Taiya and Skagway valleys—1950.

Historical accident is many times the reason for the particular origin of a town. It is, however, the existence of certain basic geographic factors advantageous to the town at a specific time that determines the kind and amount of growth the town will experience, and often determines whether the town will continue to exist at all. As the number of ships and men and goods coming to the head of Lynn Canal increased, the port of Dyea proved too small to handle them. The lower Taiya Valley was very flat. Its surface was low, moist, often mucky, and greatly dissected by the braiding of the Taiya River. The tidal flats extended nearly two miles up the valley. Ships could be unloaded only at high tide, and then only by transferring cargo and men to small boats which could navigate the shallow tide water to the shore. As more and more ships came, the unloading became more difficult. A better port was needed, and Skagway was available.

Skagway was comparatively well suited to receiving large ships. The tidal flat was less than a quarter mile long and a pier was built so that ships could unload at either high or low tide, thus making transfer to smaller boats unnecessary. At first, Skagway was only the port for Dyea. Men and goods made their way to Dyea over a narrow rocky road which serpentined the four miles between the two communities (Fig. 3). But now Skagway as well as Dyea was growing and soon the large number of people in Skagway warranted extensive exploration of White Pass and its development as a route to the interior.

White Pass was seven hundred feet lower than Chilkoot Pass. From the mouth of the Skagway to the top of White Pass was three miles shorter than from Dyea to Chilkoot Pass. The grade to White Pass, however, was greater than that to Chilkoot Pass. White Pass was, therefore, the more difficult to cross, but it could be crossed to the headwaters of the Yukon in less time. Since time rather than convenience was important, White Pass, rather than Chilkoot Pass, soon became the major route to the interior.

The death knell for Dyea was sounded in May of 1898 with the selection of Skagway as the starting point for a railroad to Whitehorse.[4] The favorable condition of the port of Skagway for the con-

[4] S. H. Graves, *On the White Pass Pay-Roll* (Chicago: Lakeside Press, 1908), p. 10.

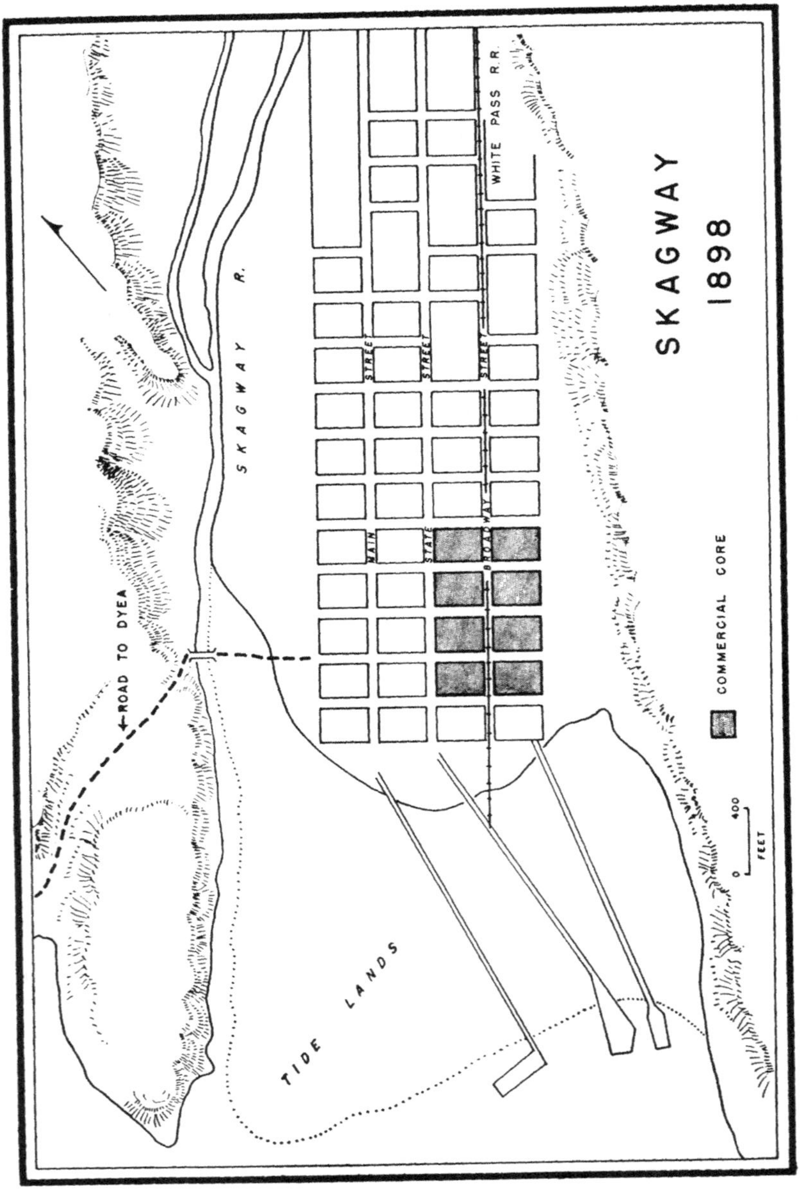

FIG. 4. Skagway—1898.

struction of a dock to handle rail traffic, the availability of land for a railyard just to the north of town, the short distance from the mouth of the Skagway to White Pass, and the lowness of the pass itself offering a minimum depth of winter snow were major considerations of the White Pass Railroad Company in choosing the route (Fig. 4). With the convenience and speed of a railroad through the mountains, nearly all of the gold seekers were now drawn to Skagway, and the town quickly became the largest and most important community in Southeastern Alaska, indeed in the entire Territory of Alaska (Fig. 5).

When the Klondike mines began to close Skagway declined. Skagway's only permanent economic asset now was the White Pass Railroad. The transient population no longer existed, and the few people who stayed in Skagway either worked for the railroad or operated a business to serve the railroad workers. The urban pattern of the town did not change, but the occupance was greatly altered. Many businesses and homes were sold or abandoned. The few people who stayed on clustered more closely around the heart of the town. Some commercial buildings were converted to homes and many others were torn down and the lumber used for building other homes. It was a period of decadence, and the area of town was ever shrinking as the population continued to decrease.

The success of the White Pass Railroad was measured by the quantity of goods needed in Yukon Territory, the railroad's only customer. The economic development of the Territory determined the success of the railroad, which in turn controlled the population of Skagway. The Klondike mines continued to produce small amounts of gold, and there were small but permanent populations at Whitehorse and Dawson. The White Pass Railroad continued to operate by supplying them. Very few men were needed to handle the small quantity of freight at Skagway.

Because of the stagnation of railroad business, Skagway continued to decline in population, and in the early 1930's reached its lowest figure, 492 (Fig. 2). The World War of 1917 and the Depression of 1929–39 aided in holding the town in the depths of economic insignificance. But in the mid-1930's the sleeping Yukon Territory awoke. Copper and asbestos mines were opened, and the ores were shipped out on the White Pass Railroad. The new mines and subsequent gain in population of the Territory created more business for the railroad

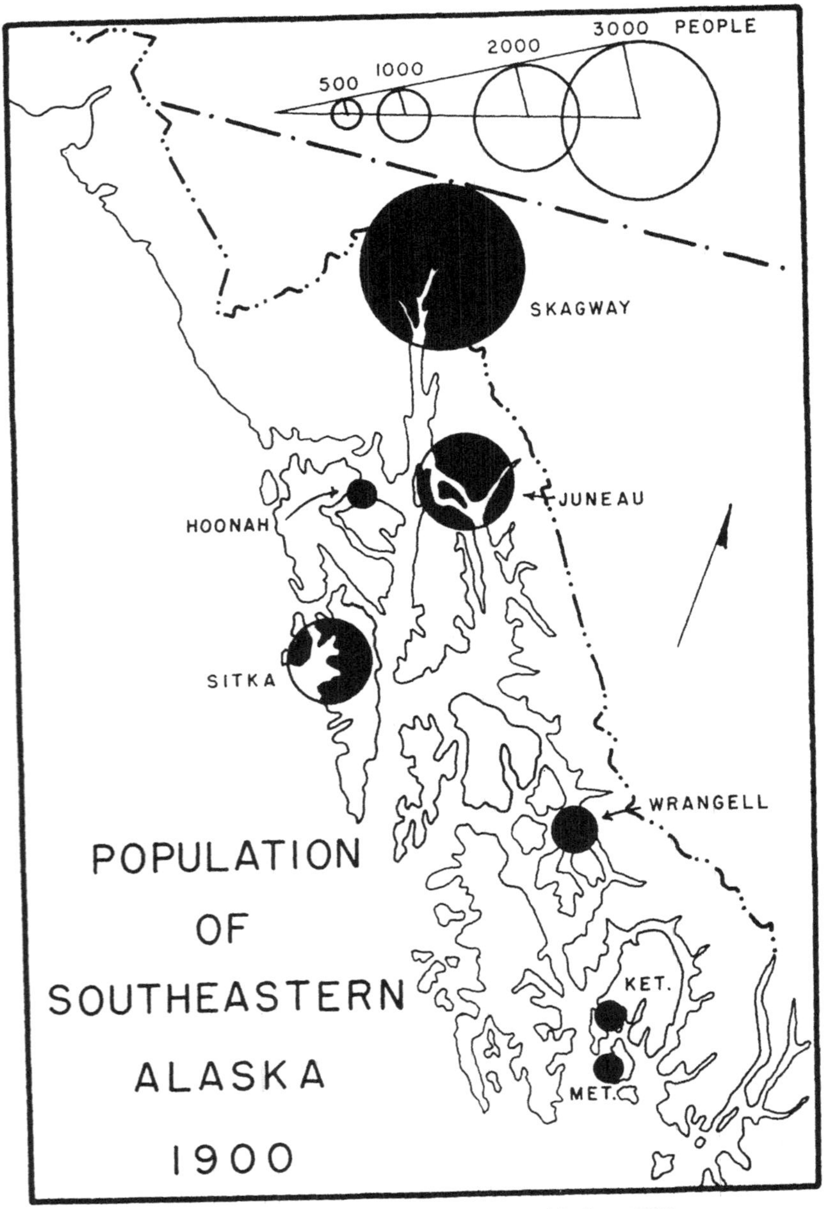

Fig. 5. Population of Southeastern Alaska—1900.

and more jobs for Skagway. The population of the town was slowly rising when the United States entered World War II. The government, fearing for the safety of Alaska and the military and naval forces stationed there, collaborated with the Canadian Government in building a highway from Dawson Creek in British Columbia to Fairbanks in Central Alaska. Until this time only the sea linked Alaska with the United States. In April of 1942 work on the road was begun.[5]

For the Yukon Territory the highway meant a means of transportation that could make feasible the exploitation of resources heretofore unused. For Skagway the highway meant much more. The town sprang to life. Mountains of material and armies of men moved through Skagway on their way to Whitehorse, a major construction center for the highway. More than three thousand soldiers and government employees were permanently stationed in Skagway to handle the traffic. Skagway was a boom town again.

The second era of prosperity for Skagway went as it had come; suddenly, but with perhaps a little more expectancy. Soon after the army had left, taking with them all that could be taken, Skagway settled into the doldrums with which she was so well acquainted. It is the economic calm that exists today. A significant variation to the pattern of settlement has followed the end of the World War II era, one that did not occur after the gold era—a tendency to disperse. Haines was tied to the Alcan Highway, but it was not until after 1948 that Skagway was connected to Haines; for in that year automobile ferry service was established between the two towns. It was now easier and less expensive to get an automobile to Skagway, which had until 1948 no land contact with other towns except by railroad, and transportation costs for shipping automobiles by rail were enormously high. The coming of the automobile to Skagway released the people to settle parts of the valley that had previously been nearly inaccessible (Fig. 6).

The occupation of Skagway during the boom era of the building of the Alcan Highway was but a stage in the history of the community. The town was again a strategic transportation center for the interior of Yukon Territory as she had been in 1898. New buildings, warehouses, rail lines, and pipe lines were constructed to meet the needs

[5] "Construction of the Alaska Highway," *The Scottish Geographical Magazine*, LVIII and LIX (1942–1943), pp. 97–99.

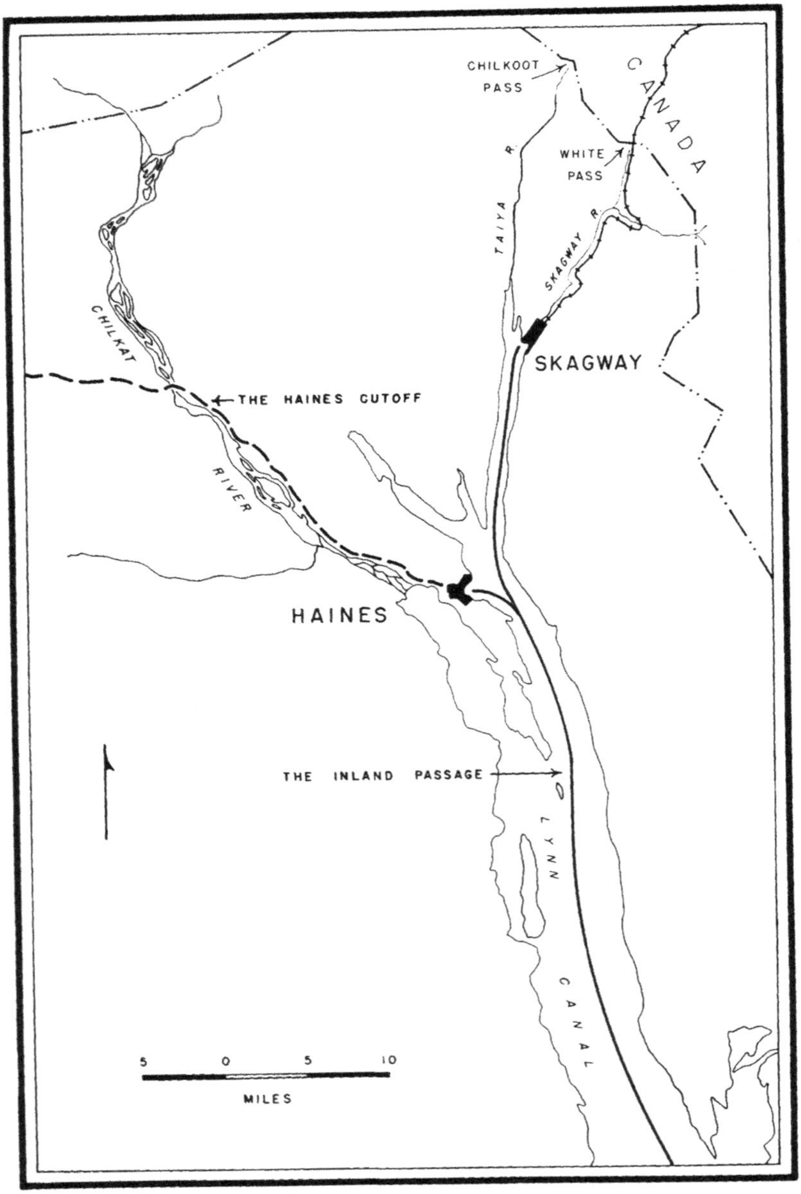

FIG. 6. The location of Haines and Skagway.

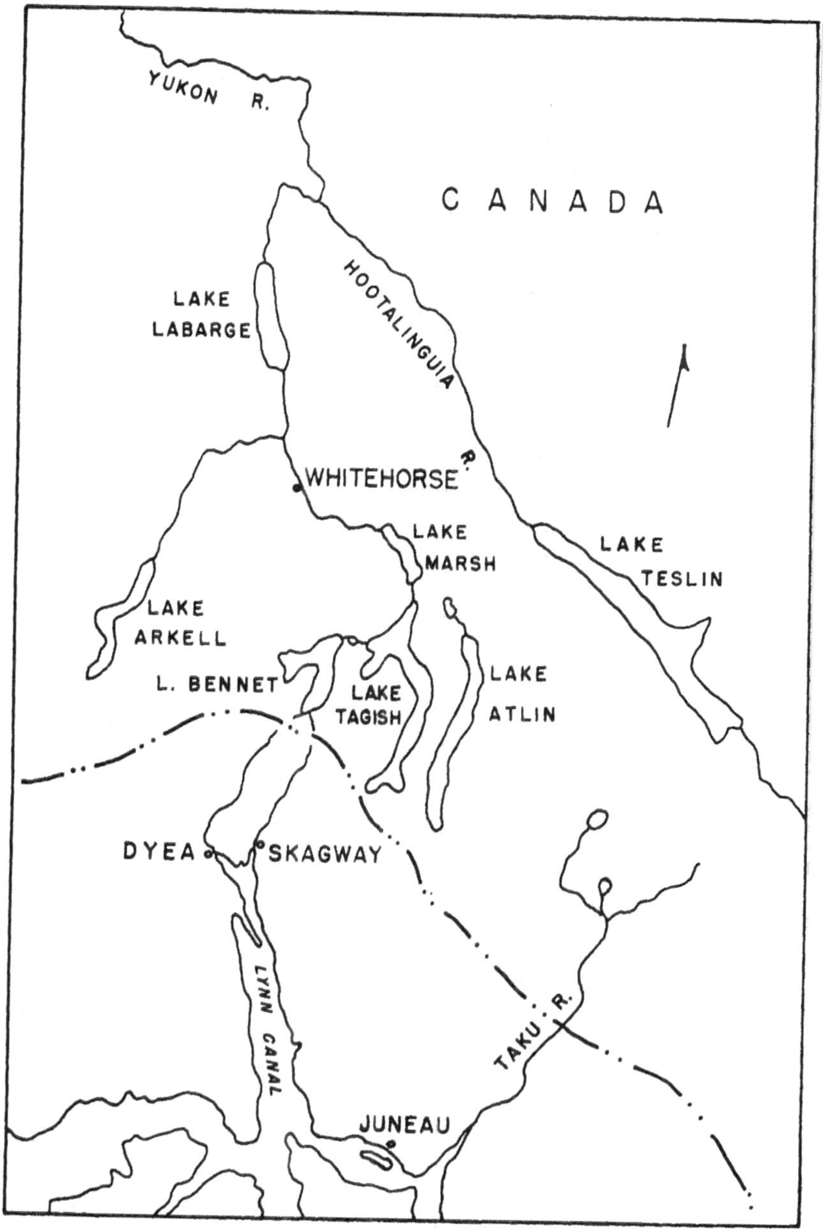

FIG. 7. Headwaters of the Yukon.

of the army in getting men and supplies over White Pass. Such construction occurred in the gold era as well. The role of Skagway during the two eras was basically the same. Perhaps the greatest difference between them was that during the first era the pattern of human occupance of Skagway was created and during the second the pattern was merely slightly altered.

In 1952 the construction of an aluminum plant was proposed for Skagway. The establishment of such a plant could create a new era of prosperity. It would be induced by advantages of the town's physical setting for the production of aluminum rather than by the significance of the location of Skagway as a port for the Upper Yukon Valley.

A number of lakes form the headwaters of the Yukon River and are less than twenty-five miles from Lynn Canal. The lakes are more than two thousand feet above the level of the canal and the water could be piped to hydroelectric plants on the coast (Fig. 7). The power created at Skagway would be low in cost, large in volume, and dependable in production. Electricity represents nearly 20 per cent of the cost of producing aluminum. Thus, Skagway would be ideally suited for aluminum reduction. Moreover, the proposed 400,000 kilowatt installation of the project could make Skagway one of the most valuable electrical and metallurgical centers in North America.[6] But the lack of coöperation with Canada, where the water is located, and the great distance that aluminum ore must be brought to Skagway, are disadvantages yet to be overcome.

A boom era of aluminum would mean the influx of many thousands of people to Skagway. But this time the nature of the population would not be the same as it was in the two previous boom eras. Aluminum production would form a more reliable basis for economic stability so that the population would also be more stable. The pattern of human occupance during such an era would most likely show greater changes than had ever occurred in the town before. A third era of prosperity could be Skagway's most important one. The physical setting of the town has not changed. But the advantages of that setting, viewed in another economic light, could indicate a future of substantial economic growth.

[6] This information is from correspondence with the Aluminum Company of America.

Papers of the Michigan Academy of Science, Arts, and Letters
Vol. XLV, 1960 (1959 Meeting)

ECONOMICS

LOCAL REVENUE NEEDS AND TAXING POWERS

CECIL M. BIRCH

University of Detroit

Detailed statistics, not available since 1942, have recently been issued. Breakdowns of local expenditures and revenues during 1957 for different types of governments and for each state have just been published as advance releases. I admit that, along with almost everyone else, I distrust estimates of trends based on two points in time that are separated not only by fifteen years from 1942 to 1957, but by social upheavals of such magnitude that there are entire centuries in the past that have not experienced equivalent changes. Nonetheless, I think the 1942–57 trend is worth something, and I propose to evaluate on the basis of this trend, modified by some subjective decisions, the adequacy in the future of local governments' tax revenues under the assumption that taxing powers do not change. This evaluation provides some perspective on the need for different taxing powers if future needs are to be met or, alternatively, some perspective on the circumstances that must develop if taxing powers do not change.

Expenditures

On the side of spending, Table I reviews the main types of expenditures made by local governments and some changes that took place between 1942 and 1957. Three important reasons—although not the only reasons—for increases in the dollar amounts are inflation, population growth, and increases in the volume of goods and services rendered. The current dollar figures of Table I show the impact of these three taken jointly. The constant (1957) dollar figures show expenditures adjusted for inflation. The second-to-last column of Table I shows the absolute increase in constant dollars for the different types of spending. Most noticeable is the increase in expenditures on education. The percentage column of Table I—the last col-

TABLE I

Local Government Expenditures in Michigan*
(in millions of dollars)

Type of Expenditure	Total Expenditures Current Dollars		Total Expenditures 1957 Dollars†		Increase or Decrease 1942–57 (constant $)	
	1942	1957	1942	1957	Absolute	Per Cent
Direct General Expenditures						
General control	26.5	55.9	59.7	55.9	*3.8*	*6.3*
Public safety	30.6	104.6	69.0	104.8	35.6	51.6
Highways	32.2	168.0	72.6	168.0	95.4	131.4
Sanitation	10.5	67.6	23.7	67.6	43.9	184.8
Health and hospitals	14.5	92.8	32.7	92.8	60.1	183.9
Public welfare	21.7	35.5	49.0	35.5	*13.5*	*27.5*
Education	87.4	614.3	197.2	614.3	417.1	211.4
Other	17.9	175.3	40.3	175.3	135.0	335.2
Total	245.2	1314.1	553.3	1314.1	760.8	137.5
Other Expenditures	61.3	175.6	138.4	175.6	37.2	26.9
Total, All Expenditures	306.5	1489.7	691.8	1489.7	797.9	115.3

* Note: Data may not add to totals due to rounding.

† Using the implicit price deflators for state and local government purchases of goods and services in the *U. S. Income and Output, A Supplement to the Survey of Current Business* (U. S. Department of Commerce) 1959.

Sources: Derived from *Governmental Finances in the United States: 1942* (Washington, 1945); *State and Local Government Finances in 1957* (Washington, 1959—Advance Release).

umn—shows that even without inflation there has been a doubling and a tripling of some kinds of expenditures.

In Table II we have expenditures on a per capita basis in current and constant dollars. The per capita constant dollar figures reflect the change in the volume of goods and services provided by local governments. It is plain that there has been a sizable increase in the per capita amount spent on education, and as a consequence education has increased in relative importance in the budgets of local governments from a position of 28.5 per cent to 41.2 per cent, as is shown in the third-last and last columns of Table II. It is plain also

TABLE II

PER CAPITA LOCAL GOVERNMENT EXPENDITURES IN MICHIGAN*
(in millions of dollars)

Type of Expenditure	Current Dollars		1957 Dollars			
	1942	1957	1942	Per Cent	1957	Per Cent
Direct General Expenditures						
General control	4.8	7.3	10.7	8.6	7.3	3.7
Public safety	5.5	13.6	12.4	9.9	13.6	7.0
Highways	5.8	21.8	13.1	10.5	21.8	11.3
Sanitation	1.9	8.8	4.3	3.5	8.8	4.5
Health and hospitals	2.6	12.0	5.9	4.7	12.0	6.2
Public welfare	3.9	4.6	8.8	7.1	4.6	2.4
Education	15.7	79.7	35.5	28.5	79.7	41.2
Other	3.2	22.8	7.3	5.9	22.8	11.8
Total	44.2	170.6	99.7	80.1	170.6	88.2
Other Expenditures	11.0	22.8	24.9	19.9	22.8	11.8
Total, All Expenditures	55.2	193.4	124.6	100.0	193.4	100.0

* Per capita obtained by dividing expenditures by population. Sources: Same as Table I.

that several items have decreased in relative importance, notably, general control, public safety, and public welfare.

How shall we project these various expenditures, say five years ahead? What assumptions shall we make? The data of Table III show the level of expenditures expected under various combinations of assumptions about population, inflation, and the trend in per capita goods and services.

Three fairly simple sets of alternatives are considered:

1. Michigan will have either the same share of U. S. population in 1964 as in 1958, namely, 4.54 per cent, or a somewhat larger proportion, i.e., 4.64 per cent. In either case it is assumed that the U. S. population will be 192 million.

2. The U. S. will not have any more inflation or, alternatively, inflation will continue but only at one-half the trend of 1942–57.

3. The trend in per capita goods and services will level off, or rise at one-half the trend of 1942–57, or rise on the same trend.

TABLE III

LOCAL GOVERNMENT EXPENDITURES IN MICHIGAN PROJECTED FOR 1964
(in millions of dollars)

Michigan Population in 1964	Inflation up to 1964	Goods and Services Per Capita in 1964		
		No change from 1957	One half 1942–57 trend	Trend of 1942–57
8,717,000	No more	1,686	1,826	1,966
	One half 1942–57 trend	1,827	1,961	2,107
8,909,000	No more	1,723	1,866	2,009
	One half 1942–57 trend	1,867	2,011	2,153

These alternatives combine to yield twelve projections. For example, the number in the upper left—$1,686 million—assumes that Michigan's share of U. S. population will be unchanged from 1958 and that there will be no more inflation and no more increase per capita in goods and services offered by local governments.

The lower right corner figure—$2,153 million—reflects the other extreme; a higher share of U. S. population, more inflation, and per capita goods and services continuing to rise as they have in the past.

For purposes of illustration a single figure is needed. In my opinion the middle figure of the last line—2,011—at least will prove to be not too high. This figure of $2,011 million assumes a larger share of U. S. population for Michigan, more inflation, and more goods and services per capita, but each of the latter two at only one-half the 1942–57 trend. Therefore $2,011 million will be taken for our purpose today as a reasonable estimate of the *need* for dollars by local governments in Michigan five years hence. Then the question is: Will local sources of revenue be as rich as these needs require?

REVENUES

We begin to answer this question in Table IV. Here are some statistics on local government revenues in 1957. The main source of revenue was, of course, the property tax, though its relative im-

TABLE IV

LOCAL GOVERNMENT REVENUES IN MICHIGAN, 1957

(in millions of dollars and percentages)

Type of Government	Taxes				Intergovernmental Revenue		Charges, Earnings, and Miscellaneous		Total	
	Property	Non-property	Total							
	$	$	$	%	$	%	$	%	$	%
Counties	114.5	0.7	115.2	45.3	101.2	39.9	36.4	14.8	253.8	100.0
Municipalities	209.3	11.5	220.8	40.2	112.9	20.5	215.7	39.3	549.6	100.0
Townships	8.8	2.0	10.8	25.1	18.5	43.0	13.7	31.9	43.0	100.0
School districts	274.6	—	274.6	46.7	272.0	46.2	41.4	7.1	588.0	100.0
Special districts	2.7	—	2.7	26.6	1.4	13.7	6.1	59.7	10.2	100.0
Total	609.9	14.2	624.1	44.2	472.4	33.4	314.5	22.4	1,411.0	100.0

Sources: Same as Table I.

TABLE V

LOCAL GOVERNMENT REVENUES IN MICHIGAN 1942, 1957, AND ESTIMATED NEED FOR 1964

(in millions of dollars)

Relative Importance of Revenue Sources in	Taxes			Inter-governmental Revenue	Charges, Earnings, and Miscellaneous	Total
	Property	Non-property	Total			
1942						
Current	181.4	3.7	185.0	91.2	33.3	309.5
Per cent	57.6	1.2	59.8	29.5	10.8	100.0
1957						
Current	609.9	14.2	624.1	472.4	314.5	1,411.0
Per cent	43.2	1.0	44.2	33.4	22.4	100.0
Estimate No. 1 for 1964 (1957 percentages applied to $2,011)	869	20	889	672	450	2,011
Estimate No. 2 for 1964 (Trend of percentages 1942–57 applied to $2,011)	734	18	752	710	549	2,011

portance varies with the type of local government. This variation suggests that it is questionable how intelligent it is to refer to the "total for local governments," but I think it is permissible, at least initially, for the purpose at hand.

For all local governments taxes were 44 per cent of revenue in 1957, intergovernmental aid was 33 per cent, and charges, earnings, and miscellaneous were 22 per cent. These percentages and the associated current dollar amounts are repeated on the third and fourth lines of Table V. The first two lines give similar data for 1942. Notable is the decline in relative importance of the property tax, and the relative increase in the charges, earnings, and miscellaneous revenue.

There are many ways to estimate revenue needed from each source in 1964. One way is to assume that the relative importance of reve-

TABLE VI

LOCAL GOVERNMENT REVENUES IN MICHIGAN: ESTIMATED NEED FOR 1964, AND PROJECTED REVENUE

(in millions of dollars)

Estimate	Taxes			Inter-governmental Revenue	Charges, Earnings, and Miscellaneous	Total
	Property	Non-property	Total			
Estimate No. 1 (from Table V)	869	20	889	672	450	2,011
Projection of Revenue (1942–57 Revenue Trend in current dollars)	810	19	829	650	446	1,925
Difference	59	1	60	22	4	86

nue sources will be the same in 1964 as in 1957. This is called here "estimate No. 1." The total for estimate No. 1 is taken from Table III—$2,011 million—and the distribution of the total is on the basis of the assumption just mentioned.

For two or three reasons, one being that I believe this assumption to be unreal, another estimate is made in Table V. Estimate No. 2 uses the same total for revenue needed ($2,011 million) but distributes it on the assumption that the 1942–57 trend in relative importance of revenue sources will continue. Unfortunately I think this also is an unreal assumption. In fact I am not sure what is a realistic assumption for this problem. The point I want to make, however, can be made by using first of all estimate No. 1, and then estimate No. 2. For this we turn to Table VI.

On the top line of Table VI is the estimated need for 1964. The total—$2,011 million—is needed on the basis of the assumptions of Table III. The distribution of the total is the same as was explained for estimate No. 1 in Table V.

The second line of Table VI is a projection of the 1942–57 trend in revenue collected. Thus we have an estimate of revenue needed in 1964, and also an estimate of revenue likely to be collected on the basis of a historical trend. The difference shown in line three is the amount the projected revenue falls short of the estimated needed revenue.

Equating Needs and Revenues: Taxing Powers

These data suggest that substantial reliance on the property tax is incompatible with future revenue needs defined in terms of future expenditure requirements. The data also indicate the extent to which we must use intergovernmental revenue and local charges if the local units of government cannot turn to nonproperty taxes. In fact, to the extent that the property tax revenue will not increase as it has in the past, then the amount involved is added to the need for more revenue from other sources.

If the property tax revenue continues to decline in relative importance (e.g., estimate No. 2 of Table V) then we face what appears to be an almost fantastic reliance upon intergovernmental revenue and local government charges and earnings. It appears that in terms of finding revenue sources for our needs at least some local governments are going to have to try nonproperty taxes. On the basis of the assumptions made here, there is not much doubt that local governments will either reduce their economic activities or else get additional sums from nonproperty taxes, charges, and state aid.

It is the estimating of revenue needs and sources, however, that brings us up against this outlook. A consideration of the powers of local governments indicates only one course of action—substantial increases in state aid. This of course means state taxes instead of local taxes. If we do not want more state aid, then a greater use must be made of local nonproperty taxes and charges for services rendered. But for this to happen, legal and other obstacles must be removed.

If the legal obstacles are cleared away, then I imagine that we surely will have local supplements to state taxes, if only because the richest local nonproperty tax bases are being used by the state, or are about to be, if an income tax is adopted. For example, I am quite sure that by 1964 the City of Detroit will have either a sales tax or an income tax.

Conclusion

In conclusion, I have glossed over, or ignored, or assumed away many problems. This disturbs me a little. For example, I am not happy with the aggregating of local government data in all the tables except Table IV, and I am not content with the passing reference to local taxing powers. But I believe that the data demonstrate that local revenue needs in the 1960's will substantially exceed current revenue from local sources. Either our needs must be modified, or state aid increased, or basic changes will have to be made in local taxing powers.

Papers of the Michigan Academy of Science, Arts, and Letters
Vol. XLV, 1960 (1959 Meeting)

THE COMPARATIVE BURDEN OF MICHIGAN'S STATE AND LOCAL TAXES ON BUSINESS[1]

DWIGHT B. YNTEMA

Hope College

This paper deals informally and briefly with the comparative burden on business of Michigan's state and local taxes. It draws upon findings from a factual study of the particular question that was initiated in the Michigan State Senate early in 1957 at the time when the House began its comprehensive review of Michigan's taxes. How should the comparative burden on business of Michigan's state and local taxes be approached? For the Senate study, it was decided to ask firms that operate in Michigan and certain other comparable states to report on their tax costs in both Michigan and the other state or states. A questionnaire requested information on taxes of each different type per employee and per $1,000 payroll. Tax figures, themselves, were not sought in the belief that requests for such information might definitely limit business coöperation.

The survey yielded returns from 100 companies that reported on comparable operations in one or more other states and Michigan. Comparability required that the firm's operations in each of two states fall in the same industry subgroup, such as "transportation equipment" or "retail trade." Most of the returns came from companies that had been sampled systematically from files of the Michigan Securities and Exchange Commission; others came almost entirely from a list of firms which, it was felt, would coöperate readily. The 100 companies provided 278 comparisons between other states and Michigan, evidencing the fact that a majority operated in more than one of the other seven states as well as in Michigan. The total includes 64 Ohio-Michigan comparisons, 47 for Illinois-Michigan, between 30 and 40 for Indiana, New Jersey, New York, and Pennsylvania, and 29 for the Wisconsin-Michigan comparison. Most of the cases are in

[1] The author served as director of the comparative tax burden study conducted by the Michigan State Senate Tax Study Committee, as established under Resolution 20, adopted February 7, 1957. All opinions and interpretations in this paper are solely those of the author.

manufacturing (28 per cent in nondurable goods and 40 per cent in durable goods), some in retail trade (20 per cent), and the rest scattered among remaining activities. Most company reports relate to fiscal year operations ended in 1956; about one third, however, are for years ended in 1957.

For comparing state and local tax burdens on business between other states and Michigan, we use herein total nonpayroll taxes. The total includes state taxes in the form of corporate franchise, sales and use, intangibles, income, business activities, and such other levies as apply in different states and as were reported by business firms. Local taxes include real and personal property taxes and other minor taxes and fees. The payroll levies, though covered in the survey, are not included in the totals being compared; these take the form of unemployment compensation taxes, costs of workmen's compensation (a sort of hybrid item), and disability insurance taxes in some of the states. Wholly excluded from the study are gasoline and other levies of the motor vehicle group, special assessments at the local level, and all taxes collected from customers or withheld from employees.

The general finding is clear. Business taxes in other states, taken either on a per employee basis or a per $1,000 payroll basis, amount usually to something like 50 per cent of the Michigan business tax figure. But let us look more carefully to the percentages which relate a company's taxes in another state to its taxes in Michigan. As noted above, we have, for example, 64 of such percentage comparisons for firms that operated in both Ohio and Michigan. To characterize the typical Ohio-to-Michigan percentage, let us take the percentage at the half-way point as to size—the median percentage. We find that the median percentage position of taxes per employee in the other state compared to Michigan ranges from as low as 33 per cent for Illinois to as high as 78 per cent for Wisconsin. Medians for the five other state comparisons fall between the 43 per cent figure for Ohio and the 53 per cent figure for Pennsylvania. On a per $1,000 basis, the low median position is at 34 per cent for Illinois and the high at 90 per cent for Wisconsin; medians for the other five states range between the 48 per cent Ohio figure and the 56 per cent figure applying to both Indiana and New York. If attention is limited to manufacturing only, the result is much the same. Wisconsin stands highest among the seven states being compared with Michigan. For the other six states, however, the tax burden is at about half of that in

Michigan. In the tier of states to the south, i.e. Ohio, Indiana, and Illinois, it falls at about 40 per cent on a per employee basis and at about 45 per cent on a per $1,000 payroll basis (see Table I).

These findings are from data collected and processed in the Michigan State Senate Tax Study. Here, we use reports of firms supplying data on taxes per employee and taxes per $1,000 payroll in one or more other states and in Michigan. Each tax figure for a firm's operations in another state was expressed as a per cent of the firm's corresponding tax figure in Michigan. The percentages were then tabulated in frequency distributions for cases of various types, e. g., taxes per employee—Illinois/Michigan—retail trade. The medians shown in the first two sections above were calculated, each from its respective frequency distribution; no median is shown where there were fewer than eight observations in a distribution. The third section gives total frequencies for each of the various distributions.

We should look at the dollars lying back of the percentages. Let us keep to manufacturing activities and again compare Michigan taxes with those in the three states to the south. We find that total state and local nonpayroll taxes per employee for the Michigan plants of the same firms that were used in calculating percentage comparisons amount to some $275 per year, as judged by median positions. For plants in Illinois, Indiana, and Ohio, the tax per employee amounts to some $105 per year. The difference is in the order of $170 per employee per year. This $170 difference in tax totals, in turn, may be divided between the part arising from differences in state taxes, about 4 out of every 10 dollars, and the part attributable to local tax differences, which accounts for the remaining 6 out of 10 dollars. Direct comparisons at the two levels suggest that state taxes per employee in Michigan are three or four times the state tax figures found in the tier of states to the south whereas local taxes are about twice as high. It may be noted that the $105 figure for total taxes per employee per year in the three states to the south represents 38 per cent of the Michigan tax total, a result in line with the median percentage figures noted earlier (see Table II). These findings come from the same source as the Table I data. In this case, however, we use medians calculated from frequency distributions of the dollar size of taxes per employee.

Some comment on the annual tax difference per employee may be useful. Instead of the $170 difference let us use $150, thus simplify-

TABLE I

STATE AND LOCAL NONPAYROLL TAXES IN OTHER SELECTED STATES AS PER CENT OF TAXES IN MICHIGAN

Tax Comparison and State	Total	Manufacturing: Non-durable Goods	Manufacturing: Durable Goods	Retail Trade	All Other
Median per cent—taxes per employee in other state as per cent of Michigan					
Illinois	33	24	42	18	—
Indiana	50	—	37	74	—
New Jersey	46	47	37	—	—
New York	51	47	54	—	—
Ohio	43	43	37	46	54
Pennsylvania	53	50	57	54	—
Wisconsin	78	66	90	—	—
Median per cent—taxes per $1,000 payroll in other state as per cent of Michigan					
Illinois	34	24	40	18	—
Indiana	56	—	50	70	—
New Jersey	52	64	40	—	—
New York	56	52	62	—	—
Ohio	48	46	47	46	63
Pennsylvania	55	52	64	50	—
Wisconsin	90	80	104	—	—
Number of firms reporting on taxes in both other state and Michigan					
Illinois	47	11	20	9	7
Indiana	33	4	14	9	6
New Jersey	31	9	14	7	1
New York	35	13	13	7	2
Ohio	64	19	26	10	9
Pennsylvania	39	12	17	8	2
Wisconsin	29	8	9	7	5

TABLE II

STATE AND LOCAL NONPAYROLL TAX TOTALS PER EMPLOYEE IN MICHIGAN AND OTHER STATES, 1956

Tax Total and State	All Firms Reporting		Reporting Firms Manufacturing: Nondurable Goods		Reporting Firms Manufacturing: Durable Goods	
	Mich.	Other State	Mich.	Other State	Mich.	Other State
Total state and local nonpayroll taxes—median position						
Illinois	$221	$ 72	$260	$ 73	$280	$100
Indiana	206	112	—	—	272	115
New Jersey	268	120	310	170	288	115
New York	260	136	290	136	296	152
Ohio	263	114	280	131	286	112
Pennsylvania	270	133	266	160	280	140
Wisconsin	233	206	320	280	300	226
Total state nonpayroll taxes—median position						
Illinois	$ 74	$ 20	$ 60	$ 20	$ 93	$ 25
Indiana	72	31	—	—	93	22
New Jersey	91	21	85	20	103	22
New York	88	80	84	104	100	90
Ohio	78	21	72	21	87	20
Pennsylvania	74	106	66	120	90	120
Wisconsin	86	71	80	40	92	90
Total local taxes—median position						
Illinois	$134	$ 57	$180	$ 60	$176	$ 80
Indiana	122	72	—	—	160	80
New Jersey	156	100	260	140	160	96
New York	157	47	194	50	170	63
Ohio	166	90	184	113	180	86
Pennsylvania	150	37	180	40	164	45
Wisconsin	124	105	220	200	170	108

ing the arithmetic and erring on the side of conservatism. And again to simplify arithmetic, let us assume a $2.50 hourly wage rate in manufacturing and a work year of 2,000 hours, i.e. 40 hours per week for 50 weeks. This gives annual earnings of $5,000. The $150 difference in taxes then is equivalent to 3 per cent of payrolls. Alternatively, it becomes 7.5 cents an hour throughout the year. It would be enough each year to meet a company's payroll for one full week and half of a second week. Compared with company profits, it could amount to 5 per cent, 10 per cent, or even much more, depending on a company's relative use of labor and the general level of company profits for the year. Such is the gain merely from a plant location somewhat beyond Michigan's southern border to the east, south, or west.

Just how the tax differential affects industry in Michigan must be a matter of speculation. If other factors were equal, the answer seems clear enough. But other factors vary widely. Capital commitments may already have been made. Access to markets may differ as well as access to the materials and services used in production. And labor supply may be crucial. In any event, there is likely to be validity in the comment that a company does not usually move its Michigan operations outside the state; rather, it is likely to expand elsewhere. In this connection, we may note the finding of Dr. Charles L. Jamison, professor emeritus, University of Michigan School of Business Administration. Had industrial growth in Michigan, he says, kept pace with growth in other states from 1947 to 1956, there would have been some 270,000 more production jobs in the state. But the actual increase represents merely one-seventh of the projected figure. Other evidence might be sought in the statements of business executives regarding location in Michigan. This I will not undertake. Suffice it to say that the tax differential usually appears as one of the reasons for locating elsewhere.

And where does this leave us? We are faced with the question of interstate rivalry, rivalry in maintaining a climate for business here as favorable as in competing states. Of course, the rivalry is limited in scope since many business activities are substantially tied to their given sites—retailing, for example. The rivalry is mainly in manufacturing. Then, with business taxes high in Michigan, we may hope that the levies will rise elsewhere, especially in the states furthest below the Michigan figure. But this may prove wishful thinking. The converse calls for lowering Michigan's taxes on business. At the state-tax level, this would require a drastic cutback, amounting to one half or two thirds of present taxes. The reduction would cancel much of the build-up in state taxes on business that has occurred since the war. At the local level, the tax disparity is ingrained. Here, it is the property tax as levied on both real and personal property that would need attention. In Illinois, Indiana, and Ohio, for example, the property tax per employee is about half as large as in Michigan. Would it be possible to cut down local property taxes and, if so, how far and how replace the loss in local revenue? Such questions raise the whole issue of taxes in Michigan, and I must defer to experts in the field.

But there is more to consider than the tax picture. When state and local finances are not in reasonable balance, expenditures as well as revenues should be reviewed. I take the liberty of remarking briefly on the former. As against the intensive study accorded taxes, study of expenditures has been grossly neglected, almost a terra incognita to the public finance student. Surely, public expenditures are neither so sacrosanct nor so elusive as to receive so little consideration when there is imbalance in public accounts.

In thinking about expenditures, I find myself wanting to classify them according to the degrees of option and personal benefit the individual may have in receiving services from government. At the one extreme would be the elemental services for protection, law and order, and so forth, as well as the modern things like compulsory public education and the welfare programs of one kind or another. Willy-nilly, the individual is a participant and, practically speaking, he has no option as to extent of participation or degree of personal benefit. At the other extreme are the services which the individual buys from public enterprises. The prices charged are supposed to be enough to cover costs, and the individual is more or less a free agent in deciding how much of each kind of service he will take. Here, prices have a twofold function. On the one hand, they provide revenue for defraying costs. And on the other, they provide for rationing of services in keeping with each individual's assessment of worth to himself. If such services were free, a mad race could be expected between suppliers and users, characterized by mounting taxes and wanton use of resources.

Between the two extremes are the various mixed cases in which general public interest is paired with personal interest and choice. Cases of this kind are likely to become increasingly important as income levels rise. They tend often to center about the individual—his security, his health, his recreation and entertainment, and his self-betterment. Government is especially likely to get a call for action in this general area when it tends to be uniquely suited for a particular task or when subsidy is sought. And what should be done about paying for and rationing the services? I myself am fearful of the unconsidered answer, based as it seems to be on the tenet that, with government doing it, services should be free and abundantly available and taxes should pay the way. But what assurance is there that free services will be used with reasonable economy? Without such assurance, little more is accomplished by the spending and taxing than a haphazard redistribution of real income, in favor usually of the have-nots and the have-lesses. But taking care of these, important as it is, can be accomplished much more efficiently through aid and assistance arrangements. This leads back to the substantial necessity of charging for the government services of the kind in question. We need a realistic system of fees—fees to help meet costs and to curb excessive and uneconomic demands. Herein lies part of the answer to present and future state and local tax problems.

Papers of the Michigan Academy of Science, Arts, and Letters
Vol. XLV, 1960 (1959 Meeting)

HISTORY

THE HIGH SCHOOL ON TRIAL: THE KALAMAZOO CASE

WILLIS F. DUNBAR

Western Michigan University

No one today questions the right and duty of public school boards to construct and maintain high schools with public funds. But less than a century ago, in a judicial case now famous in the annals of the history of American education, not only the right to levy taxes for the support of a high school but also the right of a school board to employ a paid superintendent of schools was challenged. The case was begun in what was then the Village of Kalamazoo, Michigan, and after the decision of the Circuit Court of Kalamazoo County it was carried to the Michigan Supreme Court. The decision of the Supreme Court not only decided the issue in Kalamazoo but has been cited as a precedent in similar cases originating in other states. One writer on education expresses the opinion that the "Kalamazoo decision . . . paved the way for a phenomenal growth in this new institution of democracy."[1]

Education beyond the elementary level was available in Kalamazoo for more than two decades before a public high school was opened. Judge Caleb Eldred and the Reverend Thomas W. Merrill, a Baptist home missionary, had obtained from the Territorial legislature in 1833 a charter for an institution called "Michigan and Huron Institute." It opened in Kalamazoo (then called Bronson) in 1836 and was renamed the Kalamazoo Literary Institute the following year. Also in 1837, the first state legislature enacted a law establishing the University of Michigan with such branches as the public convenience might dictate. The branches were designed to serve several purposes: to provide for the training of teachers, to teach agriculture, to provide for the education of women, and to be preparatory schools for the University. Nine of these branches were

[1] Chris A. De Young, *Introduction to American Public Education* (N.Y., 1942), p. 202.

opened, one of them in Kalamazoo. It was opened on May 1, 1838, and from that date until August 6, 1839 (when the Kalamazoo Branch was temporarily suspended), there were two schools in the little pioneer town offering secondary education. By an arrangement between the Trustees of the Literary Institute and the Regents of the University of Michigan, the two schools were merged in 1840.

Limited funds caused the Regents, in 1846, to withdraw all financial support from the branches, but the Kalamazoo school continued to operate under the control of the Trustees of the Kalamazoo Literary Institute. The principal, Dr. J. A. B. Stone, endeavored in vain for some years to persuade the legislature to give the institution the power to confer degrees. He finally was successful in 1855, when the charter of the school was amended to authorize the granting of degrees, and the name was changed to Kalamazoo College. The preparatory department, however, was retained. In the late fifties, the enrollment exceeded four hundred and was larger than that of the University of Michigan. Dr. Stone's wife, Lucinda Hinsdale Stone, was head of a female seminary connected with the College. A large building was erected for Mrs. Stone's use in 1859. At the same time, a movement was on foot in the town to establish another "female seminary" by the Presbyterians. An ambitious building program was undertaken and shortly after the Civil War the Michigan Female Seminary was opened in Kalamazoo.

A lively and continuing interest in secondary education must be inferred from this record. Dr. Stone was quoted later as saying that he often was surprised "at the interest taken in the educational work in Kalamazoo, and the readiness . . . to make what were really very generous contributions at a time when money was scarce in a new country, and wealth unknown, and the personal necessities of most persons absorbed their income."[2] Even as plans were being made to provide buildings for two female seminaries and as Kalamazoo College, with its preparatory department, was enrolling large numbers of students, a movement got under way to establish a public high school.

Public education had its beginning in Kalamazoo when School District Number 1 was organized in September, 1833. By 1850, three other districts had been formed, each with its own school board. In that year, the legislature authorized the consolidation of

[2] *History of Kalamazoo County, Michigan* (Philadelphia, 1880), p. 146.

these four districts into one village district, under control of the trustees of the village. One of the objects of this change apparently was to make possible the erection of one large school for the entire village. But differences of opinion on the proper location of such a school held up the project. Meanwhile, the village trustees appointed a Director of Schools to have general charge of the district's educational work. Two lawyers and a businessman held this position between 1851 and 1857. Early in 1857, the question of a site for the proposed new school was settled, and a five-acre tract was purchased on which a "Union School" was to be built. Then, on April 20, the trustees appointed a Kalamazoo College professor as Director of Schools. Daniel Putnam was what we would call a professional educator. On July 14, 1858, Putnam submitted to the trustees an extensive report, proposing that the schools be organized by dividing them into four departments: primary, intermediate, grammar, and high school. The plan was adopted by the trustees, and Putnam was engaged on a full-time basis, one half of his time to be spent as Superintendent of Schools and the other half as principal of the high school. On December 6, 1858, the high school department was opened on the third floor of the still uncompleted "Union School." The building was completed and dedicated on January 27, 1859.[3]

The records reveal no controversy connected with the opening of the high school, but there may have been some doubt in the minds of the president and trustees of the village concerning their powers. At any rate, the legislature, in 1859, passed an act amending the act of 1850, under which a single school district in Kalamazoo had been set up. This amendment transferred the control of the schools in Kalamazoo from the village trustees to six trustees, who were to be elected as a Board of Education. The amendment also empowered the Board of Education to classify and grade the schools (which already had been done in response to Putnam's recommendation) and to adopt courses of study and textbooks. The qualified voters were authorized to raise such sums, by taxation, as were necessary to "make the several schools of the district free of tuition in all English branches, to the resident scholars thereof."[4] This amendment to the act of 1850 was approved February 12, 1859. Two

[3] *Ibid.*, p. 240.

[4] *Session laws,* 1859, p. 339.

days later a general school act was approved by the governor. This provided that any district in the state containing more than two hundred children between the ages of four and eighteen might elect a district board of six trustees if the voters should so determine at an annual meeting. The trustees so chosen were to have "the power to classify and grade the scholars in such district and cause them to be taught in such schools or departments as they may deem expedient, to establish in said district a High School when ordered by a vote of the district at any annual meeting and to determine the qualifications for admission to such schools, and the prices to be paid for tuition in any branch taught therein, [and] to employ all teachers necessary for the several schools of the said district."[5] By a special act passed in 1861, the legislature decreed that the Kalamazoo school district was to have all the powers and privileges conferred upon school districts by general law.[6]

The question that was to be raised in the Kalamazoo case was whether the Board of Education had been empowered, under any or all of these laws, to establish a high school and to operate it with tax money. Had this question been raised immediately, it is quite possible that the courts would have found against the school board. It certainly had no such power under the general law of 1859, for that act had specifically stated that the establishment of a high school must be authorized by a vote of the district at an annual meeting, and no such vote had ever been conducted. Did the board have such power under the special act of 1859, amending the law of 1850? This enactment did authorize the Board to classify and grade the schools and to make the schools free in all English branches. This might be implied as bestowing upon the Board the right to maintain a high school, but it is curious, if that is the case, that two days later, in the general act, the lawmakers should take such pains to define exactly the conditions under which a high school might be established.

High schools were by no means uncommon by 1860. The first high school in the United States was opened in Boston in 1821, and four years later one was founded in New York City. A Massachusetts law, passed in 1827, required every town having five hundred families to maintain a high school. The first high school in Mich-

[5] *Ibid.*, p. 391.

[6] *Session laws*, 1861, p. 402.

igan was started in Detroit in 1844. Although, by 1850, there were fifty-five high schools in the nation, Detroit's was the only one in Michigan. During the 1850's, however, several leading towns in the state opened high schools. Kalamazoo was one of about eighteen such communities in 1860.[7] The best organized and equipped of these were in Adrian, Ann Arbor, Battle Creek, Coldwater, Jonesville, Monroe, Ypsilanti, and Kalamazoo, according to Daniel Putnam.[8] The Kalamazoo High School graduated its first class in 1863. At that time the average attendance was less than eighty, but of course we may assume that the Civil War, then in progress, had taken some of the young men who might have enrolled. As late as 1879, however, the enrollment was only 189, so it would be safe to estimate that the high school remained relatively small in numbers throughout the early seventies when the Kalamazoo case was tried.

In January, 1873, more than fourteen years after the Kalamazoo High School had been opened, a bill of complaint was filed in the Kalamazoo Circuit Court, asking the Court to restrain the township treasurer from collecting such portion of the school tax in the village as was intended for the support of the high school and to restrain the Board of Education from spending tax funds for this purpose. It also asked the Court to restrain the treasurer from collecting taxes to pay the salary of the Superintendent of Schools. The charge of the complaint was that there was no legal basis for such taxes, that the law of 1859 establishing graded and high schools was unconstitutional, but that if it should be found to be constitutional, its provisions had not been complied with since no vote of the district had ever authorized a high school. The petition also asserted that there was no law which authorized the Board of Education to employ a Superintendent of Schools. An injunction was granted by Judge Charles R. Brown and filed with the county clerk on January 13, 1873. The opinion of the court on the complaint was not rendered for more than a year, and we can only speculate as to how the high school and the Superintendent got along during that time without any funds.

[7] Elwood P. Cubberly, *Public Education in the United States* (Boston, 1934 ed.), p. 263.

[8] Daniel Putnam, *The Development of Primary and Secondary Public Education.* (Manuscript in Kalamazoo Public Library), p. 79.

While the decision of the Court was awaited, the issues which had been raised apparently became a matter of state-wide discussion. The Superintendent of Public Instruction, Ormael Hosford, in his annual report published in 1873, has this to say about the case:

> No little excitement has arisen in some places, from the discussion of the question of the legality of public high schools. The claim has been made that this department formed no part of the school system recognized by law, and that all taxes imposed to raise money to sustain these schools were illegal. It is claimed to be unjust to sustain such a school. The law contemplates that such schools should be self-supporting by charge of tuition. The law merely provides that every child shall be educated up to a certain point, and makes no provision for taxation for giving instruction beyond the ordinary English branches, and the school boards exceed their powers in providing high schools at public expense.
>
> It also is contended that these boards have no right to employ superintendents and pay them with public money; that they not only created the office, but had abdicated the power conferred on them by the people to the person appointed to the illegal office.
>
> I have given this full statement of the views held by not a few citizens of the state, and the ground of the opinion expressed, that the attention of the legislature might be called to the subject that they may have a definite idea of the ground of the objections made to high schools, as now established and maintained.
>
> If these statements were mere rumors or complaints made by thoughtless men, who were dissatisfied with the tax they had to pay, they would not be worthy of the least notice; but since they come from men of character and standing, who have a reputation as lawyers, and who profess to have given the subject careful consideration, I have deemed it best to call the attention of the legislature to the subject.[9]

Several writers on the history of education refer to the Kalamazoo case as a "friendly suit," inferring that it was taken to court by individuals who were only desirous of establishing once and for all the legality of tax support.[10] Archie Nevins, in his study of the case, has pretty well demolished this idea.[11] The complaint was filed with the Circuit Court by three Kalamazooans: Charles E. Stuart,

[9] *Historical Catalogue of the Kalamazoo Public Schools, 1859–1909* (Kalamazoo, 1909).

[10] Chris A. De Young, *op. cit.*, p. 202; also H. G. Good, *A History of American Education* (N.Y., 1956), p. 251.

[11] Archie P. Nevins, *The Kalamazoo Case.* (Manuscript in Kalamazoo Public Library, dated 1942.)

Theodore P. Sheldon, and Henry Breese.[12] All three men were well known in the community, and one of them had attained national prominence. Nevins points out that Sheldon and Breese were partners in the banking business and that they owned considerable real estate and business property. Sheldon had settled originally in White Pigeon where he had been deputy receiver of the land office. He retained the position when the land office was moved to Kalamazoo in 1834. He opened a private banking house in 1844, and Henry Breese had become his partner in 1862. He had served several terms as village trustee, but there is no record that he ever served on the Board of Education. There is no evidence of any civic or educational interest on the part of Henry Breese.

Charles E. Stuart was the son of a physician in upstate New York. He had apprenticed himself to a lawyer and later was admitted to the bar. He had only a common school education. In 1835, at the age of twenty-five, he had moved to Kalamazoo and within a few years had built up a lucrative law practice. He was elected to the legislature as a Democrat in 1840, and served three terms in the state house of representatives. In 1846, he was elected to Congress. Although defeated for re-election in 1848, he was returned by the voters in 1850. In 1852 he was elected by the legislature to the United States Senate. He served there from 1853 to 1859. Since the legislature was controlled by the Republicans in 1858, he was not re-elected. His prominence in Democratic politics is shown, however, by the fact that he was permanent chairman of the Democratic National Convention which met in Baltimore in 1860 and nominated Stephen A. Douglas for the presidency. Tradition has it in Kalamazoo that Stuart and Douglas were intimate friends, and that the Little Giant visited Stuart in Kalamazoo. Stuart held no office after 1859 but continued to be active in the practice of law in Kalamazoo.[13]

None of the complainants appears to have made any public statements of record beyond what appeared in their bill of complaint. It is worthy of note that Sheldon platted an addition to the Village of

[12] The latter is spelled "Brees" in the legal document, although the Kalamazoo County history, published in 1880, and other local sources render it "Breese."

[13] *History of Kalamazoo County, Michigan*, p. 120; *National Cyclopedia of American Biography* (N.Y., 1901).

Kalamazoo on June 3, 1872, and that he and one George O. Rice had previously made such a plat in 1860. Stuart entered two plats, one in 1859, the other in 1867. Thus, two of the parties to the suit were large real estate holders and stood to benefit substantially if the school tax was reduced.[14]

The decisive evidence, however, that the Kalamazoo case was not a friendly suit, is an account of a school meeting held in 1880, which Stuart addressed, and a letter which he afterward sent to the editor of the *Kalamazoo Telegraph*. Six years after the Kalamazoo case had been decided, the School Board called a public meeting to discuss the need to enlarge the high school. At an evening session, Stuart gave an extended statement of his views. These were reported in the *Kalamazoo Gazette*, then the Democratic newspaper, and an editorial appeared in the Republican *Telegraph* criticizing what he was reported to have said. Then Stuart wrote a letter to the editor of the *Telegraph* to correct what he said were misrepresentations and calumnious statements about his speech. In his letter, Stuart stated clearly his views on high schools.

According to the account in the *Gazette*, Stuart opened his speech by saying that during the past ten years, Kalamazoo, with 12,000 inhabitants, had paid about $1,250,000 in taxes. He asserted that no 12,000 people on God's earth could pay a million dollars in taxes every ten years and exist. Farmers, he claimed, moved into the Village to educate their children at public expense. He said the school tax was more than double all the other taxes. This thing, he went on to say, extended to the University of Michigan, where "they educate young men from every state in the Union and all God's creation for nothing, while we, the glorious and liberal taxpayers foot the bill." He said he had never had anything but a common school education, and he could not see that he had suffered as a consequence. In his letter to the *Telegraph*, Stuart claimed he had thought much on the subject "for the past fifteen years" and was clearly of the opinion that the high school ought to be abolished. If that cannot be done, then it should be conducted in the most economical manner possible, he advised. He concluded by asserting that the moral right of the state to tax ends with the common school. Stuart took issue with the editor of the *Telegraph*, who

[14] A copy of the decision is in the Kalamazoo Public Library.

backed the need for better high school facilities with the prediction that within a lifetime, Kalamazoo would be a city of 40,000. Stuart pointed out that the 1880 population was 11,963, including, as he put it, "676 patients at the Asylum, a source from which I presume there is not much increase expected for the next fifty years." On the basis of past growth, Stuart estimated that the population of Kalamazoo in 1920 would be about 25,000 rather than the 40,000 the editor had estimated. The actual census figure for 1920 was 48,487.

Judge Brown rendered his decision in the Kalamazoo case on February 9, 1874, some thirteen months after the complaint had been filed. Reviewing territorial and state legislation on the subject of education, he concluded that there was nothing in these acts that showed an intent to confine public education to the primary branches. He quoted the Constitution of 1850, which made it the duty of the legislature to establish a system of primary schools, but he said this could not be reasonably construed as forbidding the establishment of other schools. He brought out the important point of constitutional law that the legislature does not derive its power to legislate from the constitution. An act to be invalid, he said, must be repugnant to the constitution and violate its provisions. He also examined the constitutional provision that all instruction must be in the English language. This did not mean, he claimed, that other languages than English could not be taught but rather that precepts and information given the student must be communicated in English.

Turning to the question of whether the special and general acts of 1859 and the act of 1861 had given the Board of Education in Kalamazoo the authority to establish a high school, Judge Brown expressed the opinion that they did. He pointed out that the special act made it the duty of the district to raise by taxes such sum "as shall be necessary to make the several schools of the district free of tuition in the English branches." This clearly implied, Judge Brown said, that other than the English branches might be taught. He went on to point out that the same law authorized the treasurer of the district to collect money due from nonresidents by suit-at-law, but that it was silent as to enforcing the payment of tuition from resident pupils. He inferred, therefore, that it was not the intent of the act that resident pupils be charged tuition.

Judge Brown then considered whether the Board had acted properly in employing a superintendent of schools. He pointed out that the Constitution of 1850 had authorized school boards to employ "subordinate officers." He dismissed as untenable the claim that because the members of the Board of Education could not receive salaries they could not delegate their powers to one who did. He concluded his review of the case by recalling that the voters of the district in 1855 had voted to purchase a site for a "Union School" and that in 1856 they had voted to build such a schoolhouse. After the opening of the high school, a Board of Education had been elected. This, together with the operation of a high school for over a decade, was evidence, the judge believed, of public approval. The injunction was therefore dismissed, and costs were not awarded to either party.[14]

An appeal was taken by the complainants to the State Supreme Court. Arguments were heard on July 10 and 15, and the decision was rendered on July 21, 1874. In comparison with the long delay in the Circuit Court, the action of the Supreme Court seems remarkably fast. The Michigan Supreme Court, at that time was composed of four members: Chief Justice Benjamin F. Graves and Associate Justices Thomas M. Cooley, James V. Campbell, and Isaac P. Christiancy. The decision in the Kalamazoo case was delivered by Justice Cooley, who, though not Chief Justice, was easily the most distinguished member of the Court.[15] All the justices concurred in the decision, which sustained the opinion of the Circuit Court and thus established high schools in Michigan on a firm legal foundation.

At the outset, Justice Cooley recognized the importance of the case to the people of the state and the schools. The notion that the suit was a friendly one may have been caused by one remark Cooley made in his opinion. Said Cooley: "The suit . . . is not to be regarded as a blow purposely aimed at the schools. It can never be unimportant to know that taxation, even for the most useful or indispensable purposes, is warranted by the strict letter of the law. . . ." Although it might be inferred from this statement that the case had been started

[15] The decision is found in *Michigan Reports, 1874–75* (Lansing 1875), pp. 69–84. It is reprinted in full in Edgar W. Knight and Clifton L. Hall, *Readings in American Educational History* (N.Y., 1951), pp. 544–554.

merely to clarify the legal position of the high schools, all Cooley actually said was that it was not meant as an attack on the schools.

The great judge quickly brushed aside the question of whether the failure to take a vote of the district on the establishment of a high school meant that it had no legal foundation. Unlike Judge Brown, Cooley did not undertake to uphold the action of the Board of Education under the provisions of the laws of 1859 and 1861. "Whether this particular objection would have been worthy of serious consideration had it been made sooner, we must, after this lapse of time, wholly decline to consider," the judge said. He went on to point out that there had been a "steady concurrence" on the part of the people by electing officers, levying "large taxes," and employing high school teachers: "To require a municipal corporation, after so long an acquiescence, to defend in a merely private suit, the irregularity, not only of its own action, but even of the legislation that permitted such action to be had, could not be justified by the principles of the law, much less by those of public policy."

The larger part of Judge Cooley's opinion was devoted to the statement of the court's opinion on whether there was authority in Michigan to make high schools free by taxation levied on the people at large. After restating the claim of the complainants that instruction in the classics and modern languages is not necessary for the benefit of the many but rather is for the benefit of the few and ought to be paid for by them rather than from general taxation, Cooley proceeded to examine at some length legislation on the subject. He cited the Ordinance of 1787 and then the famous act of the Territorial legislature in 1817 establishing a "Catholepistemiad," noting that the latter envisioned a complete system of education supported by taxation. He then proceeded to examine the provisions of the Constitution of 1835, and found that it provided not only for common schools, but also for a university, "with such branches as the public convenience may hereafter demand for the promotion of literature, the arts and sciences." He also quoted from the messages of the governor and from the recommendations of the first Superintendent of Public Instruction to show that a complete system of education was contemplated. In dealing with the Constitution of 1850, he recalled that the debates in the constitutional convention dealt with a proposal that the English language and no other shall be taught in the schools. It was stated in the convention that some schools taught French, German, and Latin, and the convention then changed the wording of the clause so that it provided that instruction should be in the English language. The Constitution of 1850, Cooley stated, provided for a university. Since branches had ceased to exist by that time, he inferred that the framers of the Constitution intended the high schools to serve as preparatory schools for the university. Concluding, Cooley said "if these facts do not demonstrate clearly and conclusively a general state policy, beginning in 1817 and continuing until the adoption of our present constitution, in the direction of free schools in which education, and at their option the elements of classical education, might be brought within the reach of all children of the state, then, as it seems to us, nothing can demonstrate it." After dismissing in a single short paragraph the claim that the Board of Education exceeded its powers in employing a superintendent of schools, the court affirmed the dismissal of the bill of complaint.

Writers on the history of education invariably cite the Kalamazoo case as the decisive judicial foundation for the public high school. Grieder and Romine, in their *American Public Education* assert that Cooley's decision "influenced courts in other states, and is rightly regarded as a landmark in the development of public education."[16] Another author says the decision "established the free tax-supported high school in Michigan, and set a precedent for other states."[17] Still another writer classes the Kalamazoo case with the Dartmouth College case as one of the major judicial decisions in American educational history. He also notes that the Asheville case, decided by the North Carolina Supreme Court in 1930, upholding the right of the City of Asheville to tax itself to maintain a junior college, is parallel with the Kalamazoo case.[18] Another well-known writer in educational history cites the Kalamazoo case with the Dartmouth College case, the Oregon case (1924), the Berea College case (1908), and the Missouri case (1938) as being the leading court decisions on education in America. He adds, "This decision [in the Kalamazoo case] became the legal precedent for other states and greatly influenced the

[16] Calvin Grieder and Stephen Romine, *American Public Education* (N.Y., 1955), p. 86.

[17] Aubrey A. Douglass, *The American School System* (N.Y., 1934), p. 130.

[18] Chris A. De Young, *op. cit.*, p. 220.

development of the high school at public expense, although some states were slower than others to acknowledge the responsibility."[19] An English writer says the opposition to public support for high schools in America "was finally defeated by the famous Kalamazoo decision. . . ."[20] Cubberly says the verdict of the Supreme Court in the Kalamazoo case "was so positive that it influenced all subsequent decisions in other states."[21]

Several authors refer to the trial of similar cases in other states. A somewhat similar suit had been tried in Pennsylvania in 1851, for example, but it never was regarded as decisive. In Illinois and at least in seven other states, cases of the same kind were decided in the same way.[22] Good, in his comments on the Kalamazoo case, gives two reasons for the tendency to regard the Kalamazoo case as the leading one. First, he says, it is because Cooley argued that the high school was part of a *system* of education which the legislature intended to establish and adds, "the argument has always had a special appeal for educational historians"; second, he attributes the pre-eminence of the Kalamazoo case to the fact that none of the other decisions was written by "an equally eloquent jurist."[23]

There is no doubt that Cooley was an eminent jurist. Always a man of tremendous energy and capacity for work, even his professorship of law and his duties on the Supreme Court did not exhaust his resources. He found time to write books on the law that gained him a nation-wide reputation. He edited and wrote works of vast erudition on Blackstone, on Story's *Commentaries*, on taxation, on torts, and on constitutional law. His best-known work was his *Treatise on the Constitutional Limitations Which Rest Upon the Legislative Power of the States of the American Union*, published in 1868. In addition to this work, he also had published his edition of Blackstone prior to the Kalamazoo decision, and his edition of Story's *Commentaries* came out the same year (1874). There can be

[19] Edgar W. Knight, *Twenty Centuries of Education* (Boston, 1940), pp. 274–275.

[20] A. D. C. Peterson, *A Hundred Years of Education* (London, 1952), p. 144.

[21] Elwood P. Cubberly, *op. cit.*, p. 263.

[22] Edward H. Reisner, *Nationalism and Education Since 1789* (N.Y., 1929), p. 455 ff.; H. G. Good, *A History of American Education* (N.Y., 1956), p. 251–252.

[23] *Ibid.*

little question that the fame of the Kalamazoo case is due in large measure to the reputation of the great jurist who wrote it.

It is somewhat doubtful whether Cooley's opinion in the Kalamazoo case can be regarded as among his really great decisions. There is no searching inquiry and analysis of the nature of the constitution which one finds in some of his other ones. But the reputation of the decision is certainly well established among educational historians, and there is every reason to presume that it will continue to be cited as one of the great foundation stones of our American system of education.

PAPERS OF THE MICHIGAN ACADEMY OF SCIENCE, ARTS, AND LETTERS
VOL. XLV, 1960 (1959 MEETING)

BEGINNINGS OF THE AUSTRIAN COTTON INDUSTRY

DWIGHT CLARK LONG

The University of Michigan

THE history of cotton manufacture illustrates better than any other the progress of the industrial revolution in Austria. It began in the eighteenth century, before the advent of the new machines yet late enough to be free of the guild organization that often hampered the older textile industries. It met the typical problems of the revolution —scarcity and high cost of capital, dearth of technical skills, reluctance of labor to move from home to factory, lack of information about new machines, and inertia of officialdom. The energetic men building this new industry found at least partial solutions, aided by the government to some extent and notably by the increasingly strong demand created by cotton itself. Writers on the economic history of Europe of this period have too often either omitted Austrian developments or included them in a general way in the German picture. These deserve particular attention, however, if we are to develop a more accurate account of the expansion of the industrial revolution on the Continent. We shall follow the early development of the Austrian cotton manufacture to about 1830.

Its modest beginnings grew out of trade with the Levant. In 1719 the Austrian Oriental Company was founded, and Vienna became the center of supply of raw cotton for Germany and Switzerland.[1] The first cotton factory in Austria was a small venture near Graz in Styria about 1721.[2] In 1726 the Oriental Company opened a factory at Schwechat, near Vienna, later sold after the Company broke up in

[1] "Denkschrift über den Zustand der Baumwollspinnerey in der österreichischen Monarchie ... 1828," by a committee of the Lower Austrian spinning mills, Hofkammer Archiv, Commerz-Cammer, Fasc. 1115, No. 86 ex August 1928. Hereafter this archival source is cited as HKA, Com.-Cam.

[2] H. Hallwich, *Firma Franz Leitenberger, 1793–1893* (Prag, 1893), p. 14. Some spinning and weaving of cotton had begun among the linen workers of northern Bohemia by the mid-seventeenth century. *Ibid.*, pp. 8–9.

[3] Franz Mayer, *Die Anfänge des Handels und der Industrie in Oesterreich und die orientalische Compagnie* (Innsbruck, 1882), pp. 59, 118.

the thirties.[3] Meanwhile two or three other small plants were established, including one by the husband of Maria Theresia.[4]

This energetic Empress gave a strong impetus to the cotton industry through practical measures providing freedom, protection, and aid. Not only was the importation of most manufactured cottons prohibited, but domestic production was encouraged to supply the home market. Spinning and weaving schools were organized and new enterprises encouraged by subsidies and concessions.[5] Hand spinning and the weaving of cotton were made free industries in 1740. By 1763 the entire manufacture of cotton was opened to all, a most important advantage in its competition with other textiles.[6] Thus at this early stage the government recognized something of the economic and social potential of the new industry and gave active support to meeting its problems.

It was in the rich province of Bohemia that the new industry now made its greatest progress. To compensate for the loss of Silesia's rich trade when that province was seized by Frederick II, Maria Theresia raised a high tariff against Prussia. While this brought reprisals that cut off the flow of Bohemian textiles via Silesia to the markets of northern and eastern Europe, the effect of this diminished as new channels of trade were developed.[7] Further, the Empress declared that the cotton industry was to be firmly established in Bohemia without delay. Recognizing that many of the Bohemian nobility had experience in and facilities for industrial enterprises on their own estates, she urged them to lead in the new development. Count Kinsky, progressive head of the Bohemian board of commerce and manufactures, erected a cotton works on his estate in 1763 and his neighbor, Count Bolza, one at Josefsthal-Cosmanos. Others followed so that in 1765 officials reported over seven thousand spinners, some three hundred weavers, and twenty-four printers of cotton in Bohemia.[8] Fears of injury to the older textile crafts

[4] Hallwich, *Leitenberger*, p. 18.

[5] Stephan von Keess and W. C. W. Blumenbach, *Systematische Darstellung der neuesten Fortschritte in den Gewerben und Manufacturen . . .*, I (Vienna, 1829), pp. 120–121.

[6] Hallwich, *Leitenberger*, p. 20.

[7] On attempts to win additional markets in the south and east see Adolf Beer, "Die österreichische Handelspolitik unter Maria Theresia und Josef II," *Archiv für österreichische Geschichte*, 86 (1899), pp. 1–204.

[8] Hallwich, *Leitenberger*, p. 31.

brought a decree in 1789 that new cotton factories were to avoid regions preëmpted by the linen and woolen industries. Nevertheless, cotton cloth-making spread with great vigor. A careful estimate gives some twenty-four thousand cotton spinners in Bohemia in 1789, and a good forty thousand by the end of the century.[9]

Assured of a growing domestic market by extension of the system of prohibitions in 1784, more Austrians ventured into cotton manufacturing in spite of such obstacles as the scarcity of capital. Joseph II gave some relief in respect to the latter by his Tolerance Patent of 1781, which brought many Protestants with capital to Austria. Certain rights were granted to Jews, also.[10] Furthermore, direct government subsidies to new enterprises were granted now in considerable amounts.[11] Another problem was to secure information about the new machines and techniques, solved in part by smuggling plans and machines and by attracting skilled workers to Austria. In 1790 the Austrian ambassador to England smuggled out a spinning machine which served as a model in equipping a mill in Steiermark.[12] By a decree of January 1786 each immigrating foreign skilled cotton weaver was to receive 100 florins.[13] Such workers were assured religious tolerance, freedom from military service, and unhindered return to their homelands. Many such craftsmen shared in building the new industry.[14] In the province of Vorarlberg seasonal workers returning from France and Switzerland introduced the working of cotton in the seventies and it spread rapidly there.[15]

In this period of experimentation in machines and techniques an enterprise was established in Bohemia by a man who deserves credit for promotion of the cotton industry. Johann Josef Leitenberger is

[9] J. Schreyer, *Waarenkabinet* (Prag, 1799), pp. 167–168.

[10] Johann Slokar, *Geschichte der österreichischen Industrie und ihrer Förderung unter Franz I* (Vienna, 1914), pp. 5, 259–260.

[11] *Ibid.*, pp. 6–7.

[12] Gustav von Pacher, *Die Baumwollspinnerei in Oesterreich-Ungarn* (Vienna, 1891), p. 3. English prohibition of the migration of skilled workmen and of exportation of textile machines and plans was lifted partially in 1825 and fully in 1843.

[13] Hallwich, *Leitenberger*, p. 77.

[14] J. L. von Barth-Barthenheim, *Oesterreichische Gewerbs- und Handels-Gesetzkunde* (Vienna, 1819), I, pp. 51–52; Karl Přibram, *Geschichte der österreichischen Gewerbepolitik von 1740 bis 1860*, I (Leipzig, 1907), pp. 351–356.

[15] R. Schenkel, *Das Haus F. M. Hämmerle* (Vienna, 1900), p. 4.

a splendid example of the intelligent and energetic entrepreneur. He became a leader in the cotton industry and one of the greatest Austrian industrialists of his day. Leitenberger learned how to print and dye cotton at Augsburg in Bavaria. Marrying the daughter of a Bohemian dyer, he joined in the small family business and five years later was in full charge of the work. He introduced the Saxon single-thread wheel and the flying shuttle in Bohemia, and later improved the hand loom.[16] In 1770 he built a cotton finishing plant at Wernstadtl and to ensure an adequate supply of yarn, a spinning mill in Prag was added, the first of its kind in Bohemia.[17] In 1788 a fine new plant near Reichstadt began operations under the direction of his two sons. In 1792 the two plants engaged over nine thousand spinners and weavers and some six hundred persons in the printing works. The following year the nearby cotton mill of Count Balza at Josefsthal-Cosmanos was purchased.[18]

The elder Leitenberger introduced the first spinning machines of English design in Bohemia. Hearing of a mill in Copenhagen where a clever person might obtain drawings of English machines, he dispatched two of his mechanics to do their best. They found the English manager of the mill "unapproachable" but were fortunate in locating a Dane who had built machines for years in England. He was brought to Bohemia in 1796 and by 1799 each of the three Leitenberger plants had a spinning mill equipped with water frames and mule jennies of English design. This example of the spread of technical skills did not end there. One of the mechanics who went to Copenhagen set up a small mill near Teschen, Silesia, in 1801, assisted by his companion on that trip, and soon the latter went into business for himself, eventually founding four spinning mills.[19]

These enterprises illustrate opportunities, problems, and growth of the cotton industry under policies of Maria Theresia and Joseph II. In the brief reign of Leopold II, and for years after Francis came to the throne in 1792, however, fear of revolution, irresolute administration, and wars with the French checked economic development.

[16] Keess and Blumenbach, *Systematische Darstellung,* I, p. 121. For information on early spinning machines used in Austria see Stephan von Keess, *Darstellung . . .,* I, Part II (Vienna, 1824), pp. 83–84.

[17] Hallwich, *Leitenberger,* pp. 50–54.

[18] *Ibid.,* pp. 56, 58, 60.

[19] *Ibid.,* pp. 88–91.

But the spreading technological revolution was beginning to make itself felt in Austria, creating problems which had to be faced. War and blockade focused attention on them and emphasized the need for a new consideration of industrial policy.

Early in the new century there was a definite, if not very vigorous, attempt to ascertain proper governmental policies for business and industry. A number of questions were posed. Should the system of prohibitions be relaxed? What should be the policy toward the new enterprises that were affecting the lives of so many people? How much freedom should industry have? How much protection against machines should the state assure the hand workers? We cite but one example of the often protracted debates on such questions. In 1801 the government of Lower Austria proposed that factories be banned from Vienna and its suburbs on the grounds that they would cause a housing shortage and increase the cost of foodstuffs. While the Emperor and Chancellory supported it, the Hofkammer, charged with business matters, declared the ban would be an unwarranted restriction of freedom of industry. After eight long years of debate and delay, during which the imprudence of such a ban became ever clearer, the principle of freedom of industry was accepted as the legal basis of business policy.[20] Economic liberalism, based on physiocratic ideas and the concept of natural rights, gradually began to influence decisions, giving increased freedom to industry and trade.[21] Fundamental to the growth of large-scale industry was the fact that whereas importation of machines had been forbidden or subject to special license, it was now not only permitted but encouraged.[22]

The rising cotton industry was especially able to profit from this significant, albeit gradual, liberalization of economic policy, and in the first dozen years of the new century it made excellent progress. With twelve new spinning mills, Lower Austria became the center of the wholesale cotton yarn trade in the monarchy. The first and most notable of these mills was at Pottendorf. In 1800 Prince Josef zu Schwarzenberg, head of the Kommerz Leih-und-Wechsel Bank in

[20] HKA, Com.-Cam., F. 1115, No. 41, ex May 1825 and No. 57, ex January 1827; Slokar, *Geschichte*, pp. 25–50, 126.

[21] *Ibid.*, pp. 51–54.

[22] *Ibid.*, pp. 182–185; Barth-Barthenheim, *Gewerbs- und Handels-Gesetzkunde*, I, p. 54.

Vienna, decided that Austria should have a large English-style cotton spinning mill. Freiherr von Glave-Kolbielski, sent to England to secure information and mechanics, contracted with John Thornton of Yorkshire, a master mechanic of "extensive knowledge and untiring activity," and mechanic John Lever to build near Vienna a mill and necessary machines of English design for what was termed "a new branch of industry." A company was formed in 1802 with the bank, some of the nobility, two wholesale merchants, a government official or two, and a few others taking shares. By 1805 Pottendorf had over eighteen thousand mule-and-throstle spindles in operation, and in 1811 nearly thirty-nine thousand, making it the largest mill in Austria and one of the largest on the Continent.[23] In 1801 two English mechanics, Edward Royce and William Taylor, began building machines for new mills at Schwadorf and Teesdorf, which opened in 1802 and 1803.[24] The great dependence on English designs and technical skill is very evident in contemporary records.

A favored location in Lower Austria was the poor agricultural region south of Vienna where the inhabitants were glad to supplement their meager incomes.[25] All these mills were run by the abundant water power. Most of them were established by merchants of Vienna, whose business connections and easier access to capital were distinct advantages. Besides the new spinning mills, over twenty plants produced finished cottons in Lower Austria.

Increased importation of English machine yarn raised the question of protection in 1800. Was the government to yield to the spinners' demands or listen to the weavers, who wanted to have the superior English yarn? The spinners won a partial victory when the tariff on foreign yarn was raised from five to ten florins per *centner* and in 1804 to twenty florins. But English competition remained sharp previous to the Blockade of 1806. English spinners, Keess reports, sent in large amounts of yarn for sale at a discount of 30 per cent. Lower Austrian mills sought a compensatory premium and higher duties, but this time they got neither. The truth is that domestic production could not yet meet domestic needs.[26]

[23] *Vaterländische Blätter* . . . (Vienna, 1811), No. 90, pp. 541–542: Slokar, *Geschichte*, pp. 182–184, 280.

[24] "Denkschrift . . . 1828," HKA, Com.-Cam., F 1115, No. 86, ex August 1828.

[25] Josef J. Knolz, *Darstellung der Verfassung und Einrichtung der Baumwoll-Spinnerei-Fabriken in Nieder-Oesterreich* (Vienna, 1843), pp. 13–15.

[26] Keess, *Darstellung*, I, Part II, p. 84.

Inauguration of the Continental System in 1806 gave added protection and increased business to the Austrian cotton industry. Freedom from sharp competition of English machine yarn and the steadily growing demand for cotton wares resulted in rapid expansion and considerable prosperity over the next several years.[27] But the financial crisis and bankruptcy of the government in 1811 dealt a severe blow to all business. Depreciation of the currency cut real tariff levels and spinners demanded higher duties. In December 1811 the duty on all cotton yarn was raised from twenty to fifty florins per *centner* and a year later to one hundred florins, with importation of yarn up to No. 50 prohibited.[28] There was much smuggling, and the situation became infinitely worse with the lifting of the Blockade. In an attempt to curb the flood of English goods, customs were made payable in Convention currency in 1814. But this made the duty on yarn so high that it was evaded, and demand for the prohibited English cotton cloth was so strong that smuggling broke all bounds.[29]

During these postwar years even so capable a manager as Franz Leitenberger, Johann's son, had difficulty in saving his firm from ruin. Recognizing the necessity of adopting the best techniques, he introduced English cylinder printing and brought Carl Köchlin, a dye expert, and Jeremias Singer, a designer, from Alsace. As a result, his firm's products were able to compete successfully with those of England and France in foreign markets.[30] But curtailed production was the common lot. From 1815 to 1819 almost no new cotton factories appeared.

[27] Keess states that the product of domestic spinning mills almost doubled between 1812 and 1815. *Ibid.*, p. 85.

[28] "Denkschrift . . . 1828," HKA, Com.-Cam., F 1115, No. 86, ex August 1828. The English numbering system for yarn was followed generally. Nos. 1 to 30 were coarse and comprised the bulk of the Austrian yarn; Nos. 32 to 60 were medium; Nos. 62 to 100 were fine, and those over 100 were finest.

[29] Florins C. M., "Convention Money," were metallic florins, coined under a convention with the Elector of Bavaria. One florin C.M. was equal to about forty-eight cents. When financial reform began in 1816 the depreciated paper florin was given a value in C.M. of 40 per cent of face and was labeled "Wiener Wahrung," or "W.W."

[30] Hallwich, *Leitenberger*, pp. 108, 110. Köchlin and Singer left in 1819 to establish their own cotton printing plant. *Bericht über die erste allgemeine österreichische Gewerbsprodukten-Ausstellung im Jahre 1835* (Vienna, [1836]), p. 110.

On the other hand, because of the need for improved efficiency and quality, this period was one of increased activity in the development of machines and processes. The government had founded a polytechnical institute in Prag in 1806, and in 1815 one in Vienna. The latter, under its energetic director J. J. Prechtl, took an active part in securing and disseminating technological information.[31] Government support, in some cases financial, was given those bringing in new equipment, to Girard of Paris, for example, for his new flax-spinning machine.[32] New machines and all models were admitted duty-free. Austrian inventors did not produce many important machines but did make numerous improvements in the foreign imports, especially those for working cotton.[33]

Another change brought about by, and in turn stimulating, technological development was in the handling of patents. Under the law of 1810 decisions were in the hands of political officials rather than technical experts. The new patent law of 1820, applicable to foreigners equally with citizens, improved both procedures and patent rights, with notable results.[34] At the same time a new law on joint-stock companies facilitated investment of foreign capital.[35] The government had replaced its policy of direct financial aid to industry with that of encouragement through elimination of hindrances to its free development.

A significant result of these more favorable conditions was the increase in technical skills. Contemporary writers emphasize especially the active improvement in both quantity and quality of Austrian machine spinning, but mention other phases of production also. Keess, one of the best sources, comments that new methods and inventions of England, France, and Switzerland were introduced in the larger plants "without regard to cost and effort."[36]

Although the general economic situation remained unsatisfactory in the early twenties, signs of recovery began to appear in the cotton industry. There was a steady increase in use of cotton clothing by

[31] Slokar, *Geschichte*, p. 168.

[32] D. C. Long, "Philippe de Girard and the Introduction of Mechanical Flax Spinning in Austria," *Journal of Economic History* (1954), 21–34.

[33] See Prechtl's *Jahrbücher*, Vols. I, II, for notices of patents granted.

[34] Slokar, *Geschichte*, pp. 253–254.

[35] Anton von Krauss, *Geist der österreichischen Gesetzgebung zur Aufmunterung der Erfindungen im Fache der Industrie* (Vienna, 1838), p. 183.

[36] Keess and Blumenbach, *Systematische Darstellung* (1829), I, pp. 168–170.

all classes, a fact noted repeatedly in contemporary accounts. "The enemy cotton"—cheap and adapted to so many uses—was rightly called the chief cause of the decline of the linen industry. When the question of protection for domestic yarn was raised again in the early twenties, its resolution in lower rather than higher duties indicated the health of the industry. While production of cotton goods mounted steadily exports remained low; domestic demand was absorbing the increase.[37] Cotton was making the strongest recovery and growth of any of the textiles.

We are fortunate to have a revealing report on the Austrian cotton-spinning industry in 1828.[38] Although prepared by a committee of the Lower Austrian spinners to persuade the government to the need for more protection, it adds much to our picture of cotton spinning as well as other branches of the industry. It emphasized problems of Austrian spinning mills by comparing their conditions of production with those of English mills. Capital, the report points out, is more plentiful in England at half the Austrian interest rates. English mills have better organization and connections for securing capital. With cotton markets at his door, the English spinner pays less for raw cotton and does not need to keep capital tied up in large stocks as does the Austrian spinner. Also, Austrian spinners must finance the weavers, usually for three to six months. Often the spinners must buy the cloth and then the credit period is frequently a year or more. The English spinner has the best machinery available and at lower cost. He can specialize in certain grades of yarn and sell at wholesale in world markets while the Austrian spinner has the limited domestic market, must produce a wider range of numbers, and must sell to individual weavers as well as to factories. These important advantages are not balanced, as some might suppose, by the lower scale of wages in Austria, for in skill and productivity the English surpass Austrian workers. However, the report points out that in the last six years the average daily production per fine spindle in most mills had doubled. The committee concluded that Austrian mills still needed the existing tariff protection and a more effective control of smuggling. On the whole the report presents a rather fair view of the problems of Austrian producers in respect to capital, costs, and markets, though the facts indicate that

[37] "Denkschrift . . . 1828," HKA, Com.-Cam., F 1115, No. 86, ex August, 1828.
[38] *Ibid.*

the mills would have prospered even with somewhat reduced duties on yarn.

Statistics of spinning mills, though admittedly incomplete in the survey, show much growth since 1820. In the best mills equipment is reported as up-to-date, and this is supported by other evidence. Recovery in Lower Austria was notable. After six years without a new spinning mill, two began operations in 1820, two in 1822, three in 1823, and twelve more by the end of 1828, making a total of thirty mills with 227,426 spindles. The great majority of the spinning machines were mule jennies of 180 spindles. Pottendorf, the largest mill, had 47,460 spindles (224 mules and 62 water frames) and was second on the Continent only to the Schlumberger mill in Alsace, with 50,000 spindles. Perhaps three other mills had over 100 mules, though the average was 46.[39] Estimated production of yarn in Lower Austria doubled between 1820 and 1828.[40] Upper Austria had but one spinning mill. Vorarlberg had eleven mills with 49,884 spindles, the largest being the Escher plant at Feldkirch with 24,000 spindles.[41] Since most spinning mills in Bohemia belonged to printing works, they were much smaller, though more numerous, than in Lower Austria. Seventy-five are listed in the survey, with a total of 117,231 spindles, which is proof of vigorous growth. Moravia, although second only to Bohemia in production of cotton goods, had no machine-spinning works. Thus the total number of machine-spinning mills in the Austrian cotton industry in 1828 was reported as 111, with 394,541 spindles.[42]

While these totals are indeed small compared with those of England and even with those of France, they show clearly that by 1830 the revolution in the production of cotton yarn was well on its way in Austria.[43] The mule jenny was dominant. It was not until the thirties that the self-actor appeared. Mechanization of weaving was not so rapid. The power loom was introduced at Pottendorf in the early twenties, about the same time as in Alsace, but it spread

[39] *Ibid.*

[40] F. Dael, *Die Baumwolle und deren Verarbeitung* (Mainz, 1846), p. 35.

[41] "Denkschrift . . . 1828," HKA, Com.-Cam., F 1115, No. 86, ex August 1828. Vorarlberg and Tirol were united with the Austrian tariff area in 1826.

[42] *Ibid.*

[43] Keess and Blumenbach, *Systematische Darstellung* (1829), I, p. 120, state that the *Factoreyen*, small hand-spinning works providing yarn for factories, had almost all disappeared.

more slowly in Austria. In 1830 practically all cotton was woven on hand looms, most of it still in the weaver's home. Printing and finishing techniques were advancing, especially with the development of chemistry. Cylinder printing had been introduced and was becoming established in the larger plants of the monarchy. Power still was provided almost exclusively by humans and water. The first steam engine appeared in Austria in 1816. The first in cotton mills were installed by English mechanics in 1823 and 1824.[44] Their use spread slowly, both because of cost and the abundance of water power. Thus we find the Austrian cotton industry well established, though far from fully mechanized, by 1830, and at the threshold of technological growth and expansion which was to raise it in coming decades to the top of the national economy.

[44] H. Hallwich, *Reichenberg und Umgebung* (Reichenberg, 1874), pp. 518–519.

Papers of the Michigan Academy of Science, Arts, and Letters
Vol. XLV, 1960 (1959 Meeting)

POLITICAL SCIENCE

THE POLITICS OF MILITARY AVIATION PROCUREMENT, 1926–32

EDWIN H. RUTKOWSKI

University of Detroit

We have become accustomed to observe an annual struggle between the President and Congress over the proper composition and support for the armed services in the national security establishment. The very regularity of the debate, however, has the unfortunate effect of destroying perspective on the political process involved in this area of contention over basic policy. A case study of certain lesser-known aspects of an earlier, generally familiar dispute concerning national defense can therefore be valuable. Such a study can serve to sketch out the more specific mechanisms of interest and value that operate, in the typically confused manner we currently see, to provide for the common defense of the liberal democratic society of the United States.

The great debate in the early years after World War I concerning the future role of aviation in the defense establishment is still, thanks to recurring mass-media portrayals, well-known as one dominated by the personality and doctrines of General William ("Billy") Mitchell. Planes can sink battleships, he vigorously maintained and then dramatically demonstrated. What was more reasonable, he therefore forcefully argued, than that aviation merited an independent, pre-eminent organization among the armed forces of the country. But the most that his efforts, culminating in his sensational court-martial in late 1925, could attain toward this end was the passage by Congress of the Air Corps Act of July 2, 1926. By that act aeronautics in the army was raised from the level of an auxiliary service to the status of a corps, co-ordinate with the other branches of the War Department. But no organizational independence of traditional general staff leadership was granted.

One hopeful prospect for military aviation did nevertheless seem to be present in the Air Corps Act of 1926. The legislation provided for an expansion program that was intended to build up the equipment strength of the air corps to 1,800 planes, and to increase the army's aviation complement by 403 officers and 6,240 enlisted men in order to obtain a final personnel total of 1,650 officers and 15,000 men in the new army branch.[1] The expansion was to occur through five more or less equal increments over the same number of years, beginning with the very year of the act's passage. Such at any rate might have been the presumptions of aviation dedicated observers. Although these had viewed the act as generally contributing insignificant gains, they thought that at least in terms of what was finally included there would be from 1926 to 1931 a pattern of expansion and development toward unambiguously specific goals determining precisely the status and attention to be accorded the fledgling service in the defense establishment.

Expectations of this kind were grossly inaccurate, however, as become apparent when President Coolidge submitted his first budget estimates under the program in December 1926. His administration took full advantage of the merely permissive, rather than mandatory, character of the act. The final figures approved and submitted by Coolidge revealed the following key decisions to withdraw from the opportunity to develop the aviation arm rapidly and powerfully: the start of the five-year expansion program was delayed, the officer complement of the expansion program was not to start the first year, the enlisted-man complement was not to be obtained by adding to total army strength, and the full schedule for procurement of planes was deferred. By contrast, however, the spirit of encouragement in the act concerning support for the aviation industry was enthusiastically approved by the President.

The first indication that a restrictive interpretation might be placed on the implementation of the expansion program occurred within four months after the passage of the air corps act, when it was announced on October 1 that recruiting for the remaining nine months of the current fiscal year would be terminated from lack of funds. Additional pay costs were forcing the army to cut its enlisted-man strength to 110,940, although the original appropriations granted by Congress for the fiscal year had been scheduled to pro-

[1] U.S., *Statutes at Large*, XLIV, Part 2, 783; U.S., Congress, House, Committee on Military Affairs, *To Provide More Efficiently for National Defense*, 69th Cong., 1st Sess., 1926, H. Rept. 700, p. 11. The latter source is hereafter cited as House Committee on Military Affairs, 1926, H. Rept. 700.

vide for 118,750 men.[2] The implications of this step for the personnel expansion provisions of the aviation program were clear. Such aviation personnel increases as would come could only occur by despoiling other army units of their current strength.

This recourse could be justified despite the terms of the act. It was true that, in calling for the increases in personnel for the air corps, the act had increased the total authorized strength of the army by the same amount as the prescribed increases for the air corps. But like all such legislation, the act was an authorization only. Its provisions sketched out a legitimate scope of action for the President, but could not compel him to perform the action. If no executive can afford to disregard certain Congressional expectations under such acts, he nevertheless can administer the act over a broad range of interrelated alternatives that will disappoint those expectations materially. Moreover, such executive alteration was in this case supported by the fact that Congressional opinion itself was not of one mind on whether personnel increases for the air corps should be taken out of existing army strength, at the expense, in other words, of the other branches of the army, or whether the air corps increases should occur with an equal expansion of total army personnel. The House Military Affairs Committee report had stated that the air corps bill specifically proposed to increase the air corps "without taking this increase away from the other arms."[3] But Representative Fred M. Vinson, who acted as another Military Affairs Committee spokesman in inserting into the *Congressional Record* a many-paged, detailed explanation of the bill at the time of its final passage into law, presented just the opposite interpretation of the bill, the language of which had not been altered on this point in passing the Senate: "Authority is granted by the bill for the President to effect this increase by increasing the Regular Army or by making reallotments of the various branches."[4]

The rumors and fears concerning personnel strength materialized

[2] *Army and Navy Journal*, October 2, 1926; U.S., War Department, *Report of the Secretary of War to the President 1926* (Washington: U.S. Government Printing Office, 1926), p. 142. The latter source is hereafter cited as *Secretary of War 1926*; reports of the Secretary of War for other years will be cited in the same abbreviated form.

[3] House Committee on Military Affairs, 1926, H. Rept. 700, p. 11.

[4] U.S., *Congressional Record*, 69th Cong., 1st Sess., 1926, LXVII, Part 11, 12277.

into fact on December 11, 1926, when Coolidge submitted his budget for the fiscal year 1927–28. His request for funds for total enlisted army manpower revealed that he included insufficient increases to bring that manpower up to the 118,750 level originally planned for in the current appropriations, a level which was being dropped in actual strength to an average of 110,940 because of the depletion of funds, as already explained. Coolidge's plans for the next fiscal year would provide for 115,000 men, in an attempt to make permanent at least a part of the current drop in total strength. As to the increase of men in the air corps to begin its expansion, Coolidge simply stated in his budget message that they were to be provided for in the total army enlisted-strength figure of 115,000 men.[5] Final Congressional action rejected Coolidge's recommendation for a decreased total army manpower, when funds were later voted to restore the strength to 118,750 men. But the decision to squeeze the first air corps expansion increment into an unexpanded total army figure was left standing by the Congress.

An opportunity to take advantage of the authorization by Congress to increase total army strength was thus rejected. It is true that this disinclination was no new departure. The chief executive already had during these years ample authority, apart from the grant in the air corps act, to request increases in the strength of the army. Repeatedly over the previous years legislation stipulated that the total authorized number of enlisted men, not including Philippine scouts, should be 125,000 men, but the President had never requested, and the Congress had never granted on its own initiative, sufficient appropriations to reach that level.[6] With the passage of the air corps act, however, containing an authorization for a larger army by the amount of the increases provided in air corps personnel, strong arguments were put forth that a new temper had arisen which would make Congress willing to raise appropriations and provide for a larger total army. These arguments called attention to the fact that the chairman of the House Appropriations Committee, Repre-

[5] U.S., Bureau of the Budget, *Message of the President Transmitting the Budget for Service of Fiscal Year 1928* (Washington: U.S. Government Printing Office, 1926), p. ix. Cited hereafter as Bureau of the Budget, *Budget Message of the President 1928*.

[6] The various figures and pertinent legislation are reviewed in House Committee on Military Affairs, 1926, H. Rept. 700, p. 11, and in *Secretary of War 1926*, pp. 49–50, 141–142.

sentative Martin B. Madden, had received an estimate in April that put Congress on notice before the passage of the act as to the cost that would be involved in the adoption of the expansion program. And yet, one aviation writer reported, what did aviation supporters see but a decision by the administration not toward "placing responsibility squarely up to the Congress" but rather choosing to include additional expenditures for the air corps within the appropriations for the regular establishment.[7]

To such criticisms, however, the statement of Chairman Madden after a conference with the President directly applied. Echoing Coolidge's well-known views concerning the paramountcy of economy in government expenditures, Madden expressed his sympathy with the cuts made by the budget director in the army and navy estimates, and stated his determination to keep the administration's pledge to cut taxes by one-half billion dollars. He remarked that the United States had already spent 900 million dollars on aviation in the world war, and, if the best still was not had, only poor management was responsible. "There is no one," he stated, "more interested in the preservation of an adequate national defense than I, but I am not foolish enough to put more people on the pay roll and more taxes on the backs of the citizens."[8]

However unfortunate for total army strength the first enlisted-man increment in the air corps was, that increment was at least forthcoming in the budget estimates. The same could not be said about the officer personnel increases scheduled in the expansion program, however, but for reasons not directly related to economy. The act required that 90 per cent of air corps officers be fliers, and provided for the detailing of officers to the air corps for a year to obtain the necessary training and thus make possible their actual transfer to the air corps. But in spite of War Department encouragement, only a small number of officers were requesting such assignments.[9] The reason for this lay principally in the fact that a flying career demanded high physical qualifications, extensive training, and relatively dangerous duty. As a matter of fact, because of these considerations the aviation branch had been, long before the passage of the air corps act, consistently unable to fill the quota of officers cur-

[7] *Army and Navy Journal,* October 16, 1926.
[8] *Ibid.,* November 13, 1926.
[9] *Ibid.,* October 16, 1926.

rently allotted to it. Coolidge made this difficulty the basis of his budget message statement that no provision was present in the estimates for any of the 403 officers specified in the five-year program. "The Air Corps," he explained, "should first absorb the additional 328 officers necessary to bring its actual strength—919—up to the authorized strength—1,247 under the old law. The 328 additional officers are to be provided from the commissioned force for which provision is made in these estimates."[10] Even so, some semblance of the first of the scheduled five increments in officer personnel was preserved. As the assistant secretary of war for air, F. Trubee Davison, pointed out, the number provided by Coolidge in the estimates was of a size to "carry sufficient funds to give us more than a fifth, or at least a fifth, of the difference between the existing actual strength in commissioned personnel and the strength, authorized by the act, of 1,650 commissioned officers at the termination of the five-year period."[11]

The equipment category of the expansion program fared even less promisingly than the manpower categories in Coolidge's first budget after the passage of the air corps act. The President chose to accept the discretion accorded him in the legislation to delay by a year those parts of the program requiring expenditure of funds. Section 8 of the act stated that the total increase in personnel and equipment should be distributed over a five-year period beginning July 1, 1926, and authorized the President to submit annual estimates for the costs of the program. The year beginning in July of 1926, however, had already been supplied with appropriations when the air corps act became law on July 2, 1926.[12] The next regular presentation of budget figures to Congress was to be six months later, and these would be for the fiscal year 1928, to start six additional months later, on July 1, 1927. To bridge this gap of one year between the assigned start of the program and the next scheduled supply of expenditure funds the act provided for a supplemental estimate for the fiscal year 1928. But the merely permissive nature of the

[10] Bureau of the Budget, *Budget Message of the President 1928,* p. ix.

[11] U.S., Congress, House, Committee on Military Affairs, *Hearings, Air Corps: Progress under Five-Year Program,* 69th Cong., 2d Sess., 1927, p. 11.

[12] U.S., Congress, *An Act Making Appropriations for the Military and Non-Military Appropriations of the War Department for the Fiscal Year Ending June 30, 1927, and for Other Purposes,* Public Law 123, appvd. April 15, 1926, 69th Cong., 1st Sess., 1926 (U.S., *Statutes at Large,* XLIV, Part 2, 254).

legislation was made especially clear by the "may be submitted" wording of the act's proviso dealing with the supplemental estimate.

The chief executive chose the following course of action. He implemented the program immediately inasfar as the purely administrative changes that had to be made were concerned,[13] but declined to use the authority to submit the supplemental estimates that would avoid a delay of those parts of the program involving an expenditure of funds. He noted in his budget message that the air corps act had been passed late in the previous session of Congress, preventing an opportunity to pass appropriations at that time. He then explained why he rejected the idea that funds be now appropriated to cover the six months lost in the scheduled start. To try to catch up by submitting estimates late in the fiscal year would be to "crowd" two full yearly increments in less than a year and a half. The estimates he was submitting, therefore, were ample for carrying out in the coming fiscal year only that part of the expansion program which could be accomplished "orderly and efficiently." He was willing, however, to make one small exception in favor of a more rapid budgetary start in the expansion plan. He asked that some amounts be immediately made available for expenditure by the air corps.[14] As the chief of staff described these funds a short time later, they were to be used for stations, hangars, some new training planes, and the like, and would enable the air corps to "start in full blast next year."[15]

What the chief executive considered "orderly and efficiently" procurable in the way of new equipment for the first actual increment under the plan fell substantially short of a one-fifth addition toward the final goal of 1,800 planes. The proposed first increment was deficient by 65 bombardment planes, 20 attack planes, and 9 cargo planes in respect to the authorized first increment for these plane categories.[16]

[13] See "Steps Taken to Carry into Effect the Provisions of the Act of July 2, 1926 (Air Corps Bill)," a War Department exhibit in U.S., Congress, House, Committee on Military Affairs, *Hearings, Air Corps: Progress under Five-Year Program,* 69th Cong., 2d Sess., 1927, pp. 16–18.

[14] Bureau of the Budget, *Budget Message of the President 1928,* pp. viii–ix.

[15] U.S., Congress, House, Committee on Appropriations, *Hearings, War Department Appropriation Bill, 1928, Military Activities,* 69th Cong., 2d Sess., 1927, p. 496.

[16] *Ibid.,* p. 498.

The reaction of air corps officials was mixed. On the one hand they were able to express approval of the President's action. The assistant secretary of war for air, F. Trubee Davison, testified that he was agreeable to the postponement of bombardment plane purchases because "we are not yet certain enough of the type of bombardment plane that we ought to get."[17] The chief of the air corps, Major General Mason M. Patrick, similarly expressed unreadiness to procure this type of plane.[18] On the other hand, Davison's fuller reaction to the implementation of the program was revealed when a senator pressed him about the bomber shortages with the question, "You came pretty near getting them, did you not, in the House?" Davison's reply was that the air corps would have been glad to complete the program under contract authorization, which would have permitted the air corps to make definite promises of purchases when current tests would reveal a stage of successful development in a particular model.[19] And General Patrick was careful to spell out one important point for House Appropriations Committee members: "It must be distinctly understood that if the program is to be carried out the shortages must be made up some time and provision must be made for them in the future estimates."[20]

Moreover, as the foregoing examples reveal, such approval as was forthcoming from the air corps officials was limited to the bomber procurement plan. Concerning the attack plane shortages, Davison and Patrick both were reported to have expressed to the Military Affairs Committee a strong conviction that planes of this type were indispensable for the program at the originally appointed time.[21] But objections concerning the attack plane phases of the program, and more general objections to the executive's handling of the opening of the five-year program, were assuaged in mid-Jan-

[17] U.S., Congress, Senate, Committee on Appropriations, *Hearings, War Department Appropriation Bill, 1928*, 69th Cong., 2d Sess., 1927, p. 116.

[18] U.S., Congress, House, Committee on Appropriations, *Hearings, War Department Appropriation Bill, 1928, Military Activities*, 69th Cong., 2d Sess., 1927, pp. 501–02.

[19] U.S., Congress, Senate, Committee on Appropriations, *Hearings, War Department Appropriation Bill, 1928*, 69th Cong., 2d Sess., 1927, p. 116.

[20] U.S., Congress, House, Committee on Appropriations, *Hearings, War Department Appropriation Bill, 1928, Military Activities*, 69th Cong., 2d Sess., p. 506.

[21] *Army and Navy Journal*, January 8, 1927.

uary when Coolidge sent a special request that the funds for the attack planes be added to the procurement estimates.[22] With the support won as a result of this last-minute change, the chief executive was able to enjoy seeing his fundamental decisions regarding the start of the air corps program pass the Congress largely as he had made them. The concesssion was indeed a relatively small one and could not significantly alter the unmistakable pattern of hesitant expansion that had set in in terms of personnel and aircraft expansion, a pattern establishing precedents that boded ill for the future course of the expansion program.

The restrictive interpretation Coolidge placed on the opportunities to expand the air corps was of course not surprising in a decade characterized by a widely shared philosophy of minimal government and dominated by such events as disarmament conferences and agreements neatly dovetailing with this philosophy. The implications of these conditions for the armed services were strikingly articulated by the army's chief of staff two months after the passage of the air corps act, with its ambitious, expensive five-year program. In an address marking the opening exercise of the Army War College on September 1, 1926, at the time of the year when budget figures for the forthcoming fiscal year are under final review, Major General John L. Hines expressed the facts of life for military officers in the postwar period. "Every officer should understand the principle of the military supremacy of the dollar in time of peace," he said. He further counseled them to make no recommendations for change in the peacetime structure of the military establishment without first considering the limitations imposed by appropriations and without first understanding the procedure involved in the operation of the budget system. The last test of the analysis and recommendations the student officers would eventually make, he stressed, would be that of their practicality from the standpoint of the "sympathy you have shown for the peacetime military dollar."[23]

While Coolidge's inception of the five-year program could be interpreted as grudging, the same judgment could not be passed on

[22] U.S., Congress, House, *Communication from the President . . . in connection with the Estimates . . . for the Fiscal Year Ending June 30, 1928 under Appropriation Title, "Air Corps, Army,"* 69th Cong., 2d Sess., Jan. 12, 1927, H. Doc. No. 640.

[23] *Army and Navy Journal,* September 4, 1926.

his acceptance of the air corps act's intention to stimulate the aviation industry. In the same message in which he described his partial and delayed opening of the expansion program, Coolidge noted that the "program which the Congress has prescribed for our air forces will assure the industry continuing Federal business. . . ." He proudly stated that in the production of airplanes and accessories there was no competition between the Federal Government and private industry. He was able to wed the idea that whatever of governmental activity was after all unavoidable could be redeemed through its assistance to the private economic sphere. "It is a fortunate situation," he said, "when the needs of the Government can be met by affording an orderly stimulation of the industry upon which we depend to supply our needs."[24] Thus, just as the parsimony of funds for the aviation service had fit an economizing outlook based on an adherence to minimal government, so had a glowing appreciation of legislation intended to support industry fit a philosophy whose greatest known enthusiasms were for values associated with business life.[25]

The fundamentally restrictive cast that was placed on the implementation of the air corps expansion program at the very outset was never significantly modified by later executive or legislative action. At the end of the fifth and last increment of the expansion program the inevitable result of the consistently low budget estimates and appropriations occurred. On June 30, 1932, when according to the calendar the expansion should have been completed, the goals of 15,000 enlisted men, 1,650 officers and 1,800 serviceable airplanes had to be contrasted with an actual strength of 13,400 enlisted men, 1,254 officers, and 1,671 serviceable planes on hand or on order.[26]

What was particularly striking through the years in which the expansion program ran its course was that Congressional appropriations closely followed the recommendations of the executive in his implementation of the air corps expansion program. The fact that this implementation was weakly and restrictively pursued when compared to the original authorizations of the air corps act did not produce a successful challenge from within the legislative body to-

[24] Bureau of the Budget, *Budget Message of the President 1928,* p. ix.

[25] Arthur M. Schlesinger, Jr., *The Crisis of the Old Order, 1919–1933* (Boston: Houghton Mifflin Company, 1957), p. 57.

[26] *Secretary of War 1932,* pp. 44, 42.

ward ensuring the attainment of the goals. Misgivings and criticisms were expressed over this turn of events, but they were never sufficiently influential in forcing a revision in the pattern of year by year deficiencies in the program. A brief review will serve to point out the essentially acquiescent role of Congress in following the suggestions of the executive in the deficient budget proposals.

With regard to the enlisted-man increases that were made under the program, final Congressional appropriations consistently left standing the Presidential decisions not to increase total army strength by corresponding additions of enlisted men. Only at the outset of the program, when the budget estimates threatened to drop the size of the army's enlisted-man complement below the usual strength for these years, were any suggestions concerning this category of personnel rejected. Even in that single instance, however, the move to restore army enlisted-man strength could not be clearly interpreted as a determination to support the original intention of the expansion program. The action was argued as much from the standpoint of consistency in yearly provisioning as from a desire to have the army grow by the amount of expansion in the air corps.[27] Any doubt on this point was removed in the remaining years of the program, when enlisted men for the air corps were invariably obtained at the expense of other army branches.

In reference to the scheduled officer increments that were never once filled during the course of the expansion program, the Presidential lead again was left undisturbed. The only solution to the problem lay in increasing the total commissioned personnel of the army beyond 12,000. The executive never proposed this move. Far from vigorous opposition to this executive policy, the strongest Congressional opinion expressed in the interbellum period concerning army officers was that their number was too large relative to the efficiency that existed in the group as a whole. Such feelings were in fact so influential that by the time of the last year of the expansion program the House Appropriations Committee moved to reduce the total commissioned strength of the army through the forced retirement of 2,000 officers.[28]

A provision of the original committee proposal would have per-

[27] U.S., *Congressional Record,* 69th Cong., 2d Sess., 1927, LXIX, Part 2, 1722.

[28] U.S., Congress, House, Committee on Appropriations, *War Department Appropriation Bill, 1933,* 72nd Cong., 1st Sess., 1932, H. Rept. 1215, pp. 7–10.

mitted the chief of the air corps to disapprove of the retirement of air corps officers as he saw fit.[29] On the House floor the spokesman for the committee later accepted an amendment to exempt all air corps officers outright.[30] These special considerations in favor of the air corps, however, did not save the proposal from being rejected by the full House. Although this was the particular House of Representatives that was to help pass the first economy act of the Depression era, the general willingness to pare back expenditures did not extend to cutting army commissioned strength. The entire incident nevertheless remains a revealing indication of the fact that increasing commissioned manpower was the remotest of possibilities as a device for relieving shortages in flying personnel. If any alteration was to occur, the stronger probability was one of a cutback, the final impact of which would have made even more difficult the meeting of the heavy deficiencies in the flying-officer complement of the air corps.

In the area of airplane strength that was to be increased under the expansion program, executive and legislative action presented a somewhat more complicated picture than that concerning personnel. For the first increment, the President himself, as we have seen, thought better of his original deficient proposal for planes and sent a special message to Congress increasing the number he had at first submitted. In the second and third increments, the insistence of the Senate was effective in increasing the final sums for new aircraft over those suggested by the President and agreed to by the House. In the fourth increment the Congress followed the executive suggestion in appropriating for all of that increment, although this action did not make up for the deficiency of the past. The executive decision was also accepted by Congress in regard to the fifth and last increment, when only one half the scheduled increment was provided in an attempt to economize during the Depression. This summary of plane budgeting and appropriation might seem to indicate a strong effort by Congress to meet at least the equipment area of the expansion program's goals. A closer consideration, however, reveals that this conclusion is not warranted by the facts.

While the appropriation figures for the five-year expansion pro-

[29] *Ibid.*, p. 30.

[30] U.S., *Congressional Record*, 72nd Cong., 1st Sess., 1932, LXXV, Part 9, 10145.

TABLE I

Recommended and Appropriated Funds During the Five-Year Expansion Program*

Fiscal Year	Requested by Air Corps	Approved by Secretary of War	Approved by Bureau of the Budget	Appropriated by Congress
1928	$25,588,066†	$26,789,050	$21,396,300†	$22,191,300†
1929	36,536,028	31,604,383	25,822,268	26,435,157
1930	71,990,965‡	33,250,000‡§	31,197,429**	32,440,785**
1931	54,007,902	42,692,260	35,823,473	38,043,877
1932	54,433,599	38,390,529	31,522,295	31,479,635

* Adapted from table in Army Air Forces Historical Study No. 22, "Legislation Relating to the AAF Material Program, 1939–1945" (revised; Air Historical Office, Department of the Air Force, August, 1949; typescript), p. 162, which cites figures from chart in Budget and Fiscal Office, Army Air Forces. The figures do not include funds to pay previous contract authorization.

† The air corps requested $17,772,363 supplemental; the secretary of war approved $2,111,992, but no further action was taken.

‡ Plus $5,500,000 to care for 1929 contract authorization.

§ The secretary of war approved a supplemental request of $5,992,536, but no further action was taken.

** Plus $2,250,000 for 1929 contract authorization.

gram (Table I) do show increases over budget estimates as a reflection of the instances just cited in which Congress provided larger numbers of equipment than suggested, the increases were in fact small relative to the estimates. Moreover, not all the increases that appear in the figures were for aircraft. One-fifth of the increase in appropriations over budget estimates in the fiscal year 1930, and all of the increase in the fiscal year 1931, for instance, were for technical construction and improvements, the latter increase occurring through two deficiency appropriation acts.[31]

Even more revealing of the inadequacy of the Congressional appropriations in respect to the planning under the five-year program is a comparison of the appropriations with the expenditure figures suggested by the air corps, and even with those suggested by the

[31] U.S., Congress, House, *Conference Report, War Department Appropriation Bill, 1930*, 70th Cong., 2d Sess., 1929, H. Rept. 2619, p. 6; U.S., *Statutes at Large,* XLVI, Part 1, 1064, 1552.

War Department. Both these service-recommended sums were well above the figures supplied by the Congress. The latter's increases were at best very weak restorations of the amounts deemed necessary by the service groups for progress under the program, and took on the character of fitful but fundamentally unsuccessful protests against the nature of the executive's implementation of the program in the estimates submitted to Congress by the Bureau of the Budget.

It was thus the executive budget that was chiefly determinant of the final funds for the air corps. Before their submittal to Congress the estimates of air corps expenditures were the result of a process that was thus summarized by the assistant secretary of war for air: "Each year the War Department directive for compiling the budget has included instruction for its [the expansion program's] progressive accomplishment according to schedule. However when the final limiting figure for the army estimates is determined, it is found that the Air Corps estimates must be drastically cut if this limiting figure is not to be exceeded."[32] And within Congress these estimates were passed with relatively minor revisions, both in terms of the estimates themselves, and especially so in terms of the original service plans. The budget process of course did not necessarily compel Congress to such acceptance. The estimates could have been altered at any time by the exercise of the legislative power of the purse, although disapproval of suggested expenditures for the air corps took on the difficult form of revisions upwards, with possibilities of executive impoundings. But there was little serious danger of Congress' destroying the integrity of the budget in reference to the air corps program. This fact was clearly evident in the amount of action—actually inaction—by Congress toward filling the program's scheduled increments, especially when to the record concerning air corps appropriations for equipment was added the acquiescence in the budget's personnel planning for the air corps.

For some legislators, these developments caused considerable surprise. As one senator put it, when he learned of unmet plane quotas from a departmental spokesman:

> Well, that seems to me quite amazing. I thought when the Congress laid down a 5-year program that everybody was satisfied that would be what we

[32] *Secretary of War 1932*, p. 39.

needed, and the least that we needed. I can not understand just why this [1930 appropriation] bill should be so far short of it as you have indicated.

. . . .

My impression, Mr. Chairman, I know I was very much interested in this 5-year program, more so than in the average programs that are adopted, and the average authorizations, and I think that most of the people of the United States were, and thought that when we passed that program we were getting planes.[33]

In the House various members were ready to complain of extra-Congressional influences on the estimates submitted to them, but in doing so they inadvertently reflected their helplessness to stay the trend of the budget suggestions. Thus one representative said he saw "a great deal of resistance being offered . . . by the budget and in other quarters" to the effort to offset air corps enlisted-men increases by adding to total army strength.[34] Another representative chose far less roundabout language and attributed the lagging course of the program to "the fault of the commander-in-chief" himself.[35] When speaking of the area of deficient plane increments, an Appropriations Committee spokesman pointed out that the chief of staff had "specifically stated" that the air corps was "not short because of any lack of action on the part of Congress."[36] At least one congressman, however, was willing to point the finger of responsibility at no exclusively outside influence, but rather at an influence that could be charged against all groups involved in the implementation of the air corps expansion program. Years before the onset of the Depression, he attributed the neglect of the program to "this economy which has come to be a fetish which we worship so blindly and without discrimination."[37]

If various congressmen in the course of the program pointed to others for blame, some observers outside Congress who studied the problem afterwards were not willing to join in such exculpation. The Baker Board investigation of military aviation in 1934 found that the unsatisfactory development of the air corps expansion program was largely traceable to a lack of funds. For this condition both

[33] U.S., Congress, Senate, Committee on Appropriations, *Hearings, War Department Appropriation Bill, 1930,* 70th Cong., 2d Sess., 1929, pp. 28, 36.

[34] U.S., *Congressional Record,* 70th Cong., 2d Sess., 1929, LXX, Part 1, 1151.

[35] *Ibid.,* 1152.

[36] *Ibid.,* 1151.

[37] *Ibid.,* 69th Cong., 2d Sess., 1927, LXVIII, Part 2, 1899.

the Budget Bureau and the Congress were responsible, the board said, but then argued further that the greater share of the accountability rested on the shoulders of the legislative institution: "Since however, the Congress is not bound by the restrictions promulgated to the War Department by the Bureau of the Budget, the ultimate responsibility must rest in Congress." At another point in its report the board was even more forceful: "In general the program prescribed by the Congress was initially delayed and subjected to continuous interference and postponement by the failure of the Congress to appropriate funds for its development."[38]

Similar outside criticism in the same year came from the chief of staff of that time, General Douglas A. MacArthur. He too placed responsibility for insufficient funds on the Bureau of the Budget and on Congress, emphasizing especially the role of Congress: "The law was passed by Congress. They themselves said it should be done in five years. They specified that that is what we should have and they did not do it. That is all." In strongly didactic terms he further explained to the legislators that "because the Budget does not send the [earlier planning] figures up here, that does not relieve the Congress of the United States from raising and maintaining armies, and they had themselves established the five-year program. It does not relieve their responsibility that the Budget may have curtailed before the figures reached here. You are not bound by Budget figures. The Constitution places the final responsibility not upon the budget, not upon the War Department, but upon one group alone, and that is the Congress of the United States."[39]

Why did not the Congress follow through with its own legislative will and blueprint, as expressed in the Air Corps Act of 1926, and supply sufficient funds in the necessary categories to realize the goals of the expansion program? The feelings within Congress that others were responsible, and the opinion outside Congress that that body was itself responsible, border on oversimplified analyses in the light of what is known about the nature of that branch of the governmental system.

[38] U.S., Special Committee on Army Air Corps, *Final Report of War Department Special Committee on Army Air Corps* (Washington: U. S. Government Printing Office, 1934), pp. 53, 7.

[39] U.S., Congress, Senate, Committee on Appropriations, *Hearings, War Department Appropriation Bill, 1935,* 73rd Cong., 2d Sess., 1934, p. 26.

From one point of view, the Congress did not act to restore the program over the adverse budget decisions because of the very nature of the representation in Congress. The localism, decentralized choice, and diffused responsiveness that have been noted to be responsible for the destruction of governmental policy as incorporated in the budget[40] operate also in a reverse manner. If such forces can disintegrate policy, they have no compensating capability of integrating diversified actions into a policy. Even if it should be objected that the passage of the air corps act was a triumph over such centrifugal forces and thus serves as a successful demonstration of Congressional policy-making, there is the further fact that such a demonstration was required on a sustained basis if the success of the air corps expansion program was to be realized.

Expressed in terms of group analysis, the public is not one but many. Successful legislation in a free society is therefore a product of an extensive alliance of varied groups, whose supporting interests are already organized actually or are simply manifest as potentially so. Frequently the combination behind the passage of an authorization act is the result of a long campaign that is finally catalyzed by critical, dramatic events, as for example Mitchell's continuous postwar publicity and the series of air disasters in mid-1925, which brought on the final outburst by the popular hero and resulted in his court-martial. But an alliance so sensationally summoned into being cannot be assured continued existence to the later times when appropriation acts must be passed to implement the original authorization law.[41] It was precisely such Congressional behavior that explains the long series of events in which the expansion program in the air corps act was energetically passed but then allowed gradually to fail to meet its goals.

From a broader point of view, the failure of the air corps expansion program takes on a perspective that explains both the chief executive's restrictions on full implementation and Congress's acquiescence in the restrictions. During more than half its course, in pre-Depression years, the fate of the expansion program was sealed by a Presidential leadership that stressed economy, minimal gov-

[40] Arthur W. Macmahon, "Congressional Oversight of Administration: The Power of the Purse, I," *Political Science Quarterly*, LVIII (June, 1943), 161.

[41] David B. Truman, *The Governmental Process* (New York: Alfred A. Knopf, 1951), pp. 51, 362.

ernment, small defense establishments, and disarmament. But such characteristics were not exclusively the possession of the particular administration of the period. They were, in turn, an expression of deep currents of similar attitude and opinion found in Congress, in broad segments of unofficial popular leadership, and in the majority of the populace itself. The prevailing depreciatory outlook on government and defense institutions was, for all its triteness of expression, a national mood, a climate of opinion.[42] As a consequence of these deeply rooted conditions in the interbellum period generally, arguments in favor of any type of increased support for the defense establishment met a reception that has thus been aptly summarized: "Congressmen—and Presidents too—normally responsive to any vigorously expressed wishes of constituents, did not by their speeches or by their votes in those years demonstrate any pronounced change of heart toward a strong defense policy, nor do the records show that they were unseated at ensuing elections because of their lethargy on the rearmament question."[43] If culpability for the failure of the air corps expansion program must be assigned, singling out the chief executive, his budget office, the War Department, or the Congress are at best partial answers. The entire diffused range of policy influences in the governmental process of a free society must be considered.

One central problem remains if the foregoing conclusion is valid. What of the possibly objective requirement of the interbellum period for a strong aviation service to cope with the demands of national security? More generally, how are such requirements to be satisfied in a liberal democratic society whose operative interests and values are unwilling to support levels of armed might adequate for the very preservation of the society? Two answers to these questions are possible concerning the years just reviewed. Either within the pluralistic, competitive group life of the country no

[42] Cf. Mark S. Watson, *Chief of Staff: Prewar Plans and Preparations* (Washington: Historical Division, Department of the Army, 1950), Chapter II. An explicit rejection of this sociological interpretation, in favor of placing the blame for the neglect of air power specifically on the War Department and the Bureau of the Budget, can be found in Army Air Forces Historical Study No. 22, "Legislation Relating to the AAF Materiel Program, 1939–1945" (revised; Air Historical Office, Department of the Air Force, August, 1949; typescript), pp. 44–45, 118.

[43] Watson, p. 17.

broad, largely consensual interest existed in favor of constructing a new, expensive weapon system that had earlier been judged a necessary addition to the armed forces helping to make that life secure; or no effective political leadership was present to summon such an interest, which may have been present only in a potentially organizable form, into supporting action for adequate air-service strength.

The first possibility is unnecessarily despairing and must be rejected in the light of the demonstrated sufficiency of national willingness to back air forces in the wars that preceded and followed the period under study. The second alternative therefore remains the fruitful one, but it opens up an entirely new problem, that of the difficult, almost terminologically contradictory, role of the creativeness required in the political leadership of a democracy. Suffice it to say here that the case study just presented reveals that such leadership was the critical factor in the contention over air strength in the middle years of the two-decade armistice, and that the discontent and activity of Congressional and other groups concerning the neglect of military aviation were not effective in surmounting the policy of the chief executive establishing minimal support for this armed service.

PAPERS OF THE MICHIGAN ACADEMY OF SCIENCE, ARTS, AND LETTERS
VOL. XLV, 1960 (1959 MEETING)

JOHN HAMILTON AND THE REVITALIZATION OF THE REPUBLICAN PARTY, 1936–40

KARL A. LAMB

The University of Michigan

REPUBLICANS could gain a measure of comfort, after the election of 1932, by explaining their defeat as the result of the temporary association of the Hoover administration with the suffering of the Depression. By 1936, however, even the most rabid partisan could not escape the conclusion that a certain positive response on the part of the electorate had been elicited by the Democratic party.

The Republican National Committee met in Chicago on December 16, 1936. It was, under the circumstances, hardly a cheerful gathering. Even for an organization with a tradition of proxy representation, it was not a well-attended meeting. Of the 104 members of the committee, fifty-nine were present, and another twenty-four were represented by proxies.[1] Although the minimum function of the committee was "to serve as a kind of umbilical cord from one convention to the next," the committeemen, surveying the devastation of their party, could not fail to be struck by the need for some way of reminding the public that an opposition party did, in fact, exist.[2] John D. M. Hamilton, the incumbent National Chairman, felt that this need should be met by establishing the headquarters of the National Committee on a permanent basis.

Hamilton had just completed his duties as the campaign manager for his fellow Kansan, Alf Landon. While the campaign itself had been singularly unsuccessful, Hamilton had been a conspicuous figure in it because of his great energy and enthusiasm. In an attempt to bolster local organizations, he had undertaken a campaign tour unusual in its extent for one who had no previous national reputation as a party spokesman and was not himself a candidate.

[1] Typescript record of the Proceedings of the Republican National Committee, December 17, 1936, p. 8, files of the Republican National Committee.

[2] The phrase is from Philip S. Wilder, "The National Committee of the Party in Power" (unpublished paper prepared for presentation at the Convention of the American Political Science Association, September, 1958), p. 1.

At the Chicago meeting, Hamilton submitted his resignation, for he felt that his position was untenable, in view of the electoral disaster which the party had suffered, unless the entire National Committee should endorse him. After considerable debate, in which it was stated that only the committeemen from Maine and Vermont could claim to have outperformed him, Hamilton's resignation was rejected. Hamilton then proposed the establishment of a national organization that would function the year around. He wanted to work full time at the job; as a professional politician he would establish a professional organization. Meeting in closed session, the Executive Committee voted him a salary of "$15,000.00, with a fixed allowance of $10,000.00 for the expenses of the Chairman in Washington."[3] Hamilton thus became the first full-time, salaried, Republican Chairman. When Senator Borah was told that Hamilton was to receive $25,000.00, he retorted, "That, as I understand it, is the customary salary of receivers."[4]

Hamilton was accepting a position that has always been ill-defined in the American party system; what he could make of it would depend uniquely upon his own abilities. The pages of American history are far from overburdened with the exploits of opposition political leaders, unless they are victorious in a succeeding election, and minority party chairmen have seldom won even a footnote. The reasons for the obscurity of the office are not hard to discover. After a party is defeated in a Presidential election, its leadership is dissipated. Space is allocated to "party spokesmen" in the news media because of their past or potential importance, or because of the sheer force of their pronouncements, rather than because of their official position within the party. Although the Chairman is the only officially appointed leader of the national party, he will, if he wishes to make pronouncements on policy, find his authority challenged by that of the last Presidential candidate, former Presidents who have borne the party label, and the party's leadership in Congress.[5]

[3] Minutes of the Meeting of the Executive Committee of the Republican National Committee, December 17, 1936, files of the Republican National Committee.

[4] *New York Times*, December 19, 1936.

[5] Ralph Goldman makes an important distinction between the Presidential Chairman and the Residual Chairman. The Presidential Chairman draws real authority within the party from his close association with the White House and —in the past, at least—his status as a cabinet member. The Residual, or mi-

Lacking an authoritative voice in party policy, the minority Chairman must, in order to make his tenure of the office meaningful, seek to exert his influence in other spheres. This is rendered difficult by the fact that allegiance will be won in the minority party by those who seem likely to become the leaders of a winning faction—either by themselves winning, or being closely associated with one who wins, the Presidential nomination. Such a leader is unlikely to be the National Chairman. No party Chairman has gone on to win the nomination for the post of chief executive. Unlike the Parliamentary system, the leader of the American opposition is not an apprentice President. As a result, the minority party Chairman must turn his attention to matters of party organization, propaganda, and finance. Indeed, his success may be closely tied to his ability as a funds-raiser.[6]

It was in just these fields that John Hamilton made a lasting contribution to the Republican party. His leadership was accepted by the party organization, and the changes made in its structure under his direction mark the beginning of the modern organization of the national Republican party. Republicans who remained loyal in spite of Landon's defeat tended to be of orthodox opinion, and the very success of his leadership was in part due to the fact that Hamilton made no attempt to guide the party in new ideological paths. By 1940, Hamilton headed an organization well prepared, in a technical sense, to battle for the Republican presidential candidate. An efficient organization, however, is only one of the essentials of party victory.

Mr. Hamilton's known attitudes on certain issues of party policy were viewed with disfavor by some of Willkie's nonparty supporters. Since Mr. Willkie felt that this source of support was more important than a harmonious relationship with the established party organization, he chose to displace Hamilton. Willkie, in fact, acted as though he could depend on the loyalty of the party organization

---

nority party, Chairman has no such source of authority. Goldman concludes that the position of the Residual Chairman may be the "strategic factor" in strengthening the American two-party system. Ralph Goldman, "Party Chairman and Party Faction, 1789–1900" (unpublished doctoral thesis, University of Chicago, 1951).

[6] Goldman also concludes that the success of the Residual Chairman may be tied to the state of his party's finances. *Ibid.*, pp. 274–275.

as a matter of course. But policy attitudes are not easily separable from organizational and personal loyalty, and the removal of Hamilton placed in jeopardy his real achievement within the party organization.

At the time of Hamilton's re-election to the post of National Chairman, his conservatism was well known. The situation of the party at that time was so desolate that its members were not prepared to consider new alternatives to the New Deal solutions of recognizable problems. Keeping the party alive was the compelling problem, and this explains the appeal of Hamilton's offer to establish a permanent and energetic national headquarters.

Questions of policy were, in fact, raised by a lone voice in the National Committee meeting. Hamilton Fish, Jr., Republican Congressman from President Roosevelt's home district, protested against the appointment. Fish, who had been a follower of Theodore Roosevelt's Progressives, represented by proxy the National Committeeman from Alaska. He argued that under the Hamilton leadership there would be no hope of uniting the conservative and liberal wings of the party—or, indeed, of winning support from a very large section of the nation.

"You must liberalize your leadership and policies," he said, "or the Republican party dies, and you must bear the responsibility yourself."[7]

When the motion was put to a vote, Fish abstained, as he could not be sure of the feeling of the delegate he represented. That there was no serious movement in the party to challenge Hamilton's leadership had been evidenced by Senator McNary's telegram to the convention endorsing him.[8] McNary was an acknowledged Senate leader, and he was from the West, which was, at that time, considered to be the home of the more liberal elements of the party. Perhaps the real reason that no contest developed for the post was that it was an unenviable position; the state of the party was such that the Chairmanship did not seem a prize worth fighting for.

Alf Landon was not disturbed by the differences on policy, and he later recalled his reasons for supporting Hamilton:

> John and I never did agree, particularly, on our political philosophy. But

[7] Typescript record of the Proceedings of the Republican National Committee, December 17, 1936, p. 51.

[8] *New York Times*, December 17, 1936.

I felt that, with the collapse of the Republican organization all over the country, that we needed a good cavalry leader, a Jeb Stuart, as it were, as I described him, to round over the country and pep up the state organizations.[9]

BUILDING THE ORGANIZATION

In order to bolster support for his plan of making the Republican headquarters staff a full-time and professional one, Hamilton conceived the idea of going to England to study the workings of the Conservative party. He spent two weeks in London in the summer of 1937, interviewed extensively nearly the entire staff of the Conservative and Unionist Central Office, and studied the basic literature of the party. The report of his trip was duplicated and sent to each member of the National Committee.[10] It was an objective report; there was no recommendation as to what specific features of the Conservative organization would be useful to the Republican party. Almost his central theme, however, was that "it is a primary principle [of the Conservative party] that politics is not a seasonal or spasmodic interest" and that "the direction must be in the hands of permanent paid employees."

The report gave considerable attention to the research and publicity departments, with an analysis of the separate publications; to party finances; to the social aspect of party functions; to women's activities; and to youth organizations. Hamilton's program was no doubt influenced by his trip to England, although few specific changes could be related to the experience.[11] It may have had an important effect on the members of the National Committee. The Conservative party had great prestige among Republicans, for it had returned to power in spite of depression conditions. The need for a permanent organization should have been obvious, particularly in view of the devastating propaganda that had been produced by Charles Michelson in the permanent Democratic headquarters es-

[9] Recorded interview with Mr. Landon, August 28, 1958.

[10] John Hamilton, *Memorandum on English Conservative Party Organization* (mimeographed, London, June, 1937, 19 pp.).

[11] Hamilton recalled that the report "was submitted in its entirety to the entire committee and had an enthusiastic reception. I have no recollection now that any specific action was taken based on the Report, but it did serve as a guide to our activities during the remaining three years of my administration and it did lift the morale of the staff at headquarters, who felt that if I were to stay in they could look upon organizational politics as a career." Letter to the author, February 26, 1957.

tablished by Chairman Raskob in 1929. But the idea was certainly reinforced by Hamilton's report.

Hamilton's activities within the party were many and varied, and their logic does not lie in their chronology. Thus, while the following account separates his work into convenient categories, they were all, to a degree, carried on simultaneously.

The first job was to restore the lines of communication between the national and local organizations, which had been allowed to fall into disuse during the easy supremacy of Harding and Coolidge. When he became the National Committeeman from Kansas, Hamilton learned that

> the office of the National Committee did not even have a list of the county chairmen. Moreover, he found that in many counties there were no chairmen and that in several areas heretofore considered Republican there was no organization whatsoever.[12]

Contact with the local organizations was first effected with a small newspaper called *The Republican Record*, but this was supplanted after August, 1937, with a more personal series called *Republican Progress Letters*. Such publications were not considered a substitute for personal contact, however. During 1937 and 1938, Hamilton visited thirty-seven states, where he spoke to diverse Republican groups, from fund-raising dinners to city Republican women's clubs. During the campaign of 1938, he spoke at rallies and other gatherings at which the audience was not limited to party regulars, but his primary purpose was not that of the campaign speaker, although his oratory had lost none of the vigor that had characterized it in 1936. His mission was to contact the local organizations, to familiarize them with the work of the party on a national level, stimulate their local activity, and, above all, to add impetus to their efforts to raise money, which would be shared by the national and local organizations. Hamilton's central purpose was to build an organization that would have the best chance of placing its candidate in the White House in 1940. He cultivated a personal friendship with local Republican leaders, which may have been quite independent of attitudes on party policy or the question of who the candidate should be, although it is possible that such

[12] Malcolm Moos, *The Republicans: A History of Their Party* (New York, Random House, 1956), p. 401.

friendships were based upon an unspoken assumption of similar policy attitudes.

The effort to raise funds was notably successful. A finance division was created under the direction of Carlton G. Ketchum. It was the forerunner of the present Republican National Finance Committee. For the first time, money collecting for the National Committee and the two Congressional campaign committees was centralized in a single agency.[13]

Following the implications of his report on the Conservative party that finance should be a continuing activity rather than one concentrated around the campaign, and to develop a feeling of identification on the part of the contributor with the work of the party, a program of "Sustaining Memberships" was established. Each contributor was given a membership card, signed by Hamilton, and pledges of continuing support were solicited. In 1937, 4,738 Sustaining Members pledged stated specific sums to be donated in 1938, 1939, and 1940. In 1937, contributions totaled $189,742. By November of 1938, 95.6 per cent of the pledges for that year had been collected, totaling $202,343.72. By then there were 6,367 Sustaining Members.[14] Making a virtue of necessity, Director Ketchum pointed out that, with the dearth of Republican office holders, the party was more than ever before in the hands of laymen and uniquely represented grass-roots opinion.[15] The result of the various efforts to raise funds was that the 1936 party deficit of $900,000 was paid off, and the treasury was comfortably in the black by 1940.

With a base of party solvency established, Hamilton considered means of giving committee staff members some security of tenure. So convinced was he of the necessity of developing a kind of professional staff politician that he toyed with the idea of establishing

[13] Carlton G. Ketchum, "Political Financing, 1937 Model," *Public Opinion Quarterly*, II, 1 (January, 1938), p. 138.

[14] These figures are from Treasurer Charles Goodspeed's report to the Committee as given in Publicity Release G 175, November 29, 1939, files of the Republican National Committee. The membership fee idea was not new, although it had been applied only during Presidential campaigns. Goldman cites in the correspondence of Nicholas Murray Butler evidence of a sustaining membership plan employed in the Hughes campaign of 1916. Goldman, *op. cit.*, p. 679, n. 1.

[15] Ketchum, *op. cit.*, pp. 136–137.

a pension plan for the regular employees of the national headquarters. The scheme was not presented formally to the National Committee, however.[16]

Another of Hamilton's organizational efforts was to strengthen the Young Republican organizations. When he had first come to the National Committee in 1935 (he served for eighteen months as General Counsel of the party before resigning to manage Landon's preconvention campaign) Hamilton had suggested that a national youth organization should be established. The idea was at that time deferred for later consideration. In September of 1935, a Young Republican division was established within the National Committee. On November 10, 1935, a national meeting of twenty-one state Young Republican organizations was held in Des Moines, Iowa; the result was the formation of the Young Republican National Federation, an entity separate from the National Committee. The Chairman then elected (J. Kenneth Bradley, thirty-five-year-old State Senator from Connecticut) was also the Director of the new division of the National Committee.[17] His main task was to co-ordinate the activities of the two groups. A national monthly magazine called *The Young Republican* was issued for the first time in February of 1936. It reported on the activities of the local organizations in regular columns, while its feature articles delineated the evils of the New Deal and maintained that the Republican party was the home of "true liberalism." The accent was on youth for youth's sake; opposition to the Old Guard was on the basis of age, not philosophy. The attitude of the national leadership seems to have been to enlist the energy of youth to do battle for an unchanged Republican creed. The magazine suspended publication after Landon's defeat, but it was in business again by July of 1937. The National Committee aided the Young Republicans by publicizing and financing an oratorical contest during the 1938 cam-

[16] Mr. Hamilton told the author of this idea. A pension plan for the "90 regular employees" of the Committee was instituted in 1956, "beating the social-welfare minded Democrats to the punch." *Washington Post,* July 20, 1956.

[17] *The Young Republican* (magazine), February, 1936, p. 12. Bradley became Chairman of the State Central Committee and National Committeeman from Connecticut. The Young Republican organization was further integrated into the official framework of the party in 1948. See Hugh A. Bone, *Party Committees and National Politics* (Seattle, University of Washington Press, 1958), p. 50.

paign. Prizes totaled $15,000, and it was also a membership drive, for entrants had to be members of local Young Republican associations.[18]

In his *Memorandum on English Conservative Party Organization,* Hamilton had stressed the work of women volunteers and the capabilities of Miss Maxse, chief of the Organization Division. In September of 1938, at a meeting in Chicago, the National Federation of Republican Women was formed. Representatives of thirteen state-wide federations were present, and twelve other states were represented by local groups. Following the organizational pattern established with the Young Republicans, Miss Marion E. Martin served a double role as a member of the Executive Committee of the National Federation and Assistant Chairman of the Republican National Committee in charge of women's activities. Miss Martin fulfilled this role very capably until 1947, and Hamilton considered appointing her to the National Committee as one of the important acts of his administration.[19]

The division of the National Committee which received the most heartening support during Hamilton's tenure was the Research Division, headed by O. Glenn Saxon, Professor of Economics at Yale. Saxon was appointed early in 1936 by the then Chairman, Henry P. Fletcher. Saxon's testimony before governmental committees caused Fletcher to seek him out, and Saxon accepted the appointment although he was by heredity a Democrat.[20] During the 1936 campaign, the decision was made to make the research establishment a permanent one. Provision was made for office space, and a library of 6,000 volumes was established. Professor Saxon arranged his teaching responsibilities so that he could spend from Thursday to Monday in Washington. He commanded a staff of forty during the campaign, and the contributions made toward research after Landon's defeat allowed him to keep the nucleus of the organization together.[21]

[18] Publicity Release G 78, August 17, 1938, files of the Republican National Committee.

[19] Hamilton interview, July 17, 1956. For the growth of the Federation since its inception, see Bone, *op. cit.*, pp. 52–54.

[20] Interview with Professor Saxon, January 18, 1957.

[21] Saxon interview; C. A. H. Thompson, "Research and the Republican Party," *Public Opinion Quarterly,* III, 2 (April, 1939), p. 306.

Favorable ground having been prepared by Hamilton's emphasis on the research activities of the Conservative party in his report from London, an effort was made to solicit funds earmarked for research. Apparently this was the first such appeal made by a major American party. Professor Saxon accompanied Hamilton to fund-raising dinners in order to describe the work of the Research Division, and the effort met with considerable success. Saxon argued that research was a very necessary activity for the party, for, as the New Deal expanded the concern of government into new aspects of the national life, government bureaus developed research divisions, and their published reports were colored to favor the administration.

Perhaps the most notable production of the research division was a booklet entitled *Promise and Performance.* Editions with the same format appeared during both the 1936 and 1938 campaigns. It was divided into various campaign issues which were presented on opposite pages. On the left-hand page were the statements of the 1932 Democratic platform, campaign promises of President Roosevelt, and several of his utterances made while Governor of New York, particularly those concerning states rights. On the opposite page were statistical demonstrations, often taken from government sources, or such authorities as the League of Nations, demonstrating the failure of New Deal programs to meet the need for which they were designed.[22] Its fault, of course, was that the booklet gave little indication of what measures might be used by Republicans to solve the same problems. Impressive statistics were furnished to bolster the case of those who were already convinced that the New Deal was an evil undertaking, but it could do little to make converts to the Republican cause. The same fault could be found with the most monumental of efforts produced by the Research Division for distribution within the party. This was called *The Republican Reporter.* Its format, a loose-leaf notebook in which new sections could be placed, as they were published to fit the developing issues of the campaign, was frankly copied from the Conservative party publication called *Hints for Speakers. The Republican Reporter* was sent to paid party subscribers, who numbered more than 2,000.[23] The material was largely drawn from

[22] Republican National Committee, *Promise and Performance: The Administration of Franklin D. Roosevelt Reveals Itself* (Washington, 1938, 64 pp.).

[23] Saxon interview; Thompson, *op. cit.*, p. 307.

secondary sources, but the treatment was exhaustive and the tone of the writing was fairly objective. Its effect was largely negative. Rather than develop alternative policies, it was concerned with establishing Democratic failures and the falsity of administration claims—such as the assumption that the 1936 dollar was the same as the "gold dollar." Professor Saxon's approach seems to have been to begin with the most conservative criticisms of the New Deal and then seek data to support them. Perhaps he should not be blamed for this. He was dealing with the same reality that was later discovered by the editors of the *Democratic Digest*: "with party workers it is more popular to smite the opposition than to stress issues of one's own party."[24]

Hamilton realized that one bloc of traditionally Republican votes had shifted dramatically to the Democrats in 1936. He felt that a sincere and massive effort must be made if the Republicans were to regain the vote of the Negro. He established a division in the National Committee concerned with the affairs of the Negro, and he obtained the services of a former secretary of Booker T. Washington to direct its activities.[25] This would seem to have been the precursor of the Minorities Division which was later an integral part of the Committee staff, although, by the 1956 campaign, its functions had been further refined into three separate parts: one dealing with Negroes, another with Jews, and a third with "ethnic groups."[26]

John Hamilton thus had an important influence upon the staff organization of the Republican National Committee, for many of its present divisions seem clearly to trace their ancestry to similar divisions set up under Hamilton's leadership. They were established to fill what were—and are—recognizable needs, if the party is to return to a majority position. But the ideas of a joint finance committee, and of increased work among youth, women, and minorities, along with the conception of a permanent and professional staff for the party headquarters, were details of organization which were not necessarily related to the content of those policy proposals which might be used by the party in an effort to win the support of the electorate. That Hamilton was able to bring about these or-

[24] Bone, *op. cit.*, p. 76.
[25] Hamilton interview.
[26] Bone, *op. cit.*, p. 61.

ganizational reforms may be due to the fact that he was able to avoid the pitfalls of policy conflict which could have given rise to a factional challenge of his leadership.

Hamilton and Party Policy

The results of the election of 1936 left no doubt that some re-examination of Republican philosophy should take place; but the very devastation the party suffered meant that no faction could claim that electoral support had been given to its own idea as to what trend a revitalized party program should take. The logic of Landon's nomination, and the confused nature of his campaign, represented an attempt at compromise between the wishes of the unreconstructed Old Guard to return to the Coolidge and Hoover policies and the desires of the more liberal wing of the party.[27] Party extremists of both right and left (by 1937, there were few who could really be placed on the extreme left) were free to claim that their own outlook should prevail in the future.

After the 1936 election, there was considerable agitation within the party for a re-examination of party philosophy; unfortunately, Mr. Hoover was the first "party spokesman" to make a concrete proposal to that effect, and he was the champion of the conservative faction. He called for a full-dress convention of delegates to the last two Republican National Conventions which would issue "a new and flaming declaration of the rights and responsibilities of free men."[28] Unfortunately for Hoover's purposes, the idea had been mentioned prematurely by Arthur Hyde, his former Secretary of Agriculture, who announced at the same time that the hope of the Republican party lay in drafting Hoover for the task of re-vitalizing it for the 1940 election.[29]

Governor Landon was not impressed by this proposal. He knew that such a large body could hardly make decisions that would give a fresh turn to party policy; it could only endorse the Constitution and the Bill of Rights, without considering what Landon felt were the political realities of the time. He claimed that to call

[27] See Basil Rauch, *The History of the New Deal, 1933–1938* (New York: Creative Age Press, 1944), p. 236.

[28] Herbert Hoover, "The Crisis and the Political Parties," *Atlantic Monthly*, September, 1937, p. 268.

[29] *New York Times*, June 27, 1938.

together the former delegates of two past conventions would only result in a meeting of "ghosts from the boneyard."[30]

Reporters saw the controversy between Landon and Hoover as a struggle between the former President and the former candidate for the "titular leadership" of the party and perhaps an attempt to gain an early advantage toward winning the Presidential nomination in 1940. John Hamilton did not become involved in the developing party conflict; he refused to support the claims of either Landon or Hoover, but he did announce that the idea had not originated with Hoover. Similar proposals had been received by the National Committee as early as March, 1937, he stated. He called a meeting of the Executive Committee of the National Committee to press for decisive action. That body voted to submit the Hoover proposal to a meeting of the entire membership of the National Committee, which was called for November. Publication of Landon's reservations concerning the proposal (he insisted on a declaration on the tariff and the gold standard, for example) made Mr. Hoover's idea untenable. But the need of some re-examination of party policy was obvious, so that a proposal was adopted calling for the formation of a committee of leading laymen to study policy proposals. Congressional leaders made it clear that they would not care to be bound by any pronouncements of the group, and when Glenn Frank, former President of the University of Wisconsin, was appointed Chairman of the new committee, he accepted their view that "the custodians of party policy in the interims between National Conventions are the Republican members in the Senate and the House when, as now, the party is not in power."[31]

The decision was the best that could have been reached. Formation of the Committee answered rank-and-file demands for an examination of party policy; both Landon and Hoover were pacified; and the National Committee was removed from a debate which could only have antagonized one or another party faction. The

[30] Letter to the author, April 4, 1957.

[31] *New York Times*, January 7, 1938. Because the Glenn Frank Committee was a semi-autonomous organization, its history is not further considered here. Its operations were, of course, a revelation of the developing thought of—and conflict between—leading Republican laymen. Its report was made ready for the National Convention of 1940 but was very largely ignored. See, by the present author, "Republican Strategy and the Congressional Election of 1938" (unpublished doctoral thesis, Bodleian Library, Oxford, 1957), chapter iv.

headquarters staff was involved only to the extent of giving research aid to the Policy Committee, which was a "grass-roots" group of laymen. Moreover, Republicans in Congress were left free to choose their own issues and policies for the election of 1938.

Party leaders in Congress are indeed important exponents of policy for the party that does not occupy the White House, and the minority Chairman will foster party discord if he plainly opposes that leadership. John Hamilton found himself in a strange position in early 1937; when the President announced his plan to reform the Supreme Court, Senators Vandenberg and McNary decided that the proposal could best be fought by complete silence from the Republican party. Democratic opposition to the proposal was considerable. The only chance for passing the bill, Vandenberg informed Hamilton by telephone, would be if the sundering Democrats should close ranks to meet a vociferous Republican opposition.[32] In accepting this "strategy of silence," Hamilton acquiesced in perhaps the most devastating admission of party impotency in the history of American politics.

Hamilton explained in the National Committee's weekly newspaper that "we are foregoing partisan advantage for the good of the nation," but many Republicans felt that the most worthy purpose of the National Committee was to seek "partisan advantage."[33] They could not know that the National Committee had become an important, but secret, center of opposition to the Court plan. Dr. Saxon's research staff prepared speeches for Democratic legislators and organized the flow of opposition witnesses before the Senate Judiciary Committee. Hamilton and Saxon also established a clearing-house for funds offered by Republican contributors to fight the scheme. According to Hamilton, the sum of $200,000 was directed to various organizations, including several church groups, which were opposed to the plan.[34]

The degree of harmony that existed between Hamilton and the Republican leaders in Congress was probably based on a funda-

[32] The day of Roosevelt's proposal, Hamilton was in Philadelphia to give an address on the duties of the opposition party. Vandenberg called from Washington to ask that he not refer to the Court proposal in his speech. Interview with Mr. Hamilton.

[33] *Republican Record*, April 12, 1937.

[34] Hamilton and Saxon interviews. These activities were, in Saxon's phrase, *sub rosa*, and no documentary evidence is available. Hamilton briefly described the effort in an article, "What the G.O.P. Needs," *Liberty*, August 18, 1945, p. 17.

mental agreement concerning party policy. Republican representation was reduced to a minimum in 1936 (sixteen Senators and ninety Congressmen). The last strongholds of the Republican faith were represented; they were the "tight little islands" of political thought, where the appeal of Roosevelt and the excitement of the New Deal did not cause at least a temporary political realignment. The Republican legislators were the hard-core party conservatives; they were also astute veterans of the legislative contest. Hamilton identified his views with theirs, and looked upon them as the natural leaders of the party.

The National Committee had always been regarded as an organization aiming almost solely at the election of a Republican President; its functions had been noticeable only during the campaigns of each quadrennium, and Republican Congressmen would naturally resent any intrusion by the Committee into what was considered their own affairs. In 1937, the National Committee offered help to the Congressional Republicans that was sorely needed. Because of the slim Republican representation at the Capitol, the Republican committee assignments were heavy, and it was difficult for the minority members to keep themselves informed regarding the issues raised in the New Deal legislation that was being introduced. In those years, Congressional committees did not have their own research staffs, so the group headed by Dr. Saxon was made available to the Republican members of Congress.

Hamilton instituted a series of week-end conferences, held at his home in Washington, designed to aid party legislators in the formulation of alternative policy proposals. Leading experts in the field to be discussed were invited; an outline of the ground to be covered was furnished by the research staff; those present attempted to reach a consensus of opinion, and the conclusions were summarized and distributed to those who had been present by the research staff. The most notable of these conferences was that on Social Security, held on January 23, 1937. The research staff prepared a memorandum on the economic effects of the Act, with quotations from various economists on the subject, including a statistical summary of the financial consequences of the Act. As a direct result of the meeting, Senator Vandenberg presented a resolution and accompanying statements on the Social Security Act.[35]

[35] Hamilton and Saxon interviews; Thompson, *op. cit.*, p. 309.

Hamilton's position of leadership was solidified by his work on many aspects of the party organization. Such efforts were not directly connected with the question of choosing a platform and candidate for 1940. Hamilton's competence found such an enthusiastic response among leading Republican contributors that many of them asked major contenders for the 1940 nomination to commit themselves to retain him as Chairman. Wendell Willkie was included in those who made this pledge. Hamilton was an early and enthusiastic convert to the Willkie candidacy, and he used his power over the arrangements for the Philadelphia Convention and his strategic talents on the floor of the Convention, in Willkie's behalf, to the extent allowed by the requirement of his position that he seem neutral. According to a Willkie biographer, Hamilton was the "one man above all others responsible for the Willkie nomination."[36]

Willkie felt that one of his most important sources of support was the *New York Herald-Tribune*, owned by Ogden Mills Reid and Helen Rogers Reid. When, prior to the 1940 conventions, President Roosevelt appointed Henry Stimson and Frank Knox, both leading Republicans, to the Cabinet, Hamilton was called upon to issue a statement pointing out that their action was a personal one, and Stimson and Knox would now owe their allegiance to the President rather than to the Republican party. The desire of Knox and Stimson for intervention in the European war was so well known, the document concluded, that "the appointments speak for themselves." The statement had been discussed and approved by Hamilton, Landon, and Hoover, among other party leaders, and it was not intended as a definition of Republican views on foreign policy. The Reids regarded the statement as an isolationist document and credited Hamilton with sole responsibility for it. They told Willkie that they could no longer continue unreservedly to support him if Hamilton remained as Chairman.[37]

In Hamilton's view, this was the most important reason—and it was the reason given by Willkie—for Willkie's decision to displace Hamilton, finally settling on Joseph Martin, Jr., then House Minority Leader, as his successor. Since Willkie later in the campaign

[36] Mary Earhart Dillon, *Wendell Willkie* (Philadelphia, J. B. Lippincott, 1952), p. 177.

[37] Recorded interview with Mr. Hamilton, April 2, 1959.

asked Hamilton to again assume the Chairmanship, Willkie's doubts concerning Hamilton must not have been permanent ones.

But Willkie was surrounded with advisers, such as Russell Davenport, who were hostile to the Republican organization as well as much of the conservative Republican philosophy. Such advisers would have considered an association between Willkie and the Old Guard of the Republican party to be an unmistakable liability. Hamilton was clearly identified with that wing. He had of necessity worked with the Old Guard; there was no other group both able and willing to contribute the funds required to keep the party alive in those years. His success in raising funds may have been due to contributors sensing that Hamilton was in sympathy with their own orthodox views.

Willkie seemed unaware of the problems that the firing of Hamilton would create within the organization. Regular Republicans viewed the action as a part of the pattern of Willkie's behavior, in that he seemed to take the support of "you Republicans" for granted and go out of his way to slight or even insult the Republican regulars in public. He was never able to choose between the Republican professionals and the amateur groups formed before the Convention who claimed credit for his nomination and the right to direct his campaign.

As a result, the national political organization nurtured by Hamilton, based on party regularity, was soon disaffected from Willkie's candidacy. Traditional Republicans probably did not vote against Willkie, but their enthusiasm for the kind of effort within the precincts that can be the foundation of party victory was considerably dampened. Whether these tactics on Willkie's part attracted an offsetting number of "independent" votes is problematical.[38]

The party was again overwhelmed at the polls, and the continuity of Hamilton's work with the National Committee was shattered, as were his efforts among party contributors. Because of the demands of wartime, the party staff would have become a caretaker operation in any case, but the repudiation of Hamilton certainly did nothing to further the fortunes of the staff which he brought into the National

[38] Adlai Stevenson faced Willkie's dilemma in 1952. His attempt to dissociate himself from the "mess in Washington" and the regular Democratic organization may have lost more votes than it won.

Committee headquarters with the intimation that this could be a lifetime career.

Conclusions

It can be argued that the fortunes of the Republican party had reached such a low ebb in 1936 that any Chairman could only have improved its standing. The claim can further be made that the establishment of a permanent and energetic national staff, along with the other changes instituted by Hamilton, was inevitable and necessary. The fact remains that most of the operating divisions of the Republican National Committee had their precursors in similar organizations established by John Hamilton.

Hamilton was one of the few chairmen of a party that had lost both the Congress and the White House whose performance of his duties worked a significant change in the party. That change was largely organizational. Under his direction the party paid off its deficit. A permanent and active national headquarters was established. Local organizations found new life and purpose. Hamilton avoided the pitfalls of party factionalism and maintained friendly relationships with party leaders in Congress. His success as party fund-raiser and national leader of the organization was probably due to his tacit acceptance of the policy assumptions of the conservative Republicans with whom he worked. But he established personal relationships and loyalties which were, in a sense, independent of policy attitudes.

The American political party is usually viewed as a loose association held together by a common desire for power. The party organization is seen as a mechanical apparatus which is largely unconcerned with party policy. But party contributors, county chairmen, and precinct workers do have a general notion of the party objectives they wish the national party to seek, and their enthusiasm for party work is a complex mixture of personal loyalties and policy attitudes. The line between organizational leadership and policy leadership is hard to determine, and the candidate who alienates or displaces established organizational leadership because he is unwilling to compromise, or even recognize, policy differences, may find that the efficiency of the organization is affected. John Hamilton rebuilt the Republican organization, but Wendell Willkie so alienated the regular organization that its full potential was never used in support of his candidacy.

Papers of the Michigan Academy of Science, Arts, and Letters
Vol. XLV, 1960 (1959 Meeting)

CANADIAN TRADE UNIONS AND THE COOPERATIVE COMMONWEALTH FEDERATION[1]

SAMUEL H. BARNES

The University of Michigan

The 1958 convention of the Canadian Labour Congress (CLC) adopted a new political program—the formation of a political party based upon trade unions, farm organizations, cooperatives, and the Cooperative Commonwealth Federation (CCF). The CCF, Canada's socialist party, would be only one of several constituent bodies. While the trade union reaction to the Conservative victory in 1958 is understandable, it is doubtful that the new party will be much more satisfactory to the trade unions than was the CCF. An examination of relations between trade unions and the CCF suggests some of the causes of the relative failure of the latter. For example, trade union declarations of support were seldom backed up with money and work. Some trade union leaders opposed the CCF. Others favored it but felt that labor did not exert enough influence within the party. In the Province of Quebec, trade unions were unable to overcome the public image of the CCF as a materialistic, Western, Protestant, and anti-French party. Finally, the rank and file of labor never enthusiastically supported the CCF. Thus a successful new party must overcome many formidable handicaps.

The Rise of the CCF and the Problem of Trade Union Representation

It is often assumed that Canadian labor shied away from political activities until the formation of the CCF. This is not correct. Throughout the last years of the nineteenth century and early years of the twentieth, there was considerable activity on the local level by trade unionists in politics. On the national level there was repeated agitation for the formation of a labor party, and the Trades and Labor Congress (TLC) finally resolved in 1906 to support labor

[1] Research for this article was accomplished in Canada in 1955–56 on a grant from the Duke University Commonwealth-Studies Center. I gratefully acknowledge this assistance, as well as the kindnesses rendered me at Duke and in Canada.

candidates to parliament and to provincial legislatures.[2] This policy remained dormant until the 1917 TLC convention, when it was resolved that the precedent of British labor be followed.[3] An independent labor party consequently was organized within two years in every province except Prince Edward Island. Aided by the agrarian protest movement, many labor candidates were elected to provincial parliaments at the end of the First World War; and two of thirty nominees were successful in the 1921 federal elections. The first national Canadian Labour party was established the same year. But the TLC did nothing to aid the party, and in 1923 abandoned even verbal support in favor of nonpartisanship.[4]

The electoral successes of labor during the 1920's were few, but various labor parties survived into the 1930's and formed the basis of the CCF. The only national trade union organization represented at the founding of the CCF was the Canadian Brotherhood of Railway Employees. The CCF was originally a federation of various provincial organizations which had sent delegates to the founding convention. Thus its British Columbia provincial section was the Socialist Party of Canada and its Manitoba section was the Independent Labor party. This situation caused considerable confusion in provinces such as Alberta which had several constituent organizations. Within ten years the original organizations in each province were absorbed into a unified structure based largely upon CCF clubs and constituency associations. Thus the possibility that in strongly industrialized provinces the trade union organizations could become the core of the CCF was avoided. Hence special arrangements had to be made to affiliate trade unions as members.

Since membership in the CCF is largely a provincial matter, the national convention for a long time was not concerned with affiliated bodies. The only trade union represented at the founding convention did not at the time affiliate. Thus for six years the CCF, a socialist party, received no direct support from trade unions. M. J. Coldwell, CCF president, has written: "The CCF does not try to be exclusively a labour party, for it aims to unite all workers, farmers, and middle and professional classes, in a great national movement;

[2] *Proceedings*, Twenty-second Annual Convention of the TLC, 1906, p. 82; hereinafter cited as *Proceedings*, TLC (or CCL).

[3] *Ibid.*, 1917, pp. 43–44.

[4] *Ibid.*, 1923, pp. 92–109.

but the size and importance of labour support will make that wing of the party an essential for success."[5]

Labor support was indeed crucial to the CCF, and the terms of trade union affiliation were a subject of controversy. Differences of opinion arose over the amount of influence the trade unions in fact did have and how much they should have. The federal system, the sectional differences within Canada, the ethnic division between the French and English, and, finally, the fact that labor alone can hardly secure a majority in the House of Commons suggests that any Canadian labor party must be able to appeal to more than one economic or sectional group. While it strongly encouraged labor support, the CCF was reluctant to permit Canadian unions to gain a dominating position in the party. In Great Britain the trade unions tend to dominate the Labour party.[6] A similar relationship in Canada would, it is believed, be disastrous for the CCF. The meager successes of the CCF in industrial areas and its success in Saskatchewan, the most agrarian of all Canadian provinces, reinforces this belief. Thus the leadership of the CCF must court the support of labor without permitting unions to dominate the party.

Despite encouragement by the party, very few trade unions affiliated with the CCF. The first was the United Mine Workers, District 26, comprising Nova Scotia and New Brunswick, which affiliated in late 1938.[7] This affiliation required the adoption of some general policy concerning future affiliations. The original CCF constitution allowed intraprovincial economic organizations to affiliate either to the national or provincial CCF organization. Interprovincial organizations had to affiliate to the national CCF. The amended constitution of 1946 permitted provincial affiliation except "where circumstances make provincial affiliations inappropriate," thereby granting greater flexibility than before.[8]

The basis of representation at CCF conventions was also settled in 1946 by allowing each affiliated union, or any economic or political organization, "one delegate for the first five hundred members and

[5] Coldwell, *Left Turn, Canada* (New York: Duell, Sloan and Pearce, 1945), p. 25.

[6] See Bernard Hennessy, "Trade Unions and the British Labor Party," *American Political Science Review*, 49 (December, 1955), pp. 1050–1066.

[7] *Report*, Sixth National Convention of the CCF, 1940, p. 5.

[8] *Ibid.*, Ninth National Convention of the CCF, 1946, p. 18.

one additional delegate for every additional five hundred members or major portion," with a maximum of ten.[9] Most provincial organizations also set lower convention delegate ratios for affiliated members than for private members. Representation of affiliated members ranges from the one delegate for each twenty-five members used in Alberta to the one to five hundred used in Nova Scotia, where the affiliated United Mine Workers would swamp the private members if a lower ratio were used.

Thus the CCF avoided trade union domination by circumscribing the representation rights of affiliated members. In 1952, forty-four trade unions were affiliated to the CCF, thirty-seven of them being CCL and seven of them TLC unions. The affiliates included a mere sixteen thousand members, more than ten thousand of whom were members of a single union, the United Mine Workers.[10] As these disappointing figures show, direct affiliation and the financial aid it would bring was not the principal means by which the labor movement supported the CCF. Only a bit more effective was the action on behalf of the CCF by unaffiliated trade unions.

Trade Unions and the CCF

The TLC.—Despite the efforts of a minority to sway it toward the CCF, the TLC, the senior Canadian labor organization, paid little attention to political action between 1923 and 1944. The CCL endorsement of the CCF in 1943 and the impressive showing of the CCF in the Ontario provincial elections of that year ensured that the subject of political action would be raised in the 1944 TLC convention. Although the convention reaffirmed its policy of nonpartisanship, it also endorsed the organization of political action committees within its affiliated unions. The convention further instructed the incoming officers to consider the establishment of a national Political Action Committee (PAC) to rally support for its postwar program.[11]

As a result, a national PAC was established before the next convention, held in 1946. The PAC was given little financial support. In the 1946 federal elections the PAC endorsed the Liberal party, using the argument that the Liberals must be allowed to fulfill the promises they had made to Canadians during the Second World

[9] *Ibid.*
[10] *Ibid.*, Twelfth National Convention of the CCF, 1952, p. 19.
[11] *Proceedings*, TLC, 1944, p. 338.

War.[12] This endorsement was the only public break in the nonpartisan tradition of the TLC from 1923 until it dissolved in 1956.

Nevertheless, a continuing minority within the TLC attempted to commit the Congress more strongly toward political action on behalf of the CCF. TLC conventions rejected all pro-CCF resolutions; however, a nonpartisan Labour League for Political Education was established in 1953. But like the PAC, the TLC political education program amounted to very little. Nor did a department of political research which was organized in 1954.[13]

A number of reasons may be cited for the stand of the TLC. It should be noted that it carefully refrained from imposing its nonpartisanship on its affiliates. They were free to support the CCF if they wished; and, in fact, in 1952 seven TLC unions were affiliated with the CCF, compared with thirty-seven CCL unions affiliated.[14] Composed of diverse unions whose members favored every party from the Labor Progressive (communist) to the Progressive Conservative, the TLC found nonpartisanship a satisfactory solution. A second factor was that the TLC seemingly had little to gain from direct political action. Its governmental relations were good. It was the oldest and largest Canadian labor congress and it would have lost prestige by following the lead of the CCL. The endorsement of the CCF by the CCL had led to no impressive results, and there seemed to be no prospects for an immediate or future CCF victory.

The CCL.—The CCF has had much closer relations with the CCL than with the TLC. The CCL, to which are affiliated most Canadian sections of CIO unions, was established in 1940 by the merger of the All-Canadian Congress of Labour with the CIO unions expelled from the TLC the year before. Almost from the beginning it differed politically from the TLC. In Canada as in the United States the CIO-CCL was identified with more aggressive organizing activities and with a more positive attitude toward political action than the older bodies, the TLC and AFL. President Mosher of

[12] *Ibid.*, 1946, pp. 278, 73–74.

[13] Interview with Leslie E. Wismer, TLC Director of Public Relations and Research, April 13, 1956, Ottawa, Ontario.

[14] *Report*, Twelfth National Convention of the CCF, 1952, p. 19. The Ontario Provincial Federation of Labor (AFL-TLC) split 133–133 on a resolution to endorse the CCF at its 1955 convention. *Globe and Mail* (Toronto), November 28, 1955.

the CCL, who was also President of the Canadian Brotherhood of Railway Employees, had been a delegate to the founding convention of the CCF. Many other leaders of the CCL, such as C. H. Millard of the United Steelworkers, were supporters of that party. As early as the 1941 CCL convention, unsuccessful efforts were made to affiliate the CCL with the CCF or to have the CCL directly enter the political field.[15] And the 1942 convention resolved that the CCL support "only those candidates who have demonstrated their friendliness toward Labour and their promise to continue such a policy if elected."[16]

In the 1943 Ontario elections, thirty-four CCF candidates were elected to the provincial parliament compared with thirty-eight for the Conservative party and fifteen for the Liberal party. The possibility of imminent CCF electoral successes caused great enthusiasm in the labor movement; many CCL leaders favorable to the CCF but doubtful of its prospects were won over to partisan political activity. The CCL convention held in September, 1943, opened in an atmosphere which was highly favorable toward political action. The convention endorsed the CCF as the political arm of labor and recommended that the CCL unions affiliate with it.[17]

The debate on the resolution, which passed by a substantial majority, reflected the various attitudes toward political action within the Congress. Communists spoke of the need for labor "unity" and protested against the opposition of the CCF to communism. Others spoke in favor of political action but objected to the method of affiliation, claiming that the CCL would not have enough influence over CCF policy. But most speakers favored the resolution.[18]

Following British practice, the CCL did not itself affiliate with the CCF; it merely endorsed the party and recommended to its member unions that they affiliate. But unlike British practice, CCL affiliates did not joint the CCF in significant numbers. Nine years after the CCL endorsed the CCF, the United Mine Workers, which had affiliated with the CCF in 1938, still accounted for more than half the affiliated membership.[19]

[15] *Proceedings*, CCL, 1941, p. 44.
[16] *Ibid.*, 1942, p. 139.
[17] *Ibid.*, 1943, pp. 53–54.
[18] See *ibid.*, pp. 54–56, for this debate.
[19] *Report*, Twelfth National Convention of the CCF, 1952, p. 19.

A national PAC was established by the CCL in March, 1944, one year after the establishment of the CIO-PAC. The CCL-PAC was composed originally of five members, but all affiliated unions were subsequently invited to appoint representatives to it. The first chairman of the PAC was a strong CCF supporter, C. H. Millard of the Steelworkers, and other members were also strongly pro-CCF. The most important labor aid given the CCF came from the PAC. Its activities indicate the value of labor support to the CCF, and also the unwillingness of many labor leaders to extend themselves for the party.

Although established in 1944, the PAC accomplished little before 1948. In that year the convention proposed that the CCL campaign for the CCF, and only communist delegates were opposed.[20] The first elections in which the PAC participated on a large scale were the 1949 federal general elections. PAC activities centered in Cape Breton Island in Nova Scotia, the major industrial centers of Ontario, and the cities of Winnipeg and Vancouver. The PAC claimed to have improved the CCF vote in areas in which it operated; elsewhere, the CCF had merely held its own. The annual PAC report complained about the poor quality of CCF candidates in some areas and advocated more union influence in their selection.[21] This was to be a recurring complaint.

The CCL took important steps in 1950 to make the PAC more effective. The executive council appointed a committee consisting of C. H. Millard (Steelworkers), chairman, Pat Conroy (CCL Secretary-Treasurer), George Burt (Canadian Director of the United Automobile Workers), and Murray Cotterill (PAC Director) to coordinate the political activities of the CCL and CCF. The PAC was also granted representation on the CCF national executive council. The PAC Council established a political action fund of one dollar per year per union member for political education and action. Of this amount, forty-five cents was returned to local unions, and fifty-five cents was retained by the national and provincial PAC's. A monthly *PAC Newsletter* also began publication.[22] During 1950, community PAC schools sponsored by local labor councils in cooperation with the national PAC office were held, usually over a weekend, in

[20] *Proceedings*, CCL, 1948, pp. 75–78.
[21] *Ibid.*, 1949, pp. 46–51.
[22] *Ibid.*, 1950, pp. 29, 47.

several industrial centers. The PAC office also supplied lecturers for PAC classes held under the auspices of the CCL Department of Education, which is separate from the PAC, in several other cities.[23]

The most important support the CCF received from labor came from the work of the PAC. Although PAC activities were separate from those of the CCF, both organizations worked toward the same goal. The major effort expended by the PAC for the CCF candidate was in the form of election publicity and volunteer aid. The "dollar per year per member" program was not realized. Perhaps the most important form of assistance was the good will disseminated by unpaid local PAC workers. PAC activities were largely limited to Cape Breton Island and the industrial centers of Ontario, Manitoba, and British Columbia. The value of the assistance varied according to the place. Some M.P.'s and M.L.A.'s attributed their elections largely to the PAC; others were scarcely aware of its existence.

Although the PAC was not outstandingly successful, most CCL and CCF leaders agreed that what was needed was more grass-roots activity of some kind. Despite more than a decade of formal declarations of support of the CCF by the CCL leadership, the rank and file of the CCL did not become solidly pro-CCF.

The CCCL.—Labor politics in the province of Quebec is greatly complicated by the existence of the Canadian and Catholic Confederation of Labor (CCCL). The CCCL is limited to that province and has traditionally been a confessional Catholic trade union organization.[24] The postwar change in the attitude of the CCCL toward political action is extremely important for an understanding of the problems of the CCF in Quebec. The CCCL traditionally took the stand that political action was not a proper function for a trade union and its constitution forbids it to connect itself with any political party. During much of its existence the CCCL has strongly criticized, first, the so-called socialist and communist leanings of some international leaders and, later, the endorsement of the CCF by the CCL.[25] Nonpartisanship in politics was a guiding maxim with long-

[23] *Ibid.*, 1951, pp. 44–45.

[24] For an analysis of the changes now going on within the CCCL see: Samuel H. Barnes, "The Evolution of Christian Trade Unionism in Quebec," to be published in the July, 1959, issue of *The Industrial and Labor Relations Review*.

[25] The early opinions of the CCCL were similar to those of the Jesuit *École Sociale Populaire*, which in 1943 strongly attacked the connection between the

time CCCL president Alfred Charpentier. A basic difference between Charpentier and the present leaders of the CCCL lies in their respective attitudes toward political action.[26]

Although the CCCL has not exerted much influence in provincial politics, there has been a considerable change in the attitude of the top leadership toward political action. While the antilabor orientation of the Duplessis provincial government had long been evident, there was no open break between the CCCL and the government until 1949. In May, 1949, the Quebec government proposed a new labor code for the province which would have greatly restricted the bargaining rights of unions and would have barred all "subversive" organizations. All the unions of the province combined to defeat the bill.[27] The Asbestos strike of 1949, when Duplessis intervened strongly on the side of the employers, ended any pretense of cooperation between the provincial government and the union.[28] The CCCL, which for thirty years had held a favored position with the Quebec government, became its most bitter opponent.

In a speech to the 1949 CCCL convention, CCCL President Picard declared that he did not favor affiliating the CCCL with any particular political party. But he added: "One must understand . . . that if a government is unjust toward a social class, where it accords privileged treatment to another, it nourishes by its deeds the idea of a class party, and often the circumstances favorable to assure its realization."[29] Picard noted the nonpolitical tradition of the CCCL, but emphasized the necessity of enacting into legislation the demands of Catholic social doctrine; and as this had not been done by the present government, the CCCL had the duty to enter politics.[30]

CCL and the CCF and the power it gave to the "Jew" Mosher and the "Irishman" Conroy. Unsigned editorial, "Menace totalitaire," *Relations*, 34 (October, 1943), pp. 253–254.

[26] Interview with Alfred Charpentier, November 25, 1955, Montreal, Quebec.

[27] W. E. Greening, "Taft-Hartleyism in Quebec," *Canadian Forum*, 29 (May, 1949), pp. 29–30. Greening said that the CCF could have been considered "subversive" under the law.

[28] For an excellent study of the background and aftermath of the strike, as well as of the strike itself, see P. E. Trudeau (ed.), *La grève de l'amiante* (Montreal: Les éditions cité libre, 1956).

[29] *Procès verbal*, CCCL, 1949, pp. 43–44.

[30] *Ibid.*, pp. 46–47.

A proposal for the formation of a Legislative Orientation Committee was rejected by the convention. It did, however, create a Central Committee of Civic Action which was to be primarily concerned with education of its members and influencing public opinion. This committee recommended the defeat of some of the most strongly antilabor candidates of the *Union Nationale* and claimed credit for much educational work and for publicizing the need for the redistribution of provincial electoral districts. The committee made it clear that it would not approve, in the name of the CCCL, any particular candidate or party. The president and secretary general of the CCCL were not free to enter politics, but *Le travail,* the CCCL newspaper, was to be used to indicate candidates who were particularly hostile to workers and unions. Yet, the 1953 convention greatly reduced the effectiveness of the committee and abolished it for all practical purposes. The principal reason was a shortage of funds. The CCCL attempts to provide its affiliates with as many services as the international unions without the same resources. It was not until February, 1955, that the CCCL Executive Board finally appointed a Director of Political Orientation; and his activities have been continually hampered by a shortage of funds.

It is significant, however, that the present leadership does not feel that there is any ideological reason why the CCCL and its leadership should not engage in any form of political activity it wishes. The old opposition to political activity based on considerations of principle no longer exists: to enter politics or not is a matter to be determined by the needs and possibilities of the moment. While no ideological reason prohibits the CCCL from forming the mass base of the CCF in Quebec, there are many practical reasons for the weaknesses of the CCF in that province.

All CCCL leaders interviewed by the author were sympathetic toward the CCF.[31] Few, however, and none in the top echelon, openly campaigned for it before 1957. The majority of the leaders felt that the CCF had done very poorly in Quebec and that it would not do much better in the near future. Hence the CCCL would gain nothing by allying itself with it. CCCL leaders do wish for labor to have a greater voice in public life. Its leadership represents

[31] The Quebec CCF in 1955 changed its name to the *Parti Social Democratique* because of the difficulty of translating "Cooperative Commonwealth Federation." It was a change of name only.

a Roman Catholic left, "advanced" in its economic and social philosophy, which is awaiting a promising opportunity to enter politics. The weakness of the Quebec left in the past is shown by Duplessis's recurring parliamentary majorities; the potential strength of the Quebec left is one of the unknown factors of Canadian politics.[32]

A contributing factor to the weakness of labor in Quebec is the division of the left. The influential newspaper *Le Devoir*, which favors extensive economic and social reforms, in the past opposed the CCF on nationalistic grounds. It supported the CCF in the 1956 provincial elections, however, and would probably support an autonomous provincial left-wing party even more strongly. The CCF in Quebec is also hurt by recurring schisms within its own ranks. In addition, the continued division of the labor movement in Quebec between the confessional and international unions further weakens the left. The projected affiliation of the CCCL with the CLC could result in a unified labor movement and renewed vigor for left-wing politics in Quebec. Finally, the identification of the CCF with Protestant and anti-French Western Canada is certainly one of the best arguments for labor's abandoning the CCF and getting a fresh start with a labor party. There is a real possibility that a political regrouping will take place in Quebec in the next few years. A new beginning with a labor party by the trade unions might have a great impact in that province.

The CLC and Political Action

Neither the policy of nonpartisanship of the TLC nor the endorsement of the CCF by the CCL was completely satisfactory to labor. Nevertheless, it was anticipated that their differences in outlook concerning political action would disrupt the founding convention of the Canadian Labour Congress (CLC) held in Toronto in April, 1956. Just as the AFL and CIO had merged in the United States, the TLC and CCL merged at that time to form the CLC.[33] Yet the provisional political policy of the new Congress—neutrality—was established with little difficulty.

[32] Walter O. Filley discusses this question in "Social Structure and Canadian Political Parties: The Quebec Case," *The Western Political Quarterly*, 9 (December, 1956), pp. 900–914.

[33] For background information on the merger see Eugene Forsey's, "The Movement towards Labour Unity in Canada," *The Canadian Journal of Economics and Political Science*, 24 (February, 1958), pp. 70–83.

A number of factors account for the willingness of the CCL delegates to sever connections with the CCF. A primary reason was the desire of most delegates to avoid a protracted struggle at the founding of the new Congress. Secondly, the major groups at the convention had agreed upon a compromise policy before the issue of political action was raised on the floor of the convention. Thirdly, CCF supporters were convinced that: (1) mere endorsement, if not accompanied by enthusiastic support, was worthless; (2) even if the CCF were endorsed, it would be by an insubstantial and hence unconvincing majority; and (3) there was a good possibility that endorsement would be rejected, an action which would be interpreted, even if wrongly, as a repudiation of the CCF.

The convention merely acknowledged that the CCF "fought consistently for Labor's legislative program inside and outside of the parliament of Canada and the provincial legislatures."[34] The convention established a political education committee and a political education department to formulate programs in political action and "to give all possible assistance to individual affiliates carrying out programs of political education or action." It also urged its affiliates to interest themselves in political affairs, to continue their former political activities, and to undertake new activities which might seem appropriate. Thus individual unions were free to retain their affiliation with the CCF. Moreover, the new political education program of the CLC was strongly pro-CCF.

There was seemingly nothing that labor could do to avert the overwhelming success of the Conservative party in the 1958 elections. Although it did relatively well compared with the other opposition parties, the CCF lost all but eight of its twenty-five seats in the federal House of Commons. The illness of the CCF national leader, the decay of its constituency organizations,[35] and the potential political strength of the newly unified labor movement encouraged a new political beginning by organized labor. The 1958 CLC convention's political resolution recognized "the need for a broadly based people's political movement which embraces the CCF, the Labour movement, farm organizations, professional people and other lib-

[34] *Canadian Labour,* 1 (May, 1956), pp. 31–32.

[35] For an analysis of the contemporary CCF organization see Leo Zakuta, "Membership in a Becalmed Protest Movement," *The Canadian Journal of Economics and Political Science,* 24 (May, 1958), pp. 190–202.

erally-minded persons interested in basic social reform and reconstruction through our Parliamentary system of government." The convention instructed the executive council "to give urgent and immediate attention to this matter by initiating discussions with the CCF, interested farm organizations and other likeminded individuals and groups, to formulate a constitution and a programme for such a political instrument...."[36] The debate on the resolution reflected various viewpoints. To some, it was a repudiation of the CCF and its years of work on behalf of labor. To others, it was merely "an excuse for not doing anything in the CCF party at the moment." One speaker favored what has been a strong minority current in the Canadian labor movement, the belief that labor should have its own party, separate from farmers and other groups. Although unstated at the convention, a major argument favoring the new party was that it would sweeten for TLC members the pill of following the lead of the CCL in political matters.

The most generally accepted construction given the resolution was that it represented an attempt to fuse new life into the Canadian left, not by repudiating the CCF but by merging it with other groups to give it a mass base which it had lacked. This was the substance of the argument of Eamon Park, Chairman of the CLC Committee on Political Education, and Stanley Knowles, National Vice-Chairman of the CCF and now Executive Vice-President of the CLC.[37]

Chairman Park added several explanatory propositions: affiliation with the party would be by the individual union, not by the Congress. Individual members would not be forced to pay party dues through the union; as in Great Britain, a member could "contract out" if he wished. The relations of the CLC itself with the new party would be patterned on those between the Trades Union Congress and the Labour party. Finally, labor would itself participate in the financing and selection of candidates of the party.[38] Each CLC member would be assessed one dollar per year, which would make available one million dollars per year, or between four and five million dollars per federal general election.[39] This financial power would give labor the influence it felt it lacked in the CCF, and

[36] *Proceedings*, CLC, 1958, pp. 45–46.
[37] The debate is in *ibid.*, pp. 45–51.
[38] *Ibid.*, p. 46.
[39] *New York Times*, November 30, 1958.

would also enrich the labor party. Trade union spokesmen disclaimed any intention of wishing to dominate the new party. But the chief arguments in favor of a new party seem to be that it would represent a new start in politics, that labor would be in on it from the beginning, that labor would help finance it, and that labor would hence have a much greater voice in it. Unless these aims are achieved by labor, the new party would, from the labor viewpoint, hardly differ from the CCF.

Conclusions

The CLC thus seems to have decided that the CCF has two basic weaknesses; it cannot be controlled by labor and it cannot win elections. The Canadian labor movement is seeking a new party based on much the same groups as the CCF but which would give labor more influence. A new beginning in politics may greatly aid labor in Quebec. In the rest of Canada it is doubtful that the change will have much impact unless a dramatic enthusiasm is generated among the rank and file. Even the anticipated financial gains could be jeopardized by the apathy of the rank and file. This general lack of interest in direct political action, often shared by labor leaders, is the principal cause of the lack of success of the CCF. Unless this is altered, the change of name and leadership will have little effect.

Papers of the Michigan Academy of Science, Arts, and Letters
Vol. XLV, 1960 (1959 Meeting)

THE CLAIMS OF THE ARAB AND THE JEW TO PALESTINE

THEODORE F. BOUSHY

Hillsdale College

Until the end of the First World War, Palestine was an indistinguishable part of the surrounding Arab world. It was—and has been for hundreds of years—thoroughly Arab. Even the Jews who were living there before the turn of the century were thoroughly Arabicized. In 1770, the Jewish population numbered about 8,000.[1] This number increased to about 55,000 in 1918. Thirty years later, Israel was born. Instead of owning 3 per cent of the land, the Jewish population now owns 80 per cent. It has forcibly caused the expulsion of about one million Arabs who are now refugees, and in their place, about two million persons from seventy different countries have been settled. This problem is plaguing Palestine.

Claims by Jews and counterclaims by the Arabs have been the basis for this problem. The claims are based on the facts of history, the Bible, many prophecies, the McMahan-Husein Correspondence, the Balfour Declaration, the Covenant of the League of Nations, the Charter of the United Nations, resolutions of the United States Congress, economic and humanitarian grounds, and the establishment of Israel as a friend of the democracies in the Middle East.

The Zionists have argued that they have possessed the land, established their kingdom by conquest, and have never forgotten Palestine. They have the right to return to it. Nahum Sokolow says, "The validity of the Jewish title to Palestine rests on the same basis as the title of any nation to any particular area of the world where it has ruled and existed for centuries.... The Jewish historical rights of the land of Israel ... must become the decisive factor in the question of Palestine."[2] E. H. Moseley answers, however,

[1] A. M. Hayamson, *Palestine: The Rebirth of Ancient People* (New York: Knopf, 1917), p. 57.

[2] Nahum Sokolow, *History of Zionism, 1600–1918* (London: Longmans Green and Co., 1919), II, p. 158.

that Israel's possession of Canaan as an everlasting inheritance is based on an indefensible title.[3]

The Arabs argue that they have a longer claim to the land. The Midianites, Ishmaelites, and Canaanites mentioned in the earlier works of the Old Testament, were of Arab stock. They were well "known in the neighborhood of Palestine for nearly two millenniums before the conquest by the followers of Mohammed in A.D. 636."[4] These people, together with the Idumeans, the Moslems, the inhabitants during the Crusades and under the Turkish occupation, as well as the Christians and Arabs who now live in Palestine, are Arabs.[5] The Arabs argue that the land was theirs before the Hebraic conquest, was theirs during the period of the kingdom, was theirs after the expulsion of the Jews, and it belongs now to them. It is theirs by occupation, it is theirs by conquest and reconquest, and it is theirs by right.

The Rev. C. F. Burney, an authority on the Old Testament, states that there is "implication which is involved in the picturing of four Israelite tribes as sons of handmaids and not full wives of Jacob. . . ." These tribes were not regarded as Israelite. The Israelites were "a part only, and not the whole of the Hebrew stock." Several tribes had already been settled in Palestine before the coming of the Israelites.[6]

The Arabs reject the Zionists' claim because those who are flocking to Palestine are not Semites but non-Semitic people, descendants of the Khazars, who appeared in European history as White Huns, White Ugrians, and White Bulgars, and whom the Byzantine annals refer to as "Turks from the East." The Khazar dynasty accepted Judaism about A.D. 740. The great historian, Ibn Fadlan, states that "in addition to Khakan, the Viceroy, the princes of Samandar, and other high officials adhered to the Jewish religion." Matislav's principality, established in Russia in 988, replaced the kingdom of Khazaria, and the Jews who were there came under the Slavic do-

[3]E. H. Moseley, *The Jew and His Destiny* (Berne: Berne Witness, 1939), p. 25.

[4]W. M. Christie, "Arabs and the Jews in Palestine," *Victoria Journal*, LXII (1930), 96–116.

[5]Philip Graves, *Palestine: The Land of Three Faiths* (London: J. Cape, 1923), pp. 21 and 23.

[6]C. F. Burney, *Israel's Settlement in Canaan* (London: Oxford University Press, 1921), pp. 54–55; Genesis 49:16.

minion.[7] Today, the descendants of these Jews are coming by thousands to Palestine.

The Zionists, supported by some zealous Christians who speak of nothing but prophecies, claim that God made an everlasting promise to Abraham and to his seed, and, since the Jews are the biological offspring of Abraham, the land belongs to them. They rely on many scriptures particularly on Genesis 17:8, which says, "I will give unto thee, and to thy seed after thee... all the land of Canaan, for an everlasting possession."

To the superficial reader, it looks as if God gave the land to the Jews. However, the words "to thy seed" inevitably include Arabs, both Moslems and Christians, who can claim descent from Abraham through his son Ishmael, the reputed father of a large number of Arab tribes, and Genesis records that Abraham became the father of many north Arabian tribes through his concubine Keturah. Moreover, when the covenant of circumcision was made with Abraham (Gen. 17) and the land was promised as "an everlasting possession," or for a long period of time—the exact meaning of the Hebrew word—it was Ishmael who was circumcised: Isaac had not then been born.

When Sarah died, Abraham negotiated for a piece of property where she might be buried. Ephram offered to give the land to Abraham, but the latter refused and finally bought it for 400 shekels. Had Abraham understood that the land was physically promised to him and to his seed, certainly he would not have paid for it. Besides, Abraham made an alliance, a treaty, a binding contract with the Amorites and others who were in the land, according to Genesis 14:13.

By right of descent, the land should belong to Esau, the first son of Isaac, had not his mother and brother deprived him of the right of inheritance. Realizing his crime, Jacob fled and for many years remained a fugitive. Even when Jacob returned, he was still feeling the sense of guilt because he dared not meet his brother but placed the children and the family, as well as the servants, in the front and he far back. When the brothers met, Jacob offered to pay his brother, but Esau refused and was reconciled to his brother.

Returning to the promises, the Arabs, Christians and Moslems alike, can easily show that the argument favors them. For, accord-

[7] Sydney Harcave, *Russia* (Chicago: Lippincott Company, 1956), p. 13; see *Encyclopedia Britannica,* 11th and 12th editions; Theodore F. Boushy, "Palestine," Macon *Telegraph,* February 4, 1948.

ing to Jeremiah 31:31–34, God promised that a new covenant will be made with Israel. This new covenant is contrasted with the old in Galatians 3:17–18 by the great Apostle Paul, who was a Jew of the Tribe of Benjamin. Likewise, Peter, another apostle of Jesus, tells the Christians that they were not the people of God in the past but that now they have become a holy nation and a peculiar people.

Jesus the Son of God is of course the greatest authority. When the Jews boasted "Abraham is our father," he said unto them, "If ye were Abraham's children, ye would do the work of Abraham, but ye are of your father the devil" (John 8:44). These biblical narratives indicate to us that the promises were made to Abraham and to his seed, who is Christ. They are spiritual rather than physical promises.

The Arabs bring as witnesses certain Orthodox Jews, who understand that the promises are spiritual. Protesting against Zionism on June 16, 1897, the Association of Rabbis in Germany declared "that the attempts of the Zionists to found a Jewish national state in Palestine are contrary to the Messianic promises of Judaism as laid down in Holy Writ and in the later religious authorities."[8] In 1911, Professor Warburg spoke to the Seventh Congress, which discussed the problem of a national home. Warburg said, "The right of the Jews to the land by reason of their having possessed it two thousand years years ago is not a sufficient claim."[9]

The leaders of American reform have looked upon Zionism as "the negation of the best hope and promise of Judaism, as wilful abjuration of the role traced for Judaism by the greatest of the prophets."[10] Many of the Jews have separated themselves from the Zionists and have nothing to do with them.[11]

Another favorable argument of the Zionists and those who adhere to their philosophy is that the return of the Jews is according to the prophecy as found in Deut. 30:1–9, and 28:25–67. Israel Zangwill, a great leader among the Jews, says, "That it should be imagined

[8] J. H. Gottheil, *Movement in Judaism: Zionism* (New York: Jewish Publication of America, 1914), pp. 115–116.

[9] *Ibid.*, p. 154.

[10] Ahad-Ha-Am (Asher Ginzberg), *Selected Essays* (Philadelphia, 1912), p. 200.

[11] Jacob Gartenhaus, *The Rebirth of a Nation* (Nashville: Broadman, 1936), p. 120.

there is any direct allusion in the Bible to the present situation, is one of the oddest results of the petrification of Israel in the pages of the Bible. It sometimes seems as if nobody reads the Bible except the inmates of lunatic asylums or those qualifying for residence. . . . But the outstanding fact about the prophecies is that they had all come true five centuries before the Christian era. . . ." Modern Zionism is a "modern political movement whose origin and methods differ from the mystic Zionism of the past."[12]

Jacob Gartenhaus, a former Jew now a Christian minister, says, "If God's ancient covenant still stands, it has been fortified by the Jews," and commenting on Isaiah 11:11, 12, he says, "The meaning is that they will return, and they had from different parts . . . from Egypt and other places. Many people returned under Zerubbabel and others in accordance to the prophecy of Amos 9:14."[13]

Mahatma Gandhi says, "Palestine belongs to the Arabs in the same sense that England belongs to the English and France belongs to the French. It is wrong and inhuman to impose Jews on the Arabs."[14]

The second claim of the Zionists rests on the Balfour Declaration and other political declarations. The Zionist leaders had realized that the problem of the Jews is one of national homelessness and its solution is to give the Jews a national home.[15] The background of the Declaration is the failure of Theodore Herzl to get permission from Turkey to settle Palestine or to persuade the Jews, particularly Russian Jews, to accept Uganda, to be called "New Palestine," where the Jews might settle and create their own political institutions and find a haven of salvation.[16]

In 1916, the Zionists selected Dr. Chaim Weizmann and Dr. Nahum Sokolow to negotiate with the British Government. Dr. Weizmann had in the same year placed at the disposal of the British Government a formula for the manufacture of high explosives. A declaration was drafted, which would reconstruct Palestine as a

[12] Israel Zangwill, *The Voice of Jerusalem* (London, 1921), pp. 69–70, 80.

[13] "Gandhi's Message on Jewry," *Christian Science Monitor*, March 3, 1939.

[14] Amos 9:14.

[15] Michel G. Malti, "Is Zionism a Solution?" *The Arab World*, Fall Issue, 1944.

[16] A. C. Gabelein, *The Conflict of the Ages* (New York: Gabelein, Inc., 1933), p. 150.

Jewish national state. A struggle developed in the Cabinet over the wording of the document, and the objections of non-Zionist Jews and British governmental officials and others caused a change in the phraseology in order to protect the rights of the Arabs, as well as other Jewish communities outside of Palestine.[17] The final form was incorporated in a letter sent by Lord Balfour to Rothschild on November 2, 1917, which included the famous Balfour Declaration. The Allies soon adhered to it, and President Wilson publicly approved the aims of the Zionists in August, 1918. The Declaration reads: "His Majesty's Government views with favour the establishment in Palestine of a National home for the Jewish people and will use their best endeavours to facilitate the achievement of this object, it being clearly understood that nothing shall be done which may prejudice the civil and religious rights of existing non-Jewish communities in Palestine, or the rights and political status enjoyed by Jews in any other country."

The support of a national home for the Jews in Palestine was basically a war measure, dictated by materialistic considerations of self-interest. Germany was contemplating an announcement of sympathy for Zionism and was going to make certain offers to them in order to win their support in the war. The Grand Vizier of Turkey discussed this with prominent German and neutral Zionists. This announcement was finally issued on January 5, 1918.[18] Before the Palestine Royal Commission, Lloyd George gave testimony that it was "a propaganda measure."[19] Herbert Sidebotham states that British interest was identified with Zionist interests in the Middle East,[20] and Paul L. Hanna indicates that the Declaration was made in order to "win the support of an international community whose

[17] Graves, *op. cit.*, p. 43; *Daily Mail*, January 9, 1923; De Haas, pp. 485–486; David Lloyd George, *The Truth about the Peace Treaties*, II (London, 1938), pp. 46–50; Paul Goodman, *Zionism in England* (London, 1930), pp. 34–37; Blanche E. C. Dugdale, *Arthur James Balfour* (London, 1917), II, pp. 215–216, 223, 225; Herbert Sidebotham, *British Imperial Interests in Palestine* (Litchworth, 1937), pp. 56–62.

[18] Graves, *op. cit.*, p. 43.

[19] *Great Britain Parliamentary Debates*, Command 5479 (1936–37), p. 23; George, *op. cit.*, pp. 1118–1122.

[20] Sidebotham, *op. cit.*, also his *Great Britain and Palestine* (London, 1937), pp. 17–50.

wealth and whose influence in every country of the world were deemed valuable to the Allied cause."[21]

The Declaration was published in Russia in 1919; it was not published in Palestine until 1920. However, it was revealed in 1917 by the Turkish government, which informed the Sherif of Mecca of the agreement and tried to get the Arabs to enter into a separate treaty with Turkey on the understanding that the latter would recognize the independence of the Arab countries, including Palestine. It was not published because the army protested that any mention of it would disturb the Arabs and would hinder the freedom of further military operations. Thus it seems that the British government had issued a declaration so high-handed, unwarranted, and dangerous that it was an impediment to the progress of the British army and must be suppressed at any cost.[22] Likewise, the English press "charged that the Balfour Declaration conflicts with the term of the agreement with the Sherif of Mecca." They also described the Balfour Declaration as "illicit."[23]

It is interesting to note that London Jews opposed political Zionism and published a protest in the *Times* of London on May 24, 1917, "against this theory. We earnestly protest . . . the establishment of a Jewish nationality in Palestine, founded on this theory of Jewish homelessness. . . . Jewish political nationality [is] anachronism."[24]

Cardinal Gasparri, Vatican's Secretary of State, wrote to the Council of the League of Nations objecting to the terms of the draft Mandate, and particularly to Article XIV of the Covenant of the League, as tending to subordinate the native population of Palestine to the Jews.[25]

Speaking before the Foreign Relations Committee of the House of Representatives on resolutions advocating unrestricted Jewish immigration into Palestine, Faris S. Malouf, an eminent Lebanese

[21] Paul L. Hanna, *British Policy in Palestine* (Washington: American Council of Public Affairs, 1942), p. 37.

[22] Dugdale, *op. cit.*, pp. 215, 216, 232–233; Jeffries, *The Palestine Deception* (London: Daily Mail, 1923), pp. 27–29; Jeffries, *Palestine: The Reality* (London: Longmans Green and Co., 1939), pp. 27–29.

[23] *Daily Mail*, January 11, 1923.

[24] Sokolow, *op. cit.*, p. 109.

[25] Graves, *op. cit.*, p. 77; *Great Britain Parliamentary Debates*, Command 1708 (1922); *New York Times*, May 28, 1922.

lawyer, said, "One cannot help asking what right has England to give somebody else's country to a people who were disunited, unorganized and scattered among the nations of the world, without consulting the people who are immediately concerned and who have occupied that land as its natives from time immemorial and certainly owned it and continuously inhabited it for many centuries before the advent of the ancient Hebrew, during the Hebrew or Jewish partial occupation, and ever since the Jewish expulsion, first by the Babylonians and finally by the Romans."[26]

Although the Mandate for Palestine safeguarded the religious and civil rights of all inhabitants, nevertheless, it strengthened the Jewish settlements by allowing more Jews to enter Palestine.[27] The Arabs have always held that the Mandate was null and void, because the people of Palestine were not consulted when it was made, and they did not express their wishes to have England as the mandatory power. The Mandate was a "charter of treachery to the Arab people, an indefensible Trojan horse set within the Holy Gates of Islam."[28]

Let us observe how the Declaration conflicts with the promise made to the Arabs. In order to get the Arabs on the side of the Allies, Great Britain and France had entered into negotiations with them, and after long correspondence and negotiations, known as the McMahon-Husain Correspondence, the Sherif set down the terms, and marked the boundary lines of the would-be-Arab state.

On the north, it would be bound by Mersina-Adana up to the 37th degree of latitude to which degree fall Birijik, Urfa, Mardin, Midiat, Anadia, Jezira, as far as the borders of Persia. On the east, by the frontiers of Persia up to the Gulf of Basra. On the south by the Indian Ocean, with the exception of the colony of Aden, which is excepted from these boundaries. On the west by the Red Sea up to Marsine.[29]

With a little modification, these boundaries were accepted. Great Britain was prepared to recognize and support the independence

[26] Faris S. Malouf, "Avoid War and Bloodshed in the Holy Land," *The Syrian Ark,* VIII, No. 10 (June, 1944), 1, 11–14.

[27] *Great Britain Parliamentary Debates,* Command 1785 (1923).

[28] Hayamson, *op. cit.,* pp. 35, 100; *Great Britain Parliamentary Debates,* Command 1700 (1922); H. L. Weisman, *The Future of Palestine: An Examination of the Partition Plan* (New York: Lincoln Printing Company, 1937), p. 107.

[29] Jeffries, *op. cit.,* pp. 41–42; *Great Britain Parliamentary Debates,* 1938–1939, Command 5957 (London, 1937), p. 4.

of the Arabs within the territories included in the limits and boundaries proposed by the Sherif of Mecca.[30] J. M. N. Jeffries wrote in the *Daily Mail,* 12 January 1923, that "Mr. Churchill threw his pledge and the word and honor of England into the wastepaper basket."

The Balfour Declaration was the product of greed at its worst, a piece of double-dealing because Palestine had been promised to the Arabs. Some suggest that the Balfour Declaration canceled the agreement with the Sherif. If that is so, the declarations made later to the Arabs cancel the Balfour Declaration.

When in 1917 Turkey revealed the Declaration to Husain, he immediately cabled this information to the British government and demanded an explanation. In January, 1918, he was assured that "His Britannic Majesty's Government, in agreement with the Allied powers, confirm its previous pledges respecting the recognition of the independence of the Arab countries, and that the Jewish settlement in Palestine would only be allowed in so far as would be consistent with the political and economic freedom of the Arab population."[31]

Seven influential Arab leaders demanded further clarification and, in its answer, the British government made clear the policy that in the territory comprising Arab countries, including Palestine, "the future government of these countries should be based upon the principle of the consent of the governed."[32] The Arab hope was further strengthened by the words of President Wilson at Mount Vernon on July 4, 1918, that the postwar settlement would be based upon "the free acceptance of that settlement by the people immediately concerned."[33]

The faith of the Arabs in these pledges was again strengthened on November 7, 1918, when an official communiqué was issued to the press in Palestine, Syria, and Iraq by the British Government, declaring "that France and England aimed at the final liberation of the people living under Turkish rule and would aid these people to set up national government chosen by the people themselves in the free

[30] Jeffries, *op. cit.*, 11–12; *Daily Mail,* January 12, 1923.

[31] George Antonius, *The Arab Awakening* (Philadelphia, 1939), p. 268.

[32] *Ibid.*

[33] Henry Steele Commager, *Documents of American History,* 5th ed. (New York: Appleton, 1949), No. 423.

exercise of the popular will."[34] This is known as the Anglo-French Declaration. They were also encouraged by the Covenant of the League of Nations, Article XXII, which states that the "wish of these communities, those formerly belonging to the Turkish Empire, were held to be the principal consideration in the selection of the mandate." But the Mandate took effect without considering the wishes of the Arabs, because England and France had had secret agreements prior to the war, dividing the countries of the Near East between them.[35]

The summary of the Arab argument is as follows: promises were made to the Arabs before they were made to the Jews and, on several occasions, they were reaffirmed. When the King Crane Commission made its report on August 28, 1919, the conflict in the Declaration was pointed out. Furthermore, the Arabs argue, England did not possess Palestine at the time, and it had no right to make such promises to anyone. The League of Nations went against its charter when it imposed a Jewish nationality in Palestine, the Congress of the United States had no moral nor legal rights to resolve that a national state be established in Palestine because it goes against the fourteen points of Wilson and the Atlantic Charter, and the United Nations, in bowing down to political pressure, compromised the principle of the Charter and showed a preferential treatment to the Zionist community and a potent indifference to the welfare of the area reserved for the Arab state. Although the Jews owned only about 9.4 per cent of the land, they were awarded about 60 per cent of the total area of the land, and 20 per cent more passed later into the hands of the Jews by conquest.[36]

The third argument is the economic argument, which may be termed the argument of the "have not's" against the "have's." It runs like this. The Arabs possess four major countries. This tiny Palestinian country, on the other hand, is the only place in the world legitimately and most deeply connected with the Jewish people.[37] This idea may be found in the words of Mr. Shertok.[38] Feuer says,

[34] Antonius, *op. cit.*, p. 268.

[35] U. S. Department of State, *Mandate for Palestine,* Series No. 1 (Washington: Government Printing Office, 1931), pp. 8–10.

[36] *Ibid.*

[37] Philip Hitti, "Palestinian Arabs Descended from Natives before Abraham," *The Syrian Ark,* VIII, No. 9 (May, 1944), 1, 14.

[38] Town Meeting of the Air, "What Should Be Done about Palestine Now?" *Bulletin*, XIII, No. 7 (October 28, 1947), 1–22.

"The Jews have no place else to go except Palestine, which has a solidly founded economic base for population expansion."[39] A. Hirth stresses exploitation of the wealth of "the Dead Sea, which is valued over one trillion dollars."[40] In the hands of the Jews, says Sidebotham, the country will thrive and the economic conditions of the Arabs will improve.[41]

Leo Pinsker of Odessa stresses the point. "We need nothing but . . . a piece of land which shall remain our own property, from which no foreign master can expel us."[42] The Jews of Russia and Romania had been generally denied the opportunity to owning or working land. Louis D. Brandeis brings out the fact that "Palestine is fit for the modern Jew, and the modern Jew is fit for Palestine."[43] This line of argument was further advanced by Dr. Albert Einstein and Dr. Eric Kahler before a Congressional Committee when they said that the Zionists "took over from the period of Arabian predominance deserts and rocks and barren oil and turned them into flowering farms and plantations, into forests and modern cities." But this argument is similar to the argument of Italian apologists in Tripoli (1912) and in Ethiopia (1935). "The plain truth is that the Zionist colonies are still living on charity." Between 1848 and the present time about two billion dollars have been sent by the United States, one fourth of which was donated by the government.[44]

It will be well to recall the resolution of the Zionist Congress of 1921, that "the Jewish people should live in amity with the Arabs and develop industrial respect and unity, making Palestine a flourishing community for both." The Arabs retort, "It is not the speeches made that count; it is the action."[45] They point out that when the land was purchased by the Jewish fund, all Arabs working on it had to leave. "Once the land comes under the domination of the Jews, the

[39] Leon I. Feuer, *Why a Jewish State* (New York: Richard A. Smith, 1942), pp. 71, 73, 74.

[40] A. Hiorth, "From the River of Egypt unto the Great River, the River Euphrates," *Journal of the Transactions of the Victoria Institute,* LXX (1938), 126–152.

[41] Sidebotham, *op. cit.*, p. 11.

[42] *Ibid.*

[43] Louis D. Brandeis, *The Jewish Problem: How to Solve It* (New York: Zionist Organization of America, 1919), p. 17.

[44] Hitti, *op. cit.*, p. 14.

[45] *Ibid.*

Arabs can never buy it back. It becomes mortmain and inalienable." They remember the words of Chaim Weizmann, "We want the colonies to be Jewish and to be worked by the Jews."[46]

The Arabs point out the policy of complete Jewish exclusiveness adopted by the Zionist and how the artisans lost their jobs because Jewish employers refused to engage Arab workers. The Rutenberg Electric Works, for instance, had at a given time employed 800 employees, but there were only five Arabs among them. Near Caesarea 500 Arabs were dismissed.[47] Jewish colonization caused an abnormal dislocation of Arab cultivators.[48] When queried, the Zionists remarked, "We would not initiate the work if we were compelled to employ Arab labor." The object of the Zionist "in development is to employ Jews." This deliberate boycott of Arab labor was contrary to the principles of the Mandate, and it had been a constant increasing source of danger to the country.[49]

Today, we can see the effect of the Jewish policy on one million refugees living in tents across the fence from their former homes and land and unable to reap the fruits of their labor and denied access to their former bank accounts. Eighty per cent of the land has passed into the hands of the Zionists. The division gave over to Israel approximately 5,600 square miles, but today it occupies over 8,000, or 40 per cent more than the area given by the United Nations. This territory was annexed in three stages—before the withdrawal of the British forces and the establishment of the state of Israel, during the cease fire and before the armistice, and under the armistice when it seized, remilitarized, and controlled the demilitarized zones. The United Nations has opposed this aggrandizement in different forms on some two dozen occasions.[50]

The Arabs summarize the whole situation in this way: We have not benefited from Zionism, which has ruined our economic life. We have had enough trouble fighting European imperialists. We have suffered enough from clashes of interest over our country. We are sick and tired of being pushed around by Turks, Europeans, and Zionists. We can never trust the Zionists, because they have never

[46] Zangwill, *op. cit.*, p. 72.

[47] Graves, *op. cit.*, p. 129–130.

[48] Weismann, *op. cit.*, p. 225.

[49] John Hope Simpson, *Report of Immigration, Land Settlement and Development*, Command 3686 (London, 1930), p. 26.

[50] *Ibid.*

kept their promises; they have propagandized in Europe and America against us; they have completely disregarded our rights and aspirations and have distorted the facts about Arab deeds and Arab help to the Allies during the last two wars. We conclude that "no people benefit ultimately by giving up their country or even their civil rights to another nation."[51]

The fourth argument upon which the Zionists base their claim to Palestine is the friend-of-the-west argument. That is, Israel will be a democratic state in the Middle East, safeguarding the interests of the Western powers, particularly England, against the nationalist Arabs. Many British politicians advocated the creation of a Jewish state as a line of defense of the British Empire,[52] ". . . a loyal bulwark, in contrast to the attitude of the Arab states of the Near East, which obviously all are lukewarm to England's war effort."[53] The New York *Evening Sun* of December 26, 1917 said, "Britain cannot do better for herself and for Palestine than encourage the foundation there for an autonomous Jewish State."[54] The Waterloo, Iowa, *Courier* of January 10, 1918, in a headline, "New Zion Worthy of Old Waterloo," said, "Britain will have a political object in establishing the New Zion. Many Britishers stress the fact that the Zionist movement will be an ally of British imperialism."[55] The Duluth *Herald* wrote on January 26, 1918, "According to the Zionists' plan the new state will be established on a democratic basis. The world will have deep sympathy."

This argument, that Israel is a democratic state and friend of the West, is a farce, and the Arab points to the record again. Israel is the only country in the entire Middle East in which the Communist party is recognized by law and operates freely and openly. This fact is seldom publicized in the United States, but it has been brought out in a book by Harry B. Ellis, *Israel and the Middle East*, which was hailed by the Israeli Ambassador to the United States as "a painstaking and honest attempt to examine a very complex problem."[56]

[51] Malti, *op. cit.*, 1944.

[52] Jeffries, *op. cit.*, p. 12.

[53] *Ibid.*

[54] Waterloo *Courier*, January 10, 1918.

[55] "British Policy in Palestine: A British Hebrew Necessity," *Sunday Chronicle*, April 15, 1917.

[56] Harry B. Ellis, *Israel and the Middle East* (New York: Ronald Press, 1957), p. 154.

It is interesting to note that the total number of seats the Communists and leftist parties have had in Israeli Parliament is 25 out of a total of 120 seats.[57] The two Israeli left-wing Marxist parties, Mapan and Ahdut, participated in the five-party ruling coalition and had four ministries in the sixteen-man cabinet.[58] Over 88 per cent of the rural settlements in Israel have been organized on a communal collectivist basis and governed by Marxist principles. Production and all personal things of life, according to Joseph Dunner, are possessed by the group as a whole.[59] Reporting her visit to one Kibbutz, Mrs. Eleanor Roosevelt observes that "no one receives any money for his work; indeed he has no need for money, for the community supplies him with anything from pins and tooth paste to clothes, food and medical care,"[60] while Gerald de Gaury reports that the pay for "work of a member outside the Kibbutz goes direct to the Secretary of the community and medical care."[61]

According to Harry Ellis, "Generally boys and girls are allowed to sleep in the same room until they themselves desire separate quarters and separate showers. In some kibbutzim sexual intercourse between unmarried youngsters, while not encouraged, is considered the affair of the young people themselves."[62]

Supreme Court Justice William O. Douglas asserts that the Kibbutz is probably more strictly socialistic than the collective farms in Russia,[63] while the special Study Mission of the Committee on Foreign Affairs of the United States House of Representatives described the Kibbutz as a "form of elementary communism" and judged that it "cannot be described among democratic institutions."[64]

[57] *Facts and Figures* (Israel Office of Information, 1955), p. 17.

[58] *Ibid.*, pp. 55, 56.

[59] Joseph Dunner, *The Republic of Israel* (New York: McGraw Hill, 1950), pp. 142–143.

[60] Eleanor Roosevelt, *India and the Awakening East* (New York: Harpers, 1953), p. 43.

[61] Gerald de Gaury, *The New State of Israel* (New York: Praeger, 1952), p. 206.

[62] Ellis, *op. cit.*, pp. 169–170.

[63] William O. Douglas, *Strange Lands and Friendly People* (New York: Harpers, 1951), p. 274.

[64] *House of Representatives,* Report No. 2147, pp. 30–32. The Special Study Mission of the Commission on Foreign Affairs of the U.S. House of Representatives.

It is well to note that the first Israeli diplomatic envoy was appointed to Prague; the second to Moscow;[65] that Israel was the first country in the Middle East to purchase arms from the Soviet bloc and other countries behind the Iron Curtain,[66] against which the United States made strong protests to Czechoslovakia in connection with the alleged shipment of airplanes and arms by air from that country to Israel during the armistice in Palestine;[67] that hundreds of soldiers, officers, and technicians for the Israeli army were trained in Czech army camps, and the number of those who left and are waiting to leave approach 5,000.[68]

Israel was the first country in the Middle East to proclaim a "neutralist" foreign policy. Israel-Soviet trade relations play an important role in Israel's economy.[69] A staunchly Zionist American who was President Truman's first envoy to Israel writes, "The assistance of Russia and more particularly of Czechoslovakia during the Arab-Israel war was warmly appreciated. Israel sought to maintain an officially neutral policy in the cold war between East and West. This neutrality was expressed in numerous votes at the United Nations."[70] This neutral policy is described also in the biography of Ben-Gurion.[71]

The basic principle of Israel's foreign policy was announced by the government after the first general election early in 1949. "Israel was not to join any power or group of powers against another.... Her support of the action which the United States took in Korea, in conformity with the General Assembly's resolution, was, on the other hand, balanced by her recognition of the Central People's Government of China and her refusal to endorse the American policy in Formosa."[72]

[65] *Zionist Review,* October 15, 1948.

[66] Moshe Pearlman, *The Army of Israel* (New York: Philosophical Library, 1950), p. 146.

[67] *New York Times,* December 20, 1948.

[68] *Ibid.,* December 26, 1948.

[69] David Horowitz, *State in the Making* (New York: Knopf, 1953), pp. 272–274.

[70] James G. McDonald, *My Mission in Israel* (New York: Simon & Schuster, 1951), pp. 283–284.

[71] Barnett Litvinoff, *Ben Gurion of Israel* (New York: Praeger, 1954), pp. 228–229.

[72] De Gaury, *op. cit.,* pp. 173–174.

In an article on "Israel among the Nations," written August, 1952, Ben-Gurion announced that "Israel was one of the countries which belongs to neither bloc."[73] The same policy was reiterated in 1956 and 1957.[74] But the most important of all is the fact that in 1953 Israel pledged never to join any anti-Soviet alliance or pact, and the Soviet Government accepted this as so genuine that it re-established diplomatic relations with Israel.[75] And in the beginning of the history of the Zionists in Palestine, the paper of the Mopsi party wrote, "We are ready to shower approval of the Third International by our organizational affiliation therewith.... We also emphasize the necessity of constant communication with the Russian Poel-i-Zion."[76]

Other declarations may be cited, but these suffice to show that the state of Israel is not willing to enter the camp of the West; yet all we hear is that Nasser or some other Arab state has entered the Russian orbit.

Lastly, there is the humanitarian argument. This is perhaps the greatest weapon in the hands of the Jews. The persecutions of the Jews in Europe have created sympathy for them. In the Town Meeting of the Air October 28, 1947, this factor was presented by Shertok who said, "We could not close our eyes to the plight of millions of homeless Jews in Europe. This is a challenge to all kinds of justice, particularly Christian justice. By placing the Jews in Palestine, we solve a world problem."[77] It was this motive which caused the General Assembly to bow to political pressure and, compromising the principles of the Charter and exceeding the limits of its competence, to recommend the establishment of the Zionist state. Is it right to sell the Arabs, to settle the Jews at their expense? As Roosevelt asked Mr. Shertok, "What do you think, Mr. Shertok, that a remote historical association with a kingdom which vanished two thousand years ago justifies dispossessing the Arabs of their land?"[78]

Under this guise of humanitarianism, this blind and one-sided motive, the majority of the nations of the world have disregarded

[73] David Ben Gurion, *Rebirth and Destiny of Israel* (New York: Philosophical Library, 1954), pp. 391–393.

[74] *Israel Digest,* June 15, 1956.

[75] *American Jewish Year-Book for 1954* (New York: American Book-Stratford Press, 1954), pp. 273–276.

[76] Dunner, *op. cit.*, pp. 129–130.

[77] *Town Meeting,* October 28, 1947, pp. 11–12.

[78] *Ibid.*

the principles of President Wilson, the provisions in the Covenant of the League of Nations, in the Atlantic Charter, and in the Charter of the United Nations regarding self-determination and consent of the governed.

The claims of the Jews and the counterclaims of the Arabs have been presented. But presentation of the claims does not solve the problem. The present situation cannot go on forever. The United Nations which divided Palestine should demand the solution of the problem according to the Protocol of 12 May 1949. Both Arabs and Jews should accept the boundary lines as set forth in the division and the protocols, which both had signed; both should withdraw from Jerusalem according to the plans of the United Nations and establish the City of Salem as a symbol of peace and love; and both should let bygones be bygones and live together as they lived in the centuries past, in neighborliness and in unity. Of course, compensation for homes and lands should be made to those entitled to it. Finally, the Jews, the Moslems, and the Christians should sit down together and devise a plan of union or federation which would benefit both Arab and Jew. The friendship of the Arabs is much more valuable to the Jews than a small area of a few thousand square miles of land.

LITERATURE CITED

AHAD-HA-AM (ASHER GINZBERG). Selected Essays, Philadelphia, 1912.

ANTONIUS, GEORGE. The Arab Awakening. Philadelphia, 1939.

BEN-GURION, DAVID. Rebirth and Destiny of Israel. New York: Philosophical Library, 1954.

BOUSHY, THEODORE F. "Palestine: The Smoldering Crisis," Macon Telegraph, Feb. 4, 1948.

BRANDEIS, LOUIS D. The Jewish Problem: How to Solve It. New York: Zionist Organization of America, 1919.

"British Policy in Palestine—A British Hebrew Necessity," Sunday Chronicle, April 15, 1917.

BURNEY, C. F. Israel's Settlement in Canaan. London: Oxford Press, 1921.

Christian Science Monitor, March 3, 1939.

CHRISTIE, W. M. "Arabs and the Jews in Palestine," Victoria Jour. LXII (1930), 96–116.

COMMAGER, HENRY STEELE, ed. Documents of American History. New York: Appleton, 1944.

DE GAURY, GERALD. The New State of Israel. New York: Praeger, 1952.

DOUGLAS, WILLIAM O. Strange Lands and Friendly People. New York: Harpers, 1951.

DUGDALE, BLANCHE E. C. Arthur James Balfour. 2 vols. London, 1937.

Duluth Herald, January 26, 1918.

DUNNER, JOSEPH. The Republic of Israel. New York: McGraw Hill, 1950.
ELLIS, HARRY B. Israel and the Middle East. New York: Ronald Press Co., 1957.
State of Israel. Facts and Figures. New York: Israel Office of Information, 1955.
FEUER, LEON I. Why a Jewish State. New York: Richard A. Smith, 1942.
GAEBELEIN, A. C. The Conflict of the Ages. New York: Gaebelein, 1933.
"Ghandi's Message on Jewry," Christian Science Monitor, March 3, 1939.
GARTENHAUS, JACOB. The Rebirth of a Nation: Zionism in History and Prophecy. Nashville: Broadman Press, 1936.
GEORGE, DAVID LLOYD. The Truth about the Peace Treaties. 2 vols. London: 1938.
GOODMAN, PAUL. Zionism in England, 1890–1929. London, 1930.
GOTTHEIL, J. H. Movement in Judaism: Zionism. New York: The Jewish Publication of America, 1914.
Great Britain Parliamentary Debates, 1936–37, Command 5479; 1922, Command 1700; 1928–29, Command 3229.
GRAVES, PHILIP. Palestine: The Land of Three Faiths. London: J. Cape, 1923.
HANNA, PAUL L. British Policy in Palestine. Washington: American Council of Public Affairs, 1942.
HARCAVE, SYDNEY. Russia. Chicago: Lippincott, 1956.
HAYAMSON, A. M. Palestine: The Rebirth of an Ancient People.
HIORTH, A. "From the River of Egypt unto the Great River, the River Euphrates," Jour. Trans. Victoria Inst., LXX (1938), 126–152.
HITTI, PHILIP K. "Palestinian Arabs Descended from Natives Before Abraham," Syrian Ark, VIII, No. 9 (May, 1944), 1, 14.
HOROWITZ, DAVID. State in the Making. New York: Knopf, 1953.
HOWARD, HARRY M. "The King-Crane Commission," Moslem World (April, 1942).
JEFFRIES, J. M. N. The Palestine Deception. London: Daily Mail, 1923.
——. Palestine: The Reality. London: Longmans, Green & Co., 1939.
LITVINOFF, BARNETT. Ben-Gurion of Israel. New York: Praeger, 1954.
London Daily Mail, January 9, 10, 12, 1923.
MCDONALD, JAMES G. My Mission in Israel. New York: Simon & Schuster, 1951.
MALOUF, FARIS S. "Avoid War and Bloodshed in the Holy Land," Syrian Ark, VIII, No. 10 (June, 1944), 1, 11–14.
MALTI, MICHEL G. "Is Zionism a Solution?" Arab World, Fall Issue, 1944.
MOSELEY, E. H. The Jew and His Destiny. Berne: Berne Witness, 1939.
New York Evening Sun, December 26, 1917.
New York Times, December 20, 26, 1948.
Palestine Royal Commission. Summary Report No. 135. London, 1937.
PEARLMAN, MOSHE. The Army of Israel. New York: Philosophical Library, 1950.
PINSKER, LEO. Auto-Emancipation. Trans. D. S. Blondheim, New York, 1916.
Report of the Commission to Determine the Rights and Claims of the Moslems and Jews, December, 1930. London, 1931.

ROOSEVELT, ELEANOR. India and the Awakening East. New York: Harper & Bros., 1953.

SIDEBOTHAM, HERBERT. British Imperial Interests in Palestine. Litchworth, 1937.

——. Great Britain and Palestine. London, 1937.

SIMPSON, JOHN HOPE. Report of Immigration, Land Settlement and Development. Command 3686. London, 1930.

SOKOLOW, NAHUM. History of Zionism. 1600–1918. London: Longmans, Green & Co., 1919.

Town Meeting of the Air. "What Should Be Done About Palestine Now?" Bulletin, XIII, No. 7 (October 28, 1947), 1–22.

United States Department of State. Mandate for Palestine. Series No. 1. Washington: Government Printing Office, 1931.

United States House of Representatives. Report No. 2147.

Waterloo Courier, January 10, 1918.

WEISMAN, H. L. The Future of Palestine: An Examination of the Partition Plan. New York, 1937.

WILSON, WOODROW. The Public Papers of Woodrow Wilson. New York: Harper & Bros., 1925.

ZANGWILL, ISRAEL. Great Britian, Palestine and the Jews. New York: Doran, 1918.

——. The Voice of Jerusalem. New York: Macmillan Co., 1921.

Zionist Review, October 15, 1948.

Papers of the Michigan Academy of Science, Arts, and Letters
Vol. XLV, 1960 (1959 Meeting)

THE LEGAL AND POLITICAL IMPLICATIONS OF AGGRESSION AGAINST YUGOSLAVIA

J. DAVID SINGER

The University of Michigan

In his report to the U.S. Senate, the chairman of the committee studying aid to Yugoslavia characterized that country as "a perfectly natural next step for an expansionist Russia."[1] That American military planners are equally aware of this danger is quite obvious; the complexion of past NATO maneuvers in the Mediterranean, the operational plans drawn up by the Balkan powers with U.S. participation (before the Cyprus dispute helped nullify the Pact temporarily), and the recurrent concern over the way in which the Yugoslav government is utilizing aid personnel and military hardware, all suggest that those responsible for naval and military planning are seriously concerned over the possibility of such aggression. The military and strategic aspects of planning for such an eventuality, however, are significantly contingent upon the taking of prior political and diplomatic decisions, and there is little indication that the National Security Council, the Department of State's Policy Planning Board, or the White House have given adequate attention to the matter. As a consequence, Congressional and public opinion are ill-informed on the implications of a Soviet or satellite attack upon our highly prized "neutral," and ill-prepared to respond promptly and decisively in the face of such a crisis.

If such aggression *were* to occur, a staggering array of questions would arise. First, it might take the form, not of direct invasion or air attack, but internal subversion leading to the transportation of "volunteers" from Albania, Bulgaria, Romania, or Hungary. Or it might take the less dramatic form of blockade or boycott and embargo. In the former case, we would be confronted with the sort of questions which were raised by the use of Chinese "volunteers" in Korea, and the threatened repetition of that device during the Suez crisis of 1956. Yugoslav provocation, moreover, might conceivably

[1] Allan B. Kline, in U.S. Senate (85/1), Dec. 1952, *Foreign Aid Program* (July 1957), p. 1547.

give Moscow some moral, as well as strategic justification for an act of belligerency; comparison with the Formosa Straits situation in September 1958 is not without merit. Finally, were such intervention to take the form of indirect aggression which confronted the West in Lebanon and Jordan, the ambiguity would be bewildering indeed.

Beyond these questions, there are a great number of broader, strategic questions. What, for example, are the Soviet objectives in the situation? What forces have they committed or are they willing to commit? What sort of weapons are they employing? What price are they willing to pay to depose Tito or occupy his territory? Next, what form, if any, should our resistance take? Is this the place to employ "massive retaliation," or can a "limited war" meet our needs? Should nuclear weapons be employed? Tactically, or strategically? In addition, what is the general political situation? What sort of help do the victims request? Where do our allies stand, and what is the position of the neutrals? Should U.N. involvement be sought? What sort of resolution could win General Assembly approval? Is the Security Council the place to seek a decision? Should another U.N. Emergency Force be created? And finally, what is the nature of American opinion, as expressed in Congress, in the press, and by the public at large? How important do we think Yugoslav security is to us?

Clearly, these are staggering questions, and the chances of appraising them accurately at the moment of crisis are not good; thus it behooves us to consider as many of them as possible in advance. The intent of this article is to deal with a single, but highly significant, aspect of this hypothetical attack—that of the commitments we have already made to Yugoslavia and the possible avenues through which such commitments might best be brought to bear. Specifically, it is hoped to present a practical analysis of the responsibilities assumed and alternatives offered by our pattern of collective defense treaties and our membership in the United Nations.

Relevant Military Agreements

The only known direct and bilateral tie between the U.S. and Yugoslavia is through our mutual security and economic aid programs, and in this regard Tito has received about 1.5 billion dollars since his 1948 "break" with the Kremlin. Of this sum, approximately 780 million dollars has been in the category of military assistance whose employment and application is subject to the vague and ill-defined control of a small joint U.S.-Yugoslav military aid mission. One of the many sources of tension in our relations with Tito has been this very arrangement—only recently President Eisenhower felt compelled to order a temporary suspension in arms shipments.[2] While the program does, therefore, include loans, grants, materiel shipments, and some limited technical training, there are no other known commitments, and it may be assumed that this relationship, of itself, creates no military obligation for the U.S.

On turning, however, to the various regional groupings affecting the Balkan area, we find that the United States might be considerably more involved than first appearances indicate. First of all, there is the Balkan Alliance consummated at Bled, Yugoslavia, in August, 1954, whose moribund nature might be quickly altered. In this Treaty of Alliance, Political Coöperation, and Mutual Assistance, the Yugoslavs hope, with Greece and Turkey, to "ensure in the most efficacious manner, the territorial and political independence of their countries, in conformity with the principles and clauses of the United Nations Charter." And under Article 2, "the contracting parties have agreed that any armed aggression against one or several of them, at any part of their territories, shall be considered as an aggression against all the contracting parties."[3] This clause, therefore, while it imposes a clear set of obligations upon Greece and Turkey, does not have a similar effect upon the U.S.

A strict interpretation of the North Atlantic Treaty, however, might readily provide either the responsibility or the opportunity for U.S. involvement. In Article 5, the heart of NATO, the signatories "agree that an armed attack against one or more of them in Europe or North America shall be considered an attack against them all." Thus, while Yugoslavia is not a member of NATO, it may well be argued that by her alliance with two states which are members, her defense becomes properly a matter of concern for all of the other members. If the Greeks or the Turks were to view an attack upon Yugoslavia as an attack upon themselves (as they are required to do

[2] For text of basic agreement, see State Department Press Release 1016, 14 November, 1951.

[3] Amos J. Peaslee, *International Governmental Organizations: Constitutional Documents,* Vol. I (The Hague: Martinus Nijhoff, 1956), pp. 73–74.

under the Balkan Pact), does it not follow that the NATO powers must view this indirect "attack by definition" on Greece and Turkey as an attack upon themselves within the meaning of Article 5? In the Senate Foreign Relations Committee's report on NATO, no specific reference to this interpretation is made, nor is there any evidence to suggest that this aspect of its relation with other treaties arose. The inference is that if such an interpretation is made, so much the better in terms of NATO's deterrent effect; at the same time, it absolves the U.S. of any precise legal obligation.

If this literal reading of the Balkan and North Atlantic treaties is valid, we then find ourselves faced with an ever-widening, rapidly expanding set of military commitments in the face of our hypothetical attack on Yugoslavia. NATO would provide for the involvement of the SEATO powers via the dual membership of France, Britain, and the U.S.; thus the Philippines, Thailand, Pakistan, Australia, and New Zealand might become involved. Likewise, U.S. membership in the ANZUS Pact would reinforce the obligation of Australia and New Zealand, and U.S. membership in OAS might be interpreted as involving the Latin-American states as well. Finally, by virtue of their association with Britain and Turkey in the Baghdad Pact, Iran, Pakistan, and Iraq (which is technically still a member) might become equally embroiled. It is important, however, to reiterate that this ultimate involvement of forty-two states is highly dependent upon a thoroughly literal, and relatively unlikely—but at the same time, not impossible—interpretation of the galaxy of collective defense treaties referred to. Such an interpretation would certainly strengthen the position of those who argue that such military undertakings, contrary to their avowed objective, serve to do nothing more than guarantee the transformation of a limited conflict into a global holocaust.

The obligations assumed by signatories to regional defense groupings within the non-Soviet world, moreover, are highly subjective. For example, that paragraph of the Balkan Pact obliging its signatories to "render assistance" and take "immediately all measures, including the use of armed force," ends in a thoroughly ambiguous phrase: "which they shall deem necessary for efficacious defense." The SEATO requirement that each state will "act to meet the common danger" is followed by the limitation: "in accordance with its constitutional processes." And the North Atlantic Treaty reiterates the classic clause found in the Balkan Pact; its obligation that each signatory "will assist the Party or Parties so attacked" is closely followed with this phrase: "by such action as it deems necessary." Indeed, in consenting to ratification of this Treaty, the Senate Foreign Relations Committee expended considerable effort to "call particular attention to the phrase." Its report emphasizes that "each Party would have to decide, in the light of the circumstances surrounding the case." Furthermore, the committee stressed that Article 5 "clearly does not commit any of the Parties to declare war." All of these qualifications, it might be added, stand in strange contrast to the prior statement that "Article 5 is the heart of the treaty" and that "the solemn acceptance of this principle by all the Parties should have a powerful deterring effect on any would-be aggressor. . . ."[4]

Applicability of the U.N. Charter

In addition to the direct and indirect obligations which the U.S. may have undertaken through the above-mentioned collective defense treaties, its membership in the United Nations might also imply certain commitments. By adherence to the Charter of the world organization, several additional avenues are presented by which the U.S. may be required, or permitted, to provide direct military assistance to our hypothetical victim of armed attack. First, there is the general proviso of Article 2, paragraph 5, obliging all Members to "give the United Nations every assistance in any action it takes in accordance with the present Charter." Second, as a permanent member of the Security Council, the U.S. shares that organ's "primary responsibility for the maintenance of international peace and security." The question then arises as to how, and under what circumstances, the U.N. (and thus the U.S.) might become involved in a Balkan conflict.

Security Council action.—Under Chapter VII of the Charter, it devolves upon the Security Council both to "determine the existence of any threat to the peace, breach of the peace, or act of aggression," and to "decide what measures shall be taken . . . to maintain or restore international peace and security."[5] This Chapter goes on to

[4] U.S. Senate, (84/2) Subcommittee on Disarmament, *Disarmament and Security: A Collection of Documents,* pp. 880–883.

[5] Goodrich and Hambro, *Charter of the United Nations: Commentary and Documents* (Boston: World Peace Foundation, 1949), Art. 39, p. 593.

delineate the various actions which the Council may take, ranging from "partial interruption of economic relations" to "action by air, sea, or land forces." And while the Council may "decide" to use force and all Members are "required to carry out the decisions" of this organ, it is important to bear in mind that both the determination of any threat to, or breach of, the peace as well as remedial action, are subject to the veto. That is, both require the affirmative vote of at least seven of the eleven Council members, including all of the five permanent members. If it is assumed, therefore, that the attack by satellite troops or "volunteers" has been planned (or at least acquiesced in) by the USSR, it follows that a Soviet veto would negate any effort of the Council to take either determinative or remedial action.[6] With the Security Council paralyzed by a veto, the U.S. and its Western allies (again assuming that they so desire) would then have recourse to the General Assembly.

Assembly action.—The provisions of Article 11, empowering the Assembly to "discuss any questions relating to the maintenance of international peace and security" and to "make recommendations with regard to such questions" would seem to provide an adequate substitute for Council action, provided two thirds of the eighty-two Members concur. Article 12, however, places a serious obstacle in the path of Assembly action because of its stipulation that the larger body "shall not make any recommendation" while the Council is "exercising . . . the functions assigned to it . . . unless the Security Council so requests." As is well known, had it not been for the voluntary absence of the Soviet delegate from the Council meetings of 25 and 27 July 1950, that organ would have probably been incapacitated by his single negative vote. Thus some new strategy was called for to circumvent the veto in the event of any future crisis.

During "general debate" at the Assembly's September meeting, Secretary of State Acheson proposed, on behalf of the U.S., that certain changes in that organ's Rules of Procedure be made to "increase the effectiveness of U.N. action against aggression." The "Uniting for Peace" Resolution, which embodied these proposals, was ultimately accepted by a 52–5–2 vote at the Assembly's session of 3 November, over vigorous Soviet protests. Its major provision

[6] It should not be assumed, however, that Soviet instigation or acquiescence is *inevitable*. Several factors suggest the possibility, albeit remote, of Soviet neutrality or outright disapproval, in the event of such an attack.

is to permit a special Assembly session on twenty-four hours' notice, at the request of any seven members of the Council or a simple majority of the entire membership, if the Council, owing to lack of Big Five unanimity, fails to take affirmative action. At such a meeting, the Assembly is permitted to "consider the matter immediately with a view to making appropriate recommendations to Members for collective measures, including . . . the use of armed force."

Not until the fall of 1956 did the celebrated resolution meet its first test.[7] A U.S. proposal calling upon Israel and Egypt "immediately to cease fire" was vetoed by France and Britain at the Security Council meeting of 30 October, thus setting the stage for a Yugoslav motion (Doc. S/3719) invoking the 1950 resolution. This Council motion, noting that "lack of unanimity" had "prevented it from exercising its primary responsibility for the maintenance of international peace and security," was approved by a vote of 7–2–2, and the first emergency special session of the U.N. Assembly was convened on 1 November. During this eleven-day session, the Assembly, by the necessary two-thirds (as required in Article 18) vote, passed a series of resolutions designed to halt the Middle Eastern war, terminating in the creation of the U.N. Emergency Force.

One procedural aspect of the Uniting for Peace resolution, however, is seldom discussed, one which brings into question the very need for that resolution. As indicated above, the Assembly is precluded from making recommendations only when the Council is "exercising" its functions. This latter organ has tended to interpret Article 12 to mean that as long as a question is on its agenda, even if it has been vetoed, the Assembly may not make recommendations concerning it. But a more reasonable interpretation suggests that once a veto has prevented action, the Council is no longer "exercising" its functions. This view has been endorsed by a handful of observers, including the Australian delegate to San Francisco, Herbert Evatt, who argues that "the Council should be actually exercising its functions of conciliation or settlement . . . by enforcement measures, before the Assembly is blocked from making recommendations."[8] It is interesting to note that, despite this quite logical interpretation of the Charter, as well as the obviously substantive

[7] Unless the Assembly resolution of 1 February 1951, labeling Red China as an aggressor, be so designated.

[8] *The United Nations* (Cambridge: Harvard University Press, 1948), p. 84.

nature of the vetoed U.S. motion in the Council, Britain and France argued that the calling of an emergency special session was out of order.

Thus, it is clear that by action of either the Security Council or the General Assembly, the U.S. is given either the opportunity or the responsibility for taking affirmative action in the event of military attack upon Yugoslavia. In addition to action as a result of Council decision or Assembly recommendation, U.N. membership provides two additional bases upon which the U.S. may act in meeting such a hypothetical invasion.

Article 51 and the right of self-defense.—In Article 2, paragraph 4, we find the admonition that "all Members shall refrain, in their international relations, from the threat or use of force," but it would be misleading to interpret this caveat too literally. For in Article 51, it is unequivocally stated that "nothing in the present Charter shall impair the inherent right of individual or collective self-defense if an armed attack occurs against a Member," before the Council has acted. Admittedly, there are two specific restrictions here; first, the right of collective self-defense may be invoked only in the presence of an armed attack, and second, such right of action terminates when the Council "has taken the measures necessary to maintain international peace and security." But it should be borne in mind that the U.S. (or any power desiring to invoke the collective self-defense privilege) is ultimately the one to decide whether or not an armed attack has occurred, and whether or not the Council has taken the "necessary" measures; an affirmative or a negative finding, respectively, taken unilaterally by the affected power, should provide ample justification for whatever action its interests dictate.[9]

In order to make the application of the collective self-defense principle more meaningful, in terms both of deterrent and protective objectives, it has been found necessary to institutionalize the right by use of the terms of Chapter VIII of the Charter.

Article 52 and regional arrangements.—As suggested above, the inherent right of collective self-defense found expression in the creation of several regional groupings, including those discussed in the first section of this article. Despite the acknowledged indivisibility of peace, the traditional preference of states for limited and specific

[9] Clearly, the right of *individual* self-defense may only be invoked by the state which has been directly attacked.

commitments made it likely that the San Francisco Conference would produce a collective security system aimed at striking a balance between the needs of the world as a whole and the perceived needs of the separate states, and it is in Chapter VIII that we find this ambivalence expressed. The heart of this chapter is Article 52, which states that "nothing in the present Charter precludes the existence of regional arrangements or agencies for dealing with such matters relating to the maintenance of international peace and security as are appropriate for regional action." To demonstrate the peaceful and defensive nature of these regional arrangements, the members of each of the pacts discussed above seem to have gone out of their way to accept the three general restrictions imposed by the Charter. The first takes note of and reaffirms the primacy of the Security Council as a peace-keeping agency. The second accepts the stipulation that the Council be notified of any military action taken, and the third expresses willingness to cease such action upon the Council's taking of the "necessary measures." As noted above, these restrictions on collective self-defense, found in Article 51, are recognized (explicitly or implicitly) in all of the known alliances bearing upon a Yugoslavian invasion: the Balkan Pact, NATO, the Baghdad Pact, OAS, SEATO, ANZUS, and even the Warsaw Pact. In addition to these three limitations, there is a fourth, which applies only to regional arrangements (Chapter VIII) and not to self-defense measures taken under Chapter VII. This requires that the Security Council be kept "fully informed of activities undertaken, or in contemplation, under regional arrangements or by regional agencies" (Article 54). It is interesting to note that none of these pacts specifically invokes Chapter VIII and thus avoids the attendant requirement of Article 54; the NATO powers have even gone so far as to explicitly claim exemption from this provision. For example, the U.S. Senate report referred to earlier goes into considerable detail on the question of reporting to the Security Council. In making its distinction, the committee categorically stated that "Article 54 can in no way be interpreted as applying to plans for collective defense under Article 51."

Conclusion

The preceding discussion indicates, it would seem, that ample justification exists for U.S. intervention on behalf of Yugoslavia.

Even without a declaration of the type found in the Truman or Eisenhower "Doctrines," we have a legal and moral foundation composed of both the U.N. approach and the regional collective defense systems created under the U.N. system. Therefore, the question which we will have to answer becomes essentially one of policy. If the Yugoslavs are attacked, we will be confronted not so much with a legal or even a moral issue, so much as with a political-strategic one. It is hoped that this article has satisfactorily dealt with the legal aspect, and the moral one would seem to be self-evident, regardless of our likes or dislikes for Tito and his brand of Communism.

Thus, in the final analysis it would seem that our military and civilian leaders must address themselves primarily to the political and strategic issue, and it cannot be urged too strongly that this be done at once. By making our position clear now, we may well avoid having to face the crisis later. But whether it works out that way or not, it would be prudent to be prepared for the worst. If such an attack is analyzed in advance we may succeed in mitigating the humiliating indecisiveness of Indo-China and Hungary, and the costly blunders of Korea and Suez.

PAPERS OF THE MICHIGAN ACADEMY OF SCIENCE, ARTS, AND LETTERS
VOL. XLV, 1960 (1959 MEETING)

SOCIOLOGY

RELIGIOUS GROUPS AND SOCIAL CLASS—SOME INCONSISTENCIES

GLENN M. VERNON

Central Michigan University

THE transition from a sacred to a secular society which the United States, along with many other societies, has been experiencing has produced many inconsistencies and contradictions. The role of religion has been undergoing marked changes in our society, and while the general overall change in religion has been somewhat consistent with the basic underlying changes in other institutions,[1] in many respects the harmony is far from complete. This paper reports a study which explored some of these inconsistencies. More specifically, it was concerned with the manner in which a sample of college students visualized or defined the current interrelationships between our religious structure and our social class structure, including the value definitions held by the students concerning certain aspects thereof.

Participants in the study were 207 students enrolled in sociology classes at Central Michigan University during the fall of 1957. The sample is not representative of the student body as a whole or of any larger group. Thus, any generalizations made from this data would be highly tenuous. The fact that the respondents were all sociology students should be noted, since this probably had an influence upon their responses. Of all students on campus, they would be the ones most alert to the existing relationships between social class and religion. At least one third of the respondents, however, were freshmen who had not studied this specific relationship.

Inconsistencies concerning social class are widespread in the United States. For instance, when one studies social class, it is not uncommon to find respondents who report that they "just don't believe in social class." Such statements might be expected in a so-

[1] See, for instance, Thomas Ford Hoult, *The Sociology of Religion* (New York, 1958), chap. 2.

ciety which holds to be self-evident the premise that all men are created equal—even though we also strongly endorse a contrary premise that a man should be able to climb from a humble birth to a higher level. In a similar manner, we might also expect that in a culture where the dominant religions emphasize a fatherhood of God and a brotherhood of all mankind there might be a reluctance on the part of the individuals to recognize that there is a relationship between social class position and religion. Since our culture indicates that we are all equal in the sight of God, we might easily overlook or minimize the fact that all religions may not be equal in the sight of individuals at different class levels. Although we are all well aware of the religious diversity in the United States, we may not also be aware that this diversity is related to social class position.

My first objective in this study, then, was to determine how well aware the respondents were of the fact that there is a relationship between social class and religious identification, and to discover whether their perceptions as to specific relationships corresponded accurately with the relationships found in objective studies.

The behavioral scientist has provided us with ample evidence to indicate that there is a relationship between the social class position of an individual and his religious behavior.[2] One such study, that done by the Office of Public Opinion Research,[3] I have used in my analysis. With a nation-wide sample in 1946–47, they were able to demonstrate that the percentage of membership drawn from the various class levels (upper, middle, and lower) was not the same for all of the eight religious groups they considered. Rather, a discernible pattern emerged which permitted ranking the groups from high to low, according to the per cent of the religious group from each class level. Starting with the one with the largest percentage of upper-class members, they ranked the religious groups as follows: Episcopalian, Congregational, Presbyterian, Jewish, Methodist, Lutheran, Roman Catholic, and Baptist.

I made no effort to duplicate the above-mentioned study. The concern was, rather, to see how consistent the perception was with the relationship already found to exist by this and other studies.

[2] See Hoult, chap. 11, or J. Milton Yinger, *Religion, Society and the Individual* (New York, 1957), chap. 7.

[3] Taken from *Information Service,* Federal Council of Churches of Christ in America, Vol. 27, No. 20, Part 2 (May 15, 1948).

More specifically, the students were provided with a list of thirteen religious groups: Baptist, Christian Science, Congregational, Disciples of Christ, Episcopal, Jehovah's Witnesses, Jewish, Lutheran, Methodist, Mormon (Latter Day Saints), Presbyterian, Roman Catholic, and Seventh-Day Adventist. The list includes the ones used in the nation-wide study referred to above, thus permitting comparisons. The list was kept small for convenience' sake. The respondents' perception as to the relationship between religion and social class was taken from their answers to the following questions: "Circle the name of the religious group which you think would have the greatest appeal to upper class individuals." Then followed the names of the thirteen religious groups enumerated above. This was followed by identical questions for the middle, working, and lower classes. It should be noted that the question does not ask for answers concerning any specific community or geographical area, but concerns class and religion in general. The respondents were permitted to circle more than one name if they felt that more than one group would have equal appeal at a given class level.

This procedure was followed for all except one class of freshmen. For comparison, this class was asked the same questions but was not given the names of the religious groups and was asked to provide the names of the groups they associated with each social class.

The answers to the questions are shown in Table I. An analysis of the first part of this tabulation indicates that the students' perception of the pattern of appeal which the different religious groups have at different class levels is, in general, consistent with that which was found in the previously cited study. The pattern obtained from the respondents who were not given the names of the thirteen groups is shown under the heading "No Names Provided." This pattern does not differ markedly from that obtained when the respondents were provided the names. The Jewish group, Mormons, Disciples of Christ, Jehovah's Witnesses, and the Seventh Day Adventists all were mentioned less than when these names were provided, but the patterns were still similar. As might be expected, the respondents rated the appeal of their own religious group somewhat higher than the rating provided by the overall group, as shown in the last section of Table I.

It was observed while tabulating the data that occasionally a student would rate his own church as having equal appeal for per-

TABLE I

Appeal Which Different Religious Groups Are Believed to Have for Individuals at Various Class Levels

	Upper		Middle		Working		Lower	
	No.	Per Cent	No.	Per Cent	No.	Per Cent	No.	Per Cent
Names Provided								
Episcopalian (1)*	54	67	19	23	7	9	1	1
Jewish (4)*	5	40	8	53	1	7	0	0
Presbyterian (3)*	70	49	54	38	15	11	3	2
Congregational (2)*	34	37	31	33	20	21	8	9
Methodist (5)*	35	21	85	52	35	21	10	6
Lutheran (6)*	3	4	44	60	23	32	3	4
Christian Science	6	26	5	22	7	30	5	22
Roman Catholic (7)*	33	16	43	22	64	32	59	30
Baptist (8)*	3	2	28	21	68	50	36	27
Mormon	0	0	2	12	7	44	7	44
Disciples of Christ	0	0	3	6	17	36	27	58
Jehovah's Witnesses	1	2	1	2	7	13	43	83
Seventh Day Adventists	0	0	1	2	14	30	32	68
No Names Provided								
Episcopalian	8	67	4	33	0	0	0	0
Jewish	4	100	0	0	0	0	0	0
Presbyterian	19	56	12	35	3	9	0	0
Congregational	2	15	8	61	2	15	1	9
Methodist	10	19	25	46	15	28	4	7
Lutheran	1	8	6	46	3	23	3	23
Christian Science	0	0	1	17	3	50	2	33
Roman Catholic	22	24	18	16	16	14	13	11
Baptist	1	3	7	18	18	47	12	32
Mormon	1	25	0	0	0	0	3	75
Disciples of Christ	0	0	0	0	0	0	0	0
Jehovah's Witnesses	0	0	0	0	0	0	1	100
Seventh Day Adventists	0	0	0	0	0	0	4	100
Rating of Appeal of Own Religious Group								
Episcopalian	7	58	5	42	0	0	0	0
Presbyterian	10	34	19	59	2	6	0	0
Congregational	7	50	5	36	2	14	0	0
Methodist	11	19	36	61	10	17	2	2
Lutheran	0	0	7	78	2	22	0	0
Christian Science	0	0	1	100	0	0	0	0
Roman Catholic	14	17	28	35	23	28	16	20
Baptist	0	0	13	46	10	36	5	18
Mormon	1	25	1	25	1	25	1	25
Disciples of Christ	0	0	1	100	0	0	0	0

* Rank order of churches in the study by the Office of Public Opinion Research.

TABLE II

Number of Respondents Who Felt Their Own Religious Group Had Greatest Appeal

	At every class level	At three class levels
Roman Catholic	12	7
Methodist	3	2
Baptist	1	3
Mormon	1	1
Congregational	0	1
Presbyterian	0	1

sons at every class level or as having a somewhat universal appeal. A check of these students revealed that this pattern was most frequent for the respondents who identified themselves as Roman Catholic and secondly for the Methodists, as indicated in the following tabulation.

Thus, it was found that the students' perception of the appeal which the different religious groups have for members from different social classes corresponds fairly accurately with the distribution of religious members according to class. The perception, however, tends to be distorted in favor of the person's own religious group. In this respect, then, the inconsistencies are not marked.

A second part of the questionnaire elicited answers as to how extensively the student had thought about the relationship between these two factors before he was exposed to the questionnaire. It was found that only 2 per cent of them reported extensive thought. Fifteen per cent checked "rather extensively," 39 per cent "not too deeply," 21 per cent in "only a vague manner," and 23 per cent indicated that they really hadn't considered this relationship.

Thus, it would seem that the respondents had not concerned themselves extensively with such questions. This might be interpreted in different ways. It might suggest, for instance, that they are really not aware of such relationships. Such an interpretation, however, is contradicted by the findings of this study, which suggest not only an awareness of such a relationship but also a fairly accurate perception of it.

A more adequate explanation stems from our religious emphasis upon the equality of mankind. It may also be a reflection of the fact that in America the institution of religion is separated not only from the state but, in the minds of some at least, from other aspects of culture. It may be felt that religion should not be related to our class structure, and thus there may be an unwillingness on the part of our people to concern themselves with the question. Whatever the reason, evidence from this study suggests that our society apparently does not stimulate much thought among college students about the relationship between socio-economic class and religion. Thus, a convenient way to escape the apparent contradiction is to avoid it. This inference is supported by penciled notations on the margin of two of the questionnaires: "I don't know—have never considered the question and don't think it's important." "Never taken that much interest in other people's religion."

The respondents were also questioned concerning their evaluations of the *religious* morality of certain aspects of our class structure, each respondent allowed to define "being religiously moral" in whatever way he chose. When asked whether the existence of social classes in a society is religiously moral, about one fourth indicated that they were uncertain on the issue and one fourth stated that the existence of social classes is definitely not religiously moral. Thus, only one half of the students endorsed the morality of having social classes in a society.

Thus, while the respondents have a fairly accurate conception as to the relationship which exists between religious affiliation and social class position, they appear to avoid thinking much about this relationship, even though fully one half of them are either uncertain that even the existence of social classes is religiously moral or definitely feel that it is not.

An additional inconsistency centers around the question as to whether or not it is religiously moral to try to improve one's social class position. This great American dream received religious endorsement by the respondents in that a big majority of them (78 per cent) indicated that such efforts were religiously moral. Only 12 per cent rejected this as being religiously moral, and 10 per cent were undecided. It would appear that some, at least, define the presence of social classes as immoral, but define efforts to change one's social position as moral. The existence of social classes is immoral, but the use of this immoral structure is moral.

The picture is somewhat further confused when we look at the responses to the question as to whether it is religiously justifiable for a person to change his religious affiliation as he changes his class position. Slightly better than half of the respondents (54 per cent) rejected this idea, with 11 per cent being undecided. Only 35 per cent indicated approval. In this respect, the Roman Catholic respondents presented a somewhat atypical pattern. Of the thirty-nine Roman Catholic respondents who answered the question, only two indicated acceptance of the premise that it is religiously moral for an individual who is able to improve his social class position to change his religious affiliation and join a group which he feels is more harmonious with his new position.

This high rejection of changing religions as class position changes probably reflects a feeling that such religious changes would not represent sincere religious conversions, which, in turn, suggests a failure to recognize that the type of religious dogma about which a person can be sincerely convinced is related to his class position. This may be a reflection of the fact that the respondents really hadn't given much consideration to the ramifications of their perceived relationship between class and religion.

Thus, in summary, evidence from this study indicates that the question as to the religious morality of certain aspects of our class structure is somewhat confused and contradictory. The students participating did recognize that religious patterns are related to our class structure, and they have a fairly accurate perception as to what the relationship is, although they do not think extensively about this relationship—possibly stemming from the fact that they do not want to think extensively about it. Only approximately half of them feel that the existence of social classes is religiously moral, yet three fourths of them feel that it is religiously moral to try to improve one's class position, although many of them feel that it is not moral to "improve" one's religious position accordingly. This study, then, would support Hoult when he claims that "perhaps the only social phenomenon that is more complicated, varied and seemingly contradictory than stratification itself, is the religious reaction to it."[4]

[4] Hoult, p. 278.

Papers of the Michigan Academy of Science, Arts, and Letters
Vol. XLV, 1960 (1959 Meeting)

PSYCHOLOGY

FURTHER STUDIES OF TURN-SIGNALING BEHAVIOR[1]

ABRAM M. BARCH

Michigan State University

The behavior of drivers on the highway is important for those concerned about traffic and safety as well as for students of human behavior. Nevertheless, relatively little has been done in this field. This is due, in part, to the difficulty of obtaining measures of driver behavior that reflect variations in circumstances and people and yet remain stable enough for studies experimentally varying the physical or social environment of the driver.

In a previous study[2], my colleagues and I chose the relative frequency of signaling a turn by the driver of an automobile (turn-signaling) for intensive study on the assumption that it could provide such a measure of driver behavior. It could be observed in the field with little if any distortion of the natural event. It required a minimum of equipment, yet displayed a high degree of reliability of measurement. Turn-signaling percentage was found to be a sensitive indicator of both situational and personal characteristics. It varied with type of intersection, direction of turn, sex of driver, and the signaling of the preceding car.[2] A partial check of stability disclosed no significant differences related to time of day. Furthermore, no significant change in signaling percentage was found for an intersection checked three months later.

On the other hand, we did not find that signaling percentage was affected by a closely following car. This seemed contradictory to the other findings. Therefore, I reconsidered the method of testing

[1] This research was supported by the Highway Traffic Safety Center, Michigan State University. The assistance of John Nagle, Richard Passine, Clyde Randall, Thomas Trabasso, and John Ufford in the collection and analysis of the data is gratefully acknowledged.

[2] Signaling behavior of the preceding car was initially reported as having no effect (2); however, a reanalysis using all the data in combination found a weak but significant effect (1).

for this effect. It seemed that the relationship might have been obscured by using only two categories in the analysis: within 100 ft. of the "observed" car when it made its turn, and not within 100 ft. One of my primary purposes in this study was a closer analysis of the influence of the following car on turn-signaling.

Another major purpose was to find the stability of turn-signaling percentage for periods up to two years or more. Moreover, to permit the collection of other information, the distance that the turn signal is made from the intersection (signaling distance) was not studied in the previous research. It is an obvious additional measure. Effective knowledge of and control over signaling distance could lead to increased control over traffic flow. Also, it might provide corroborative evidence for the relationships implied by the signaling percentage measure. A hypothesis that seemed likely was that the higher the signaling percentage, the farther from the intersection the signal would be given. The third major purpose, then, was to study the relation between signaling percentage and signaling distance.

METHOD

General procedure.—I carried out observations at four different intersections in the Greater Lansing (Michigan) area, making three sets of observations (a total of 42 hours)—in June 1957, in July–August 1957, and in November 1958. Each set had a different major observer. All observations were made during daylight hours; none was obtained during periods of rain or fog. Reliability checks for the first two sets of observations were made by comparing per cent of agreement found with one of the major observers of the previous study. As before, per cent of agreement for number of cars signaling, cars not signaling, and direction of turn was 99 per cent or better. Per cent of agreement for judgments of signaling distance ranged from the middle eighties to middle nineties.

Observation sites and specific procedure.—A full description of Site 2-A is given in the previous report.[2] It was one of the legs of the secondary two-lane road at a two-way stop, at the edge of the M.S.U. campus. In addition to noting direction of turn and turn signal, observations were also made of: (*a*) the distance the signal was made prior to intersection; and (*b*) presence of a following car and its relative distance from the observed car.

[2] *Ibid.*

Signaling distance from the intersection was coded into four categories: 0–50 ft., 50–100 ft., 100–175 ft., 175 ft. to the next intersection. The distances were chosen so as to utilize some natural landmarks and thus facilitate estimation.

Determination of following car distance was somewhat more complex. The observer stood at a fixed point, 175 ft. from the intersection. When the "observed" vehicle reached this point, the location of the first following car was noted. If the first following car was within 75 ft. of the "observed" car, it was recorded as "following"; if the first following car was more than 1,075 ft. or if no following car was seen, the situation was recorded as "nonfollowing." The relative spacing of cars might change between the time the "observed" car reached the fixed point of observation and when it completed its turn. If the following car entered a different zone than the one it had occupied at the time of fixed observation, a "change" was noted. No "change" pairs were considered in the statistical analysis. Observation of a given car always began when it was at least one full block (250 ft.) from the intersection under study.

Observations were made in June 1957 and in November 1958. Between these two periods, the characteristics of the intersecting road had been changed from a two-lane road into a divided four-lane road. In this change, the length of the block just before the intersection was shortened. Due to this change, signaling distance from the intersection in these two sets of observations did not represent the same sections of roadway.[3]

Site 6 was the intersection of a four-lane divided expressway (widened to six lanes at the intersection) with a two-lane street (widened to three lanes at the intersection) within the city of Lansing and controlled by multiple stop signals (2). Site 6-A observations were made on the expressway; Site 6-B observations were made on the street. Observations were made during the morning rush (6–9 A.M.) on the south and east leg, and in the middle to late afternoon on the north and east leg.

The use of the south leg in the morning and the north leg in the afternoon was due to the low volume of traffic on these legs at other times. Previous results had not found any differences between op-

[3] Observational details on all sites are on file at the Highway Traffic Safety Center, M.S.U.

posing legs that were structurally similar at the same or at different times of day. Signaling distance from the intersection was recorded in units of 50 ft. Not all turning cars were recorded since the observation procedure required that a car be under observation for at least one block prior to the intersection and that only one car be observed at a time.

Site 6 observations were made in July–August 1957. This intersection was previously studied in October–November 1956.

Site 8, the north leg of the primary road at the intersection of two two-lane roads (Farm Lane and Shaw—west), on the M.S.U. campus at a two-way stop, was observed in July–August 1957 and in November 1958. The road at this point is upgrade toward the intersection with the right-hand road the more heavily traveled leg of the secondary road. Between the two periods of observation, a number of changes occurred. The intersecting road immediately to the north of Site 8 and to its right, which had been blocked, became an entrance to a large parking lot. The right-hand leg at Site 8 became one-way away from the intersection.

Site 9 was about 300 ft. to the south of Site 8, and was the same type of intersection as Site 8 except that one leg (the observed leg) was one-way into the intersection. All observations were made in November 1958.

Signaling distance was obtained in units of 50 ft. from the intersection for both Sites 8 and 9. In addition observations were also made on following car distance. The fixed point of observation was 175 ft. from the intersection.

When the "observed" car came abreast of this point, the position of the first following car was recorded in one of four categories. The size of the categories varied slightly from Site 8 to Site 9 to permit use of natural landmarks. They were behind the fixed point of observation approximately 0–180 ft., 180 ft.–425 ft., 425 ft.–600 ft., 600 ft., and beyond. All cases were discarded in which the following car changed its categorical location between the time of fixed observation and the time the "observed" car completed its turn with the exception of those following cars which had been in the first following zone and passed the fixed point of observation before the "observed" car completed its turn.

Results and Discussion

Stability of turn-signaling percentage.—Table I presents in summary form the signaling percentages for right and left turns at Sites 2-A, 6-A, 6-B, and 8 at various times from July 1956 to November 1958. In only one comparison was a significant chi-square found, that for left turns at Site 6-A. On further analysis this difference was found to be due to the difference in signaling percentage for left turns between morning and afternoon signaling for the same period of observation—August 1957. Signaling percentage was not different between the 1956 and 1957 observations when compared for the same time of day (afternoon). It will be recalled that morning and afternoon observations made at this site in 1957 used different legs to compensate for low volume. Thus, the only exception to the finding of prolonged stability of turn-signaling percentage involves

TABLE I

Signaling Percentage at Selected Sites Over a Two-Year Period

Site	Observation Date	Right Turns % Sig.*	Right Turns Cars Turning	Left Turns % Sig.*	Left Turns Cars Turning
2-A	July 1956	41	1341	66	207
	June 1957	42	1000	60	377
	Nov. 1958	43	154	67	114
		$\chi^2 = 0.2, p > .05$		$\chi^2 = 3.1, p > .05$	
6-A	Nov. 1956	58	576	56	1,304
	Aug. Morn. 1957	52	82	38	657
	Aug. Aft. 1957	59	135	51	331
		$\chi^2 = 2.0, p > .05$		$\chi^2 = 57.2, p < .01$	
6-B	Nov. 1956	37	548	73	875
	Aug. Morn. 1957	41	231	75	69
	Aug. Aft. 1957	34	238	74	337
		$\chi^2 = 2.4, p > .05$		$\chi^2 = 0.3, p > .05$	
8	Aug. 1957	58	1220	—	—
	Nov. 1958	58	1304	—	—
		$\chi^2 = 0.0, p > .05$			

* % Signaling.

TABLE II

EFFECT OF FOLLOWING CAR DISTANCE ON SIGNALING PERCENTAGE OF "OBSERVED" CAR

	Right Turns		Left Turns	
	% Sig.	Cars Turning	% Sig.	Cars Turning
Site 2-A				
Within 75 ft.	58	139	78	27
More than 1,075 ft.	35	140	42	24
	$\chi^2 = 14.3, p < .01$		$\chi^2 = 5.5, p < .05$	
Site 8				
Within 190 ft.	61	696	87	94
190–430 ft.	57	297	88	34
430 ft.-	52	311	68	50
	$\chi^2 = 7.0, p < .05$		$\chi^2 = 6.6, p < .05$	
Site 9				
Within 175 ft.	—	—	70	83
175–425 ft.	—	—	69	35
425 ft.-	—	—	40	60
			$\chi^2 = 14.4, p < .01$	

a variation in time of day and specific leg observed. It is likely that this exception also involves an altered sex-ratio of drivers.

Influence of a following car.—Table II summarizes all comparisons for the influence of a following car. In each comparison, there was a significant increase in signaling percentage when the "observed" car was closely followed by another car. The influence of the following car held for following distances up to about 400 ft. Thus, the previous negative finding, as surmised, was probably due to the inadequacy of the method used to measure this effect.

Signaling distance—Table III presents the median signaling distances, signaling distance distributions, and signaling percentages for right and left turns at the various sites.

We initially assumed a relationship between signaling distance and signaling percentage: namely, the higher the signaling percentage, the farther from the intersection the signal would be given. To hold site characteristics constant, a comparison was made of right and left turn signaling distances at Site 2-A, using 1957 ob-

TABLE III

SIGNALING DISTANCES AT THE VARIOUS SITES

Site		Median Sig. Distance (ft.)	Signaling Distance from Intersection (50 ft. units) 1	2	3	4	5	6	7	% Sig.
2-A	RT†	76	145	125	(149)*	—	—	—	—	42
(1957)	LT	43	79	34	(24)*	—	—	—	—	66
2-A	RT	127	8	20	(38)*	—	—	—	—	43
(1958)	LT	105	30	7	(39)*	—	—	—	—	67
6-A	RT	113	13	30	8	18	8	9	4	52
(Morn)	LT	115	48	64	48	29	20	19	24	38
6-A	RT	143	47	28	30	38	31	13	4	59
(Aft)	LT	129	37	35	20	19	21	19	16	50
6-B	RT	112	44	35	40	18	9	20	11	38
	LT	91	110	49	33	30	39	25	15	74
8	RT	172	10	64	192	198	131	63	46	58
(1957)	LT	—	—	—	—	—	—	—	—	—
8	RT	168	24	40	68	140	40	29	21	58
(1958)	LT	168	9	6	10	35	7	4	4	82
9	LT	121	—	(51)**	—	21	—	(14)**	—	58

† RT (Right Turn), LT (Left Turn).

* Includes all signals given more than 100 ft. from intersection.

** First three and last three categories grouped.

servations. At this site left turns are signaled significantly more often than right turns. The two signaling distance distributions were significantly different by a chi-square test, but the left turns tended to be signaled closer to the intersection than the right turns. Similar results were found at Site 6-B and at Site 6-A (afternoon). On the other hand, the median signaling distance for right and left turns at Site 6-A (morning) are almost identical, and these signaling distance distributions do not differ significantly. The same is true for the right and left turn signaling distances at Site 8 (1958 observations). Thus, left turns were either signaled closer to the intersection than right turns or at about the same point irrespective of any difference in signaling percentage.

This suggested separate consideration of right and left turns in relation to signaling percentage. Also, if signaling percentage was the major determiner of signaling distance, different sites with similar signaling percentage should have similar distance distributions. Therefore, the distributions of right turn signaling distances were compared for two sites with near identical signaling percentage—Site 6-A (afternoon) right turns at 59 per cent and Site 8 right turns at 58 per cent (1957). The distributions were significantly different. A similar significant difference was found when Site 8 right turns were compared with Site 6-A morning right turns with a 53 per cent signaling percentage. Further comparisons showed that the right turn signaling distances were significantly different from morning to afternoon at Site 6-A even though their signaling percentages were not. On the other hand, left turn signaling distances at Site 6-A did not differ significantly from morning to afternoon even though, as noted in Table I, their signaling percentages did.

From this and other comparisons it became obvious that signaling distance and signaling percentage were not related in any simple fashion. In an attempt to further study this relationship, it became apparent that the use of chi-square as our statistical test ignores some of the information in the data. Signaling distance can be regarded as an ordered dimension rather than as one composed of independent categories.

With this view, the Kolmogorov-Smirnov Two-Sample test (the K-S test) was applied to all comparisons of signaling distance. This test takes ordering into account and is considered as powerful as the chi-square test for small samples and nearly as powerful for large samples (3). In most of the comparisons the K-S test failed to yield a significant difference at the .05 level whereas the chi-square test did at the .01 level or beyond. The reverse was never true.

If the evidence for the power of the K-S test is dependable, the inference from these results is that in a number of instances the differences between signaling distance distributions is not one that has a cumulative or ordering aspect, but that unique characteristics peculiar to 50 ft. segments of the roadway per se determine signaling distance.

The sensitive nature of the signaling distance distribution is also shown in the follow-up studies. Despite some changes in the physical characteristics near Site 2-A and Site 8, no change occurred in signaling percentage. However, the chi-square test of signaling distance distributions at Site 2-A (both right and left turns) in

June 1957 and November 1958 were highly significant (.01 level) as were the right turn signaling distance distributions obtained at Site 8 in June–August 1957 and November 1958. The K-S test showed significant changes in signaling distance at Site 2-A but not at Site 8.

Both the chi-square and K-S test showed that a closely following car (Site 2-A) significantly affected signaling distance. The "observed" car tended to signal farther from the intersection in the presence of a closely following car. On the other hand, the presence of a following car had no effect on signaling distance for right or left turns at Site 8 and for left turns at Site 9.

In general, signaling distance appears to be a very sensitive measure of traffic behavior reacting to relatively slight changes in situation. Although it could be observed with a reasonable level of reliability, there is no guarantee that it itself is a reliable measure of traffic behavior. The evidence suggests that it is not.

Summary and Conclusions

I carried out observations of turn-signaling behavior at four different intersections in the greater Lansing (Michigan) area, recording direction of turn, presence of signal, distance from intersection signal given, and relative distance of following cars. The relative frequency of signaling a turn was found quite stable, showing little change for periods up to more than two years. Signaling percentage was significantly increased by the presence of a car following within about 400 ft. There was no apparent relationship between signaling percentage and signaling distance, and the measure of signaling distance was unstable over a period of time.

LITERATURE CITED

1. Barch, A. M., D. Trumbo, and J. Nagle, 1957. Social Setting and Conformity to a Legal Requirement. Journ. Abnorm. Soc. Psychol., 55: 396–398.
2. ——, J. Nagle, and D. Trumbo, 1958. Situational Characteristics and Turn-Signalling Behavior. Highway Res. Bd. Bull. No. 172: 95–103.
3. Siegel, S. 1956. Nonparametric Statistics for the Behavioral Sciences. McGraw-Hill Book Company, New York.

PAPERS OF THE MICHIGAN ACADEMY OF SCIENCE, ARTS, AND LETTERS
VOL. XLV, 1960 (1959 MEETING)

FREUD AND ROGERS: A COMPARATIVE STUDY OF TWO PSYCHOLOGICAL SYSTEMS

BEN O. RUBENSTEIN

Wayne State University

THE following report draws preliminary conclusions from the raw material at hand. In subsequent papers I hope to make more detailed presentations of Rogers' system, his philosophical roots and influences, and his particular philosophy of science. I have examined all of Rogers' books, articles, and pamphlets, and they have been considerably enriched and illuminated by direct discussions with him. Although Rogers' nondirective system has been examined many times, we might find some value in an examination of Rogers' significance as a phenomena on the contemporary American psychological scene. Initially, I am curious about the great appeal of the very positive optimistic expectations of nondirective therapy, the modicum of technical needs, and the emphasis upon its scientific validity. In my comparison I do not place Freud and Rogers in an equal relationship. History has already placed Freud as a great man. History has yet to assess Rogers, and I feel I am justified in using Freud as the larger and more familiar backdrop against which we can examine Rogers.

I propose to compare systematically the ideas and methodology of both men and to use their life histories as a starting point. Study reveals that both men place different values and meanings upon their early lives. It strikes me as significant that Rogers developed a system that is not concerned with genetics and that Freud created one which rests entirely upon the meaning of his early life. In fact, I am impressed with the striking differences in the lives of both men. Freud developed a theory that was based upon his own early experiences and directed his interest toward sexual curiosity, the role of the father, and sibling rivalry. He chose to use himself as his own clinical material to formulate hypotheses and conjectures. Conversely Rogers is, wittingly or unwittingly, very careful to separate his own early years from the part they may have played in the formulation of his theories and hypotheses. He emphasizes the fact that he arrived at them solely from the assessment of objective evidence. Although it is not within the province of this thesis to draw inferences about the character of both psychologists, a meaningful comparison of their ideas must consider the basis for their different views about genetic psychology. Perhaps we can make our first significant comparison at this juncture.

I feel that Rogers never resolved a dichotomy which prevailed from his earliest years. There are two Rogers, the dreamer and the practical man. I do not wish to suggest that Rogers is unaware of these aspects. He visualized the purpose of science as providing verification for the hunches that grow out of subjective "experiencing." We find Rogers' family as being so concerned with the young dreamer that the father sought to bring him into touch with practical affairs through an extended trip. His autobiography tells of the parents' rigid materialistic and moralistic orientation and their reaction to Rogers' adolescent religious experience in which he changed his faith and his later marital and theological plans. Likewise we are told of both his academic interest in cataloguing and his practical interest in agricultural experiments. Freud wrote about the significance of a religious experience.[1] He regarded it as a recrudescence of the oedipal conflict and displacement of the struggle with the father to religion followed by a complete submission to the new father, God. It is precisely this type of explanation which was and is abhorrent to Rogers. He has never displayed any interest in causation but only in discovery of order in nature. Freud, on the other hand, had an overpowering need in his youth to understand the riddles of the world and to contribute something to their solution.

In another sense we find Rogers interested in helping humanity and Freud disclaiming any craving in his childhood to succor suffering mankind. We must apply the word "compartmentalized" to Rogers: he wears the costume of either the dreamer or the practical man, and they remain unsynthesized. For example, as the objective researcher he cannot in the process visualize any subjectivity. If he is engaged in unrestrained subjective "experiencing," the scientist is not on the scene. Freud never conceived of such a dichotomy; he was coolly scientific in his observations of his own phantasies. He could not imagine a mind in which all parts were not interacting

[1] Sigmund Freud. "A Religious Experience," *Collected Papers*, Vol. V (London: Hogarth Press, 1950), p. 246.

simultaneously. A fully functioning mind was one which possessed simultaneous knowledge of its disparate parts. We have arrived at the first definitive difference in the ideas of the two men, that of their different reactions to their own beginnings. Rogers disclaims the influence of the past upon his present and further feels that his scientific findings are entirely uninfluenced by his past. Freud, with equal clarity and emphasis, sees himself and his thought as a product of his past and feels that the two are not in conflict.

We discover that Rogers, whose background consisted of a fire and brimstone religiosity and a family atmosphere of hard work and biting sarcasm, developed with irreducible optimism a unique psychological system resting upon the need for unconditional acceptance. It contained a therapy which sought to cleanse the patient of all introjected values (conditions of worth) and the buoyant belief that everyone who so desired could be helped to become fully-functioning. Freud, whose background was not demanding and contained adequate love, pessimistically fashioned a system based upon instinctual conflict, the influence of the past, immediate and archaic, and the belief that therapeutic results were somewhat questionable in face of the implacability of the unconscious. Rogers turned his face from the past and saw an unlimited future; Freud studied the meaning of the past and cast a quick, dubious eye ahead.

Intriguing as it might be, this study is not the place to examine what existed in Rogers' roots that caused him to reject all biological interest and curiosity about man's body. We must mark this fact as the second difference between the two men. While Rogers amassed his collection of moths and noted the difference in yield in relation to amounts of fertilizer, Freud studied ancient history and devoted a summer to his attempt to discover the whereabouts of testes in the eel. Freud never forgot the body and its influence on the mind, while Rogers maintained a dichotomy between them.

One more point. Rogers reacted sharply to his parents' attempt to keep him tied to their faith. He dropped his study for the ministry and rejected Freudian thinking because he could not accept dogma or orthodoxy. Yet we find that he has developed a therapeutic process which he believes can answer all questions. Freud was never bound to any orthodoxy, yet similarly rejected fixed and rigid creeds and chose to fashion a therapeutic system which he felt might be of some limited help for certain neurotic difficulties.

A COMPARISON OF PHILOSOPHICAL IDEAS

I believe that the magnet which will bring order into our diffused discussion will be a philosophical one. Once we compare and understand the differences in the Freudian and Rogerian conception of the nature of man, the epistemological frame of reference, the dynamic, the genetic character and the technique of their psychologies will be understandable.

As Americans we react to the notion of men rigidly fixed to a predestined direction. As a consequence we have moved toward environmentalism and situationism. We mark the lack of interest in the complex Freudian model of personality with its intricate arrangement of drives, defenses, and derivatives. Freud's pessimism and emphasis upon constitutional factors does not well suit us. Our therapeutic aims are more ambitious than his. Our American goal is almost inspirational, as exemplified by Rogers and others.

The above statements are all true of Rogers who, diagnosis notwithstanding, stated that anyone can be helped if he so desires to become a fully-functioning person. But I do not propose to have the American culture speak for Rogers. He has made his own position sufficiently clear for comparison. Rogers emerges unequivocally as an idealist concerned with the tenuous body of the spirit.

Both of our protagonists saw man as part of nature, but Rogers saw him through the eyes of Hegel in the process of becoming part of the Absolute. In a somewhat contradictory fashion, he holds to a lesser degree an existential position. Freud saw man through the eyes of Darwin; man surviving through adaptation with no moral evaluation as to his peculiar adaptation in time and space. The self for Rogers had only one window and that window opened on the phenomenal world; all behavior at the moment was controlled by an actualizing tendency at the control of conditions of worth. All perceptions of the phenomenal world were determined by the self-concept. Freud's ego had two windows of different sizes, and he left no doubt where he placed the larger one. He saw the ego as the tip of an iceberg perched precariously: engaged with the surface environment but very much influenced and guided by the mass of id below the surface. The personality of man evolved from his conflict with his primitive sexual and aggressive drives and the fear of punishment or loss of love. The resolution of the conflict in a social setting produced man as we know him.

I suggest that McGiffert at Union Theological Seminary offered young Rogers an opportunity to integrate his new-found intellectual insights with his former religious belief under the rubric of humanitarianism. Further, McGiffert stressed the value of studying only accessible phenomena and depreciated the unscientific character of going beyond the facts. Sin was now dressed in modern garb and was defined as a consequence of imperfect development. Since Rogers believed that the ultimate nature of man is good, he further believed that he only needed to be helped to achieve his potential. A partial explanation of Rogers' optimism about therapy may be attributed to the conviction that man can only evolve in a "good" direction. In addition, we are acquainted with Rogers' conception of the enriching character of the therapeutic relationship for the therapist. Freud, conversely, felt that the patient was helped when the physician proceeded in a cool, dispassionate fashion. I think that William Heard Kilpatrick at Columbia likewise gave Rogers additional reinforcement for his own urgent need to protect individualism with the same rationale; man contains a potential for the realization of his best self if he is given choice. We find Rogers in agreement with Kilpatrick in an indifference to the past and a focus on the present existential reality. Each new situation contains novel elements which, as one reworks them, enlarge his life pattern and reorient his perspective.

Throughout the writing of all of Rogers' mentors—McGiffert, Kilpatrick, Hollingworth, and Rank—we discover the same high morality and an unshakable belief that man is in his essence good and positive. We recall that Rank not only saw man's will as positive, but in addition felt that his "will psychology" was a reaction against Freud's "making evil of the will."

Rogers' apparent later existential position may not be as contradictory to the aforementioned beliefs as they may appear at first glance. His emphasis upon the insular character of man's mind and his simultaneous rejection of systematization and education may be understood in part from McGiffert's position: there are regions which are inaccessible to human knowledge. I am suggesting that Rogers offers us a certain sequence of philosophical ideas, such as: man is good; we need only remove all interference and he will achieve his full potential. It follows, therefore, that we need not further investigate what is within, since everything we experience is good. Moreover, we cannot investigate more deeply since we can only judge observable data. (J. B. Watson comes to mind.) Finally, we have no way of knowing more since we can only learn from our own experience and not from that of any other person.

Freud too had his mentors, but the ideas held in common with them dealt with different material. He shared with Plato ideas about memory recall, the love instinct, and myths surrounding sexuality. Galton and Hartmann offered direction to his thoughts about the unconscious and free association. Together with Goethe, Helmholtz, and Brucke, he worshiped at the new shrine of the science of nature which was now his *Weltanchauung*. Goethe and Freud found common cause in their conception of nature; her perpetual springs of action and her crown of law. It was Darwin who buttressed Freud's perception of the primacy of the intellect. Darwin's thinking strengthened Freud's recognition about the fallacy of final causes and gave him encouragement for his naturalistic position about behavior and the evolutionary flavor of his constructs. Our past shapes us in the process of our evolution. Helmholtz and Fechner gave Freud a model for his physicalistic mathematical method of conceptualizing psychic forces. Therefore, I place Freud in a naturalist deterministic orbit and Rogers in an idealistic one, with a later inexplicable contradictory existentialist interest.

Science was Freud's God, and it was a warm intuitive worship, albeit one without an apparent morality. We know of his interest and identification with literary figures and themes. From these themes we learn again of Freud's conception of the essence of man. Shakespeare enriched Freud's views of man as a helpless victim of great forces, Dostoevsky illustrated, perhaps too well, Freud's vision of man's primitive, irrational unconscious feelings upsetting his well-intentioned conscious wishes. Freud thought of man as being neither good nor bad; rather a being whose direction was determined by the character of the socializing process. He did not view conscience or morality as being different from any other psychical expression and regarded conscience as having potentiality for destruction. Freud has never left us in doubt about what he felt his motivation to be for his adult, scientific pursuits. He knew he was driven by infantile ambition to be a conqueror, a veritable Hannibal. He recognized the basis for his wish to stake out a new world for conquest. The insights presented him with no apparent conflict in his scientific work but he strived to understand the relationship between the two parts.

Conceptions of Psychic Functioning

We are ready to turn to comparison of the two systems. Freud's system was composed of five aspects of the psychic apparatus, the theory of instincts, the development of the sexual function, mental qualities, and dream interpretation. We can eliminate the development of the sexual function and dream interpretation from our comparison, which has no counterpart in Rogers' system. Freud envisioned psychic function as deriving from three mental structures. The id, an unconscious agency, was the reservoir for instinctual impulses, and we have no comparable structure in Rogers' system, except as subsumed under the construct of the actualizing tendency. Can we compare the ego with Rogers' self-concept? Freud saw the ego as developing out of the id as an intermediary between the id and the outer world. It has the simultaneous task of both self-preservation and of providing for the appropriate discharge of id impulses. In the process it developed a variety of additional functions including synthesis, cognition, motility, adaption, et cetera, for an efficient manipulation of the environment for the discharge of id demands at some times and the control of them at other times. Rogers, on the other hand, defined the self as an organized conceptual gestalt composed of perceptions of "I" or "me" and perception of "I" or "me" in relationship to others with values attached to these perceptions; in a word, an individual's view of himself. It is clear that in this sense the self relates to only a part of the ego. We can visualize certain ego activities which might be compared to the self-concept. Similarities between the ego and Rogers' self-structure may be revealed by a juxtaposition of the two abstractions. The ego was conceived as a mediator between three areas: the id, the super-ego, and the outside world. Initially the outside world and the representatives of the conscience are contained in the person of the parents. They stand both in the position of gratifier and frustrater. The ego is compelled, in these vastly oversimplified circumstances, to develop an adjustment which will permit gratification of id impulses which will be consonant with super-ego requirements.

Freud contended that, in the ensuing frustrating and socializing process, the ego with the employment of a variety of defenses, could arrive at a consistent character quite early in life. The ego becomes an integrated one if the frustrations and denials in these formative years occurred under reasonably benign circumstances which permitted the child some discharge, forced some repression, encouraged sublimation, and allowed fusion of hate and love in an enduring relationship. Excessive repression forces an unconscious break-through of id impulses with the aid of various defensive mechanisms. The superego, with its roots in the unconscious, perceives these maneuvers and causes guilt and anxiety. His viewpoint may likewise be seen as an economy of scarcity; too much repression leaves little energy for the ego to carry on its affairs with the outside world. Conversely, Rogers, as do Maslow and others, rejects the energy concept and visualizes energy being evoked by situations.

Although Freud and Rogers recognized that the need for love (unconditional positive regard) by the infant determined his behavior, the two men differ about the consequences. Rogers posited three phases. In one phase the child experienced positive regard when he was aware that he had satisfied another's need for positive regard. We interpret this statement to mean that the child feels a sense of well being when he gives his mother a sense of well being. In the next phase the child is driven to give the mother affection; i.e., make her happy, although what he may do does not gratify his own organismic needs. The child develops, subsequently, a structure called a self-regard complex. This complex is the repository of those experiences for which he received affection and when he repeats them independently he no longer needs the approval of another, but can like himself for his own action. However, when the child has self-experiences and the parents react to them negatively the child can no longer esteem that experience and he loses self-regard. The child avoids that particular self-experience because of this new condition of worth. The distinction between the last two phases appears to rest upon the thesis that the child owns or had to disown the feelings arising out of the inhibition or expression of behavior.

Rogers' entire system rests upon this central thesis. Conditions of worth force selection of perceptions, and those which are denied cause an incongruence of dissociation and, finally, maladjustment. No conditions of worth would develop in the child if the mother gave him unconditional acceptance, warmth, or sympathy (unconditional positive regard), no matter what he felt. Consequently all perceptions would be accurately received and the individual would be adjusted.

The genetic position of psychoanalysis is quite different. Freud never conceived of an unconditional situation at any time; such a thought would be alien to his conception of the dynamic character of the mind. Further, Freud would have held that the mother's statements or communications to the infant regarding the acceptance of his feelings as she frustrated his wish would be ignored by him. In addition, in any case she would be considered a bad mother by the child. The infant would be compelled to wrestle with his angry feelings about her and would be compelled out of his dependent needs to impose a value judgment upon his own act. More precisely, Freud felt that children had little doubt that mothers could read their minds and would reject them after perceiving their angry thoughts. Anxiety over the possible loss of love would cause the child to temporize his needs in the following situation. Rogers regarded this particular sequence as a tragic development in infancy. Therapy was to be directed at this estrangement in man's functioning. Freud saw it as natural but not tragic since the infant, through the utilization of a variety of mechanisms of defense in the socializing process, gained control of his primitive feelings. The personality developed out of the conflict situation as the child was constantly forced to assess his reality and to determine what actions were possible. The mother may accept the child's feelings but the child is not necessarily aware of her acceptance, since he experiences his frustration. Perhaps Rogers can visualize an unconditional situation since he rejects Freud's concept of the innate primitive forces within the child and, consequently, the child's own fear of them.

Psychoanalysis differs with both the possibility of an unconditional acceptance of a child's feelings and the consequences of such acceptance. At this juncture we may be closer to the discovery of the relationship of Rogers' self-concept to Freud's ego structure. I feel we can place it not within the ego per se but at the point of interaction between ego and superego. In fact, the bulk of Rogers' system may be pin-pointed in the area between conscience and the ego. The superego is responsible for man's view of himself since he is driven constantly by guilt to compare what he actually does to what the introjected parents feel he should have done. Rogers' sole concern appears to be that of freeing man from the yoke of his conscience. He proposes to do this by exposing the individual to a therapeutic situation whose prototype is an idealized conception of the mother—child relationship. The situation would be roughly comparable to Franz Alexander's corrective emotional experience.

Freud was likewise interested in the pathological consequences of a too strict superego, but to the same degree he was also concerned with pathological ego defenses. He was aware of the primitive unconscious roots of the superego but was equally cognizant of the fact that the ego possessed equally primitive methods for restricting or inhibiting the personality.

We have discovered another area of major difference between Freud and Rogers centered around the importance of conscience and its psychic consequences. In addition, we have found that the Rogers theories of personality and therapy have at their core the belief in the primacy of the conscience and the dissociation it causes, whereas Freud visualized ego-superego conflict as only one aspect of psychic function.

A Comparison of Methodology

We can proceed to examine a basic methodological difference which stems from the aforementioned dissimilarity between the two psychologists. Rogers stated unequivocally that in therapy if the patient experienced the unconditional positive regard of the therapist a series of events would be initiated; the patient would become aware of certain previously denied experiences and include them in his self-structure.

Psychoanalysis has a somewhat different methodology. The therapist accepts the patient's productions and feelings without judgment. The patient is driven to project upon the inactive, nonjudgmental, accepting analyst a variety of unconscious needs and feelings originally attached to significant persons in his past. Frequently the patient is compelled to misinterpret the above features of the analytic situation and respond to the acceptance with a positive transference. Freud defined the positive transference as one in which the analyst is unconsciously conceived of as being that parent who will gratify the patient's forbidden infantile wishes. It is of some historical interest to mention that it was precisely this particular technical phase that made for so much distortion of and consequently disillusionment with psychoanalysis in this country. Many untutored and poorly trained therapists chose to make this initial permissive phase of treatment and the ensuing positive transference of the patient into the complete therapeutic process. They would mistake the frequent spontaneous disappearance of symptoms and the appearance of good spirits of the patient for the cure. In reality the patient's remission was due more properly to the unconsciously heightened anticipation of infantile gratification.

In what way might the positive transference ameliorate, albeit temporarily, the yoke of the conscience? Psychoanalytic theory explains the situation from a genetic position in a rather simple fashion. The child will accept frustrating conditions of worth for that situation which any child fervently desires: the complete undivided accepting attention of the important parent. There is no change as a consequence in the superego itself but a greater tolerance of it. In addition, the belief of the nondirective therapist that the patient is, in his deepest essence, a good and positive person does not threaten the patient's fear that he will need to modify his infantile unrealistic superego.

Psychoanalytic technique continues its different course. The analyst interprets the positive transference and inevitably unleashes the resistance of the patient to the possibility of change. Freud held that the mind resisted change for several important reasons. Any change implied the re-experiencing of the anxiety which caused the original neurosis and, simultaneously, implies the permanent forsaking of infantile phantasy. The resistance can be expressed in the following language. The patient unconsciously wishes to hold on to the strict conscience and asks only that he be given more love to keep the repressed aggression in check. He does not want this need interpreted since he fears that the anxiety will break out in the face of the weakened defense against his own aggression. Analysis would encourage the examination of each defense against anxiety until full knowledge is gained of the genesis of the infantile feelings and fears. Rogers' therapeutic technique does not disturb the unconscious disposition of forces since there is no interpretation of unconscious material and mechanisms.

The different views about technique are likewise expressed in the differing treatment goals. Rogers declares simply that the successfully treated patient is less defensive, more open to experience, and more creative, i.e., he accepts himself. Freud is somewhat more practical in his conception of the sought-for maturity. The individual, according to Freud, should be able to work, to love, and to achieve genital heterosexuality.

Epistemologies

Finally, I wish to compare the two different epistemologies and the character of the theories within the philosophy of science. The most immediate difference in selection of data lies in Rogers' rejection of the unconscious. This factor alone would make it scientifically impossible to compare the two systems since neither system could agree upon the very observations from which predictions could be made. For example, an adolescent boy might say that he wishes help in working out better relationships with girls. Treatment or counseling may uncover a different state of affairs. For example, the boy might be an active homosexual who consciously expects or feels that the knowledge of such activities would not be acceptable to the therapist. He may, once he discovers the Rogerian or Freudian therapist accepts his homosexuality, attempt to effect a homosexual relationship with him. The Freudian, from his genetic understanding, recognizes the homosexuality as an unconscious infantile defense against a more threatening heterosexuality. He proceeds to look for historical explanations. The Rogerian reflects back to the patient the homosexual feeling with unconditional acceptance and hopes to relieve the adolescent of whatever conscious surface guilt he has. Additionally, we found in our survey of epistemological differences that Rogers is selective of his empirical data in other significant ways, as is Freud. The Rogerian therapist in his belief that historical data and instinctual needs are unimportant, chooses to reflect back to the patient solely the contemporary self-referrent material. The psychoanalyst, from his genetic and dynamic position, chooses through his understanding of unconscious processes to make conscious the historical ways in which the patient reacted to the various psychic forces.

Conclusion

We found in our examination of the theories of both men that Rogers was primarily interested in a theory that offered accurate predictions and Freud was interested in a theory that offered adequate room for explanation. The former was guided by a wish for a theory that lent itself to operationism and to prediction. The latter wished to develop a theory that could contain statements about relations derived from empirical data that were initially sensed before they could be demonstrated. Freud's concern was that concepts might become permanently concretized before they had ample applicability. As a consequence we find that Freud left us a system that has been most fruitful in terms of explanation and expansion of psychological frontiers of knowledge. Rogers has given us a system that offers no explanation but a description of certain functional relationships. Yet operationism is an unquestionable scientific requirement. Rogers' phenomenological view has the great appeal due to its clarity. Its attractiveness appears due to the ease with which it can be converted into operational terms. Theories which are too speculative may lead to cultism. But Carnap, Hempel, and Frenkel-Brunswick tell us that operationism is not definitive in itself and contains a variety of "open end" or "surplus" meanings. Marx and Allport caution us against placing too much weight upon operationism since a lack of imagination would discourage good theories. Moreover, the central bulky problems of personality would be left untouched. Methodologically, Rogers' system is further suspect, since it rejected the variable contained in the influences of desires. It was this rejection that led us to the speculation that Rogers had remained primarily concerned with the age-old problem of the will, sin, and redemption. Bergman, the nonpartisan experimentalist, feels that psychoanalysis was scientific and was of great impact upon the burgeoning field of psychology. It was this science that developed a comprehensive theory that was based on causal mechanisms.

We have completed our journey. Following Dewey's belief that every psychology is based upon a philosophy, we have studied Rogers' origin for clues to his thought, and we have examined his thoughts within the philosophy of science, its epistemology and ontology. In viewing his system in comparison with Freud, it is hoped that we are now able to place Rogers in a larger more meaningful context.

PART III

PHILOSOPHY, LITERATURE, AND ART

PAPERS OF THE MICHIGAN ACADEMY OF SCIENCE, ARTS, AND LETTERS
VOL. XLV, 1960 (1959 MEETING)

PHILOSOPHY

AUTHORSHIP OF SIGNS

HENRY S. LEONARD
Michigan State University

WHEN may a person be said to have been the author of a sign? What is to be regarded as the sign of which he is the author? And what may be said to be the deliberate signification of that sign? One thing that adds importance to these questions is the fact that most instances of speech and of writing in a conventional language are instances of authorship. I do not insist that all of them are such. Perhaps baby's first babblings will hardly count. And perhaps several of the more uncontrolled verbal responses of older children and adults should not be counted either. But in any case, I am convinced that an analysis of the concept of authorship can eventually throw considerable light on the problems of meaning in language.

Not least among Dewey's many amazing and suggestive insights is his insistence on examining, no matter what the question may be, both the natural and the social or cultural matrix within which the facts relevant to answering the question have their place. This is so whether he is dealing with ethics, art, religion, logic, or language. The approach that I shall make to the questions I intend to put is reminiscent of this Deweyan stress on both natural and cultural factors that must be integrated in any defensible solutions to the questions. Moreover, my remarks will reflect a point of view broadly consonant with those points of view in Dewey's philosophy that are suggested by the labels "instrumentalism," "particularism," and "end in view."

On the other hand, the details of what I shall say will often be at variance with details in Dewey's philosophy. Close students of Dewey will be even more aware of this than I am. First, Dewey's approach to philosophical problems abjures formalism to a greater extent than I find congenial. Second, Dewey uses the word "language" in a much more inclusive sense than I find profitable. And third, Dewey nowhere (to my knowledge) distinguishes between sign-tokens and sign-types, although his treatment of sign-situations and symbol-situations almost perpetually suffers for the failure or refusal to make this distinction. Finally, I believe that he draws his distinction between signs (i.e., natural signs) and symbols (i.e., arbitrary or conventional signs) prematurely.

Nevertheless, I hope that if Dewey had been able to read this paper, he would have recognized in it a study broadly within the tradition fostered by his own work in such a book as *Logic, The Theory of Inquiry*. And, indeed, that very book includes his blessing of studies that will perchance deviate in detail but still pursue the broad pattern of his inquiry: "The present volume is an approach not a closed treatise. The aim it hopes to fulfill is that of being a sufficiently coherent and systematic approach to move others to undertake the long coöperative work . . . needed to test and fill in the framework which is outlined in this book."[1]

I should like, however, to make another use of this passage from Dewey. By altering only slightly its wording, it will express the spirit of the present paper, which, likewise, outlines an approach, not a finished theory. It will probably raise more problems than it settles. The aim it, also, hopes to fulfill is that of being a sufficiently coherent and suggestive approach to move others to join in the long coöperative effort needed to test and fill in a theory of signs.

Sign Reading

The signs that people make, as well as those that they interpretatively observe, are more often than not constituted of some concrete particular: a sound, a mark, a cloud, a swelling leaf bud, a contorted face. What is more natural and obvious than to speak of signs, with Dewey, as "physical existences"?[2] I have myself in another place written of "physical sign-tokens."[3] Other investigators, sometimes even the same ones, have remarked that it is the *character* of the sound, the *shape* of the work, the *type* of the cloud, the *swollen state* of the leaf bud that make these physical particulars to be signs of a *dog*, of *justice*, of *rain*, or of *spring*. Some of these investigators have gone on to insist that it is not the physical token, but rather it is the type of the token, which should be regarded as the sign.

Such great figures in the history of philosophy as Locke, Des-

[1] *Logic, The Theory of Inquiry* (New York: Henry Holt & Co., 1938), p. 40.
[2] *Ibid.*, p. 46.
[3] *Principles of Right Reason* (New York: Henry Holt & Co., 1957), pp. 92 ff.

cartes, and Kant found a principal focus of their interest in signs to be a focus upon ideas, sensations, percepts, or, more recently, sense data. Such signs are assuredly not physical existences; at least they were not so conceived. They are particulars, yet what they represent, or are signs of, seems to depend upon their characters. Hence again the question might be raised, is it the token (in this instance, nonphysical), or the type that is the sign of the thing supposedly represented?

Let me suggest that the idioms "is a sign of" and "stands for" are misleading and confusing. But another idiom, "is a sign that," promises a more fruitful clue to an effective theory of signs. For example, instead of saying, "The swollen leaf bud is a sign of spring," let us say, "That the leaf bud is swollen is a sign that spring is near at hand." When the sign situation is so expressed, it becomes apparent that neither the token, physical or otherwise, nor the type, is the sign. Instead, the sign is to be identified as the *fact* that the token belongs to the type.

Furthermore, *where* the token is *when* it has the character in question and *when* it is that it has this character, are frequently indispensable factors in the determination of what is signified. For example, to say that clouds are natural signs of rain is a useful, but misleading oversimplification. Rather, the fact that that thing is now on the western horizon *and* is a cloud is a sign that it will rain here in about a half an hour.

What I want to suggest is that sign-reading may best be viewed as a triadic relation connecting an observant interpreter to two states of affairs. I use the word "proposition" to refer to states of affairs rather than to sentences. If you can bear with me in this idiosyncracy, then I shall say that b reads that p as a sign that q means simply that b infers that q upon the more or less direct observational apprehension that p. I shall symbolize this situation by

$$R^t(b,p,q)$$

The variable "t" identifies the time at which the sign-reading occurs.

There are many problems relative to sign-reading that could, and should, be considered at this juncture. I hope that some of them will be brought out in the discussion. I shall make only two remarks. First, sign-reading is frequently inference more nearly on the order of what Russell calls "animal inference" or Whitehead calls "physiological inference" than on that which we might call "logician's inference." Second, it is very difficult to be confident that one has identified accurately what sign a given organism *is* reading in a given situation. What, for example, is the multiplicity of cues *in addition* to the shape, color, and location of that cloud—cues such as wind, time of day, or whatever—to which we are responding when we infer that it will soon rain? The psychologist's pains to eliminate uncontrolled cues in his rat experiments, and his search for undetected cues when he attempts to explain the occult cognitive powers of a medium, should have their sobering effect upon the philosopher who presumes to identify instances of sign-reading.

But the problem is not wholly a psychologist's problem. There is the *logic* of the situation that must be considered. For example, are these several cues, p_1, p_2, p_3, etc., so many different propositions *each* of which is read as a sign that q, *or* is their conjunction read as a single, but complex, sign that q? Until such questions can be worked out, no ultimately satisfactory theory of signs can be developed.

Purposeful Behavior

I may as well declare right now that I regard human beings as acting purposefully and that, furthermore, I regard productive sign-events as instances of purposeful behavior. A good deal could, and should, be said about purposeful behavior in general, but there are only a few principles to which I shall appeal, and I shall keep my general remarks confined rather closely to those general principles.

If I were to develop a formal system I might start by using,

$$\mathrm{W}^{t}(a,p,q) \tag{1}$$

to abbreviate "a works at time t at bringing it about that p in order to bring it about that q." I should also use

$$\mathrm{B}^{t}(a,p) \tag{2}$$

to abbreviate "a brings it about at time t that p." These two formulas would enable me to define "$\mathrm{D}^{t}(a,p,q)$":

$$\mathrm{D}^{t}(a,p,q) = \mathrm{Df}\ [\mathrm{W}^{t}(a,p,q)\cdot \mathrm{B}^{t}(a,p)] \tag{3}$$

"a brings it about at time t that p in order to bring it about that q" is defined as meaning "a works at time t at bringing it about that p

in order to bring it about that q, and a brings it about at time t that p." We could also define purposing:

$$P^t(a,q) =\text{Df } (\exists p)W^t(a,p,q) \qquad (4)$$

"a purposes at time t that q" is defined as "There is some proposition, p, such that a works at time t at bringing it about that p in order to bring it about that q." It will then follow that

$$W^t(a,p,q) \supset P^t(a,q) \qquad (5)$$

If a works at time t at bringing it about that p in order to bring it about that q, then a purposes at time t that q. And furthermore,

$$D^t(a\ p,q) \supset P^t(a,q) \qquad (6)$$

If a brings it about at time t that p in order to bring it about that q, then a purposes at time t that q.

Working at is a transitive relation. That is,

$$[W^t(a,p,q) \cdot W^t(a,q,r)] \supset W^t(a,p,r) \qquad (7)$$

If a works at time t at bringing it about that p in order to bring it about that q, and at q in order to bring it about that r, then a works at time t at bringing about that p in order to bring it about that r.

From Law (7) it follows that

$$[D^t(a,p,q) \cdot W^t(a,q,r)] \supset D^t(a,p,r) \qquad (8)$$

If a brings it about at time t that p in order to bring it about that q, and works at time t at bringing it about that q in order to bring it about that r, then a brings it about at time t that p in order to bring it about that r.

As Law (8) shows, a purposeful action may have more than one purpose. In point of fact, most purposeful actions have several purposes. The purposes of bringing about p, as recorded in Law (8), are the one of them, q, subservient to the other, r. But frequently, two or more purposes of a given action will be independent, one of the other.

Productive Sign-Events

So that we may see what is involved in authorship, that is, in a productive sign-event, let us examine three imaginative situations. In the first example, there is no authorship, but a situation otherwise quite like a productive sign-event. The second example involves authorship, but not what I shall call candid authorship. The third example is one of candid authorship. Practically all uses in processes of communication of conventional symbols such as those forming a language involve candid authorship.

Example 1.—I am standing, a few minutes after the hour, outside of Berkey Hall when I see a middle-aged man carrying a brief case and hurrying toward that classroom building. I infer that he is a professor who is late to a class and that he is hurrying in order to reach the class before it disperses.

Let us suppose that my inference of purposeful conduct was correct. We would then symbolize this part of the man's action and my interpretative observation of it as a substitution instance of the general form

$$\mathrm{D}^t(a,p,q)\cdot\mathrm{R}^{t+i}[b,\ \mathrm{B}^t(a,p),\ \mathrm{D}^t(a,p,q)] \qquad (9)$$

a brings it about at time t, that p (namely, that a is hurrying toward Berkey Hall) in order that q (namely, that a reaches his class before it disperses); and I, b, read at time $t + i$, that a brings it about at time t that a is hurrying toward Berkey Hall, as a sign that he brings this about in order to reach his class before it disperses.

In this example, a was engaged in purposeful behavior and, furthermore, his purposeful action was read by an observer as a sign. But I shall refuse to call a an author of a sign because it was not among his purposes in performing that significant act that it be read as a sign. The next example will illustrate this difference.

Example 2.—Just a moment later, as I continue standing outside of Berkey Hall, Professor Volstead comes into view, also hurrying toward Berkey Hall. I make an inference in this case similar to the one made in the last. But suppose that this time I had made a mistake. Suppose that in fact Professor Volstead had no class but had spotted me just before I noticed him. He was afraid that I might speak to him and he did not want me to. So he cleverly joined the hurrying stragglers, hoping I would make the inference that I did make and would consequently leave him alone.

We could symbolize Professor Volstead's purposeful action and my subsequent sign-reading as substitution instances of the following.

$$\mathrm{D}^t\{v,p,\mathrm{R}^{t+i}[b,\ \mathrm{B}^t(v,p),\ \mathrm{D}^t(v,p,q)]\}\cdot\mathrm{R}^{t+i}[b,\ \mathrm{B}^t(v,p)\cdot\mathrm{D}^t(v,p,q)] \qquad (10)$$

v brings it about at time t that p (namely, v is hurrying toward Berkey Hall) in order that I, b, read at time $t + i$ that v brings it about at time t that p as a sign that v brings it about at time t that p in order to bring it about that q (namely, v gets to his class before it disperses); *and* I do so read it at time $t + i$.

Formula (10) affirms that Professor Volstead's action has among its purposes my reading the action as a sign with a certain meaning. It is this which makes Professor Volstead's action an instance of authorship. In fact, we may use this example as a clue to the way in which the authorship of directed signs should be defined:

$$P_1{}^t(a,p,b,q) =\mathrm{Df}\ D^t\{a,p,R^{t+i}[b,\ B^t(v,p),\ q]\} \qquad (11)$$

"At time t, a produces p for a sign$_1$ to b that q," means by definition that a brings it about at time t that p in order that b reads at time $t + i$ that a brings it about at time t that p, as a sign that q.

There are two things to notice about Definition (11). First, in contrast with Formula (10), the definition abstracts completely from any limitations on the character of the intended signification of the proffered sign, $B^t(v,p)$. In Formula (10), we preserved a special feature of the illustrative example, namely that $B^t(v,p)$ was to be read as a sign of purposeful behavior. But in Definition (11), we allow, by using "q" for the intended signification that the author may be intending to signify, without restriction, whatever he will.

Second, Definition (11) does retain the feature of Formula (10) that makes $B^t(a,p)$ to be the offered sign. Another kind of example would generate another type of productive sign-event, in which p, rather than $B^t(a,p)$, would be the offered sign:

$$P_2{}^t(a,p,b,q) =\mathrm{Df}\ D^t[a,p,R^{t+i}(b,p,q)] \qquad (12)$$

"At time t, a produces p for a sign to b that q" means "a brings it about at time t that p in order that b reads at time $t + i$ that p as a sign that q." A situation illustrating this latter type of productive sign event would be provided by a detective hiding in a house at night who puts out all the lights so that the expected thief will infer that nobody is at home.

Many other types of productive sign-event could, and should, be discriminated one from the other and defined. But to explore the spectrum of possible variations would take up more time than we have left, and I want to get on to a very special and important type of case.

Example 3.—It must be confessed that although Professor Volstead was an author, he was neither honest nor candid. He was not honest in that he did not have any class toward which he was hurrying; and he was not candid in that the sign which he deliberately produced for me to read imitated a natural sign in such a way as to be mistaken for one. In contrast, consider the case of Professor Allen, who is also walking toward Berkey Hall. Since he is not hurrying, I do not think he has a class, so I call to him. He responds by waving his hand and then pointing several times first to himself and then to Berkey Hall.

These gestures I recognize as a sign; but more than that, I recognize them as a sign with an author, directed to me by their author for me to interpret. If he had used English, instead of merely gesturing, the case would have been the same. I would have recognized the sounds as a sign that he was undertaking to communicate with me. In this sense, the sign-making activity is candid, and it is candid whether or not it is honest.

Upon recognizing the noises or gestures as the product of a sign-making activity, I must, so to speak, read these same noises or gestures again, in order to find out what message Professor Allen is addressing to me. But these two readings of the same sign have a curious relationship to one another. We can see what this is if we consider the following definition of a *candid* productive sign-event:

$$^{c}\mathrm{P}_1{}^{t}(a,p,b,q) \;=\mathrm{Df}\; \mathrm{D}^{t}\{a,p,\;\; \mathrm{R}^{t+i}[b,\mathrm{B}^{t}(a,p),\mathrm{P}_1{}^{t}(a,p,b,q]\} \qquad (13)$$

"At time t, a produces p for a candid sign_1 to b that q" means by definition that "at time t, a produces p in order that b read at time $t + i$ that a brings it about at time t that p as a sign that at time t,a produces p for a sign_1 to b that q."

Some Conclusions

The symbolic representation of a candid productive sign-event in Definition (13) is so complex as to make impractical in a meeting like this many attempts at logical transformation. We can, however, contemplate it in order to bring out certain important considerations. In particular, it can help us to distinguish one from the other a number of problems or questions that are often confusedly identified one with the other. It is my opinion that many of the difficulties experienced in the analysis of sign situations arise from not confronting these original problems and making properly the distinctions they call for.

1. *What* is the sign? Is it that a brings it about that p, or is it that p? Line (13) suggests that each of these may be a sign of something or other. But after all, they interact and what each of them signifies depends on what the other signifies. Furthermore, the formula offers a misleading simplification of the situation in that it has abstracted away from *other* cues upon which both a and b depend for their effective communication. For example, it is often noticed that the meanings carried by words vary in accordance with the identity of the author and the addressee. It is not merely that words such as "I" and "you" vary in this manner. Frequently one has to know the context of events within which sign-making takes place, and be aware of purposes of the author in order to discover the meanings of his words. Now when you are engaged in conversation, you not only hear the sounds having such and such character, but you see moving lips. These lips are one sign, among others, that identify the speaker. Also, you see the direction in which he is looking. That he is looking at you is read as a sign that you are his addressee. These natural accompaniments of the uttered words are signs read by you but affecting what meaning you read into the words themselves. Contrast this situation with one in which you read a letter. Here conventional signs, the salutation and the signature, replace such natural signs as moving lips and eye-focus to identify for the receiver the author and the intended addressee of the letter. But in so identifying them, they contribute to determining what signification is to be put upon the body of his letter.

2. Suppose that a is at least in part successful in his candid sign-making, so that

$$\mathrm{R}^{t+i}[b,\ \mathrm{B}^{t}(a,p),\ \mathrm{P}_1{}^{t}(a,p,b,q)] \tag{14}$$

b reads a's bringing about of p as a sign that a brought about p for a sign_1 to b that q. Had b misunderstood the message we could record this misunderstanding by substituting "r" for "q" at the end of Formula (14). And had b grasped that there was a message but been unable to figure it out, we could record this fact by substituting

$$(\exists r)\mathrm{P}_1{}^{t}(a,p,b,r)$$

for

$$\mathrm{P}_1{}^{t}(a,p,b,q)$$

in Formula (14).

But let us suppose that b did understand the message and that Formula (14) is correct. From (14) it follows that

$$b \text{ believes that } P_1{}^t(a,p,b,q) \tag{15}$$

b believes that a produces p for a sign_1 to b that q.

But surely b's belief of (15) is not all that a wants. Beyond that, he wants b to believe that q, and he may want that for any number of still more remote purposes. Yet for it to follow from (15) that b believes that q requires the coöperation of the observer, b. For example, and at the very least, it requires that b accept a's productive sign activity as honest.

3. Let us, finally, consider briefly the way in which the present theory of productive sign events applies to illuminate an author's use of ordinary or conventional speech. Here the value for "p," for example, in Formula (15) would be a proposition to the effect that the sign-tokens uttered or written by a belong to such and such sign-types. What does this temporal fact, created by a's utterance or writing, signify? Our analysis has shown that a *would* have it signify *to* b that q. But *does* it signify that q? Undoubtedly many an addressee, b, will take it as signifying that q. This is, however, more characteristic of the young than of the older and more worldly-wise. Before a more experienced and skeptical addressee accepts p as signifying that q, he questions not only a's honesty, as already noted, but his competence as a judge of q. There will even enter into the consideration some question of a's motivation for purposing that b accept p as a sign that q.

Let me suggest one direction our thought might take. I want only to suggest this, not to pursue it. Remember the earlier mention of unremarked supplementary cues, on the presence of which we take the cloud to be a sign of rain, and the speculation that it might be the conjunction of these cues with the cloud which is properly the sign, and not each of them separately. So, it may be here, that p alone is not a sign of anything, unless it be a sign that somebody spoke. It may be that a tenable theory of signs will not develop until we stop trying to treat that p as a sign and consider some such thing as "a brought it about that p," or even something more complex, as the sign.

But to get back to the main line of our thought, quite apart from the point just made, it would seem apparent that a will not succeed in getting a worldly wise addressee, b, to accept p as a *candid* sign that q, without limiting the range of q. So long as a offers that p for a sign to b that a has such and such a purpose, he may hope to get b to move from the situation described in Formula (15) to a belief that a purposes such and so. Here *all* that is requisite is a confidence on b's part that a is honest.

But what *kinds* of purpose might a undertake to signify by p when that p is brought about as a candid sign for b? Let me limit attention to cases in which p, a proposition to the effect that such and sounds or marks have such and such a form, is imputing to the sounds or marks a form which is a *sentential* form. The conventions of language correlate *that* sentential form with a specific proposition, r. But a's purpose in uttering the sentence is to put b in some definite relation to the proposition, r. We may symbolize all of this by using the form defined in Formula (4):

$$\mathrm{P}^{t}[a, \mathrm{f}(b,r)] \tag{16}$$

What I am proposing is that the values of "q" in Formula (15), when p is brought about as a candid sign that q, are always also values of (16). In another place[4] I have argued that the sign-tokens of productive sign-events always signify a purpose of their author, and I analyzed that purpose into two factors, a *concern* and a *topic of concern*. I undertook to maintain that the tokens *expressed* their author's concern and *indicated* their author's topic of concern. But the question how one drew the line between concern and topic of concern was left inadequately treated. In still another place,[5] I have argued that all purposes are analyzable into a *primitive* concern (to make some proposition true) and a *primitive* topic of concern (the proposition that one would make true), but that when the purpose is one appropriately signified by speech or discourse, the analysis of the purpose might also be made at another point: into an *ultimate* concern and an *ultimate* topic of concern. What I proposed was that sign-tokens *expressed* the ultimate concern and *indicated* the ultimate topic of concern. I there applied the theory to show that declaratives, interrogatives, and imperatives having the same subjects and predicates expressed different ultimate concerns with the same ulti-

[4] *Ibid.*, pp. 137–145.

[5] Interrogatives, Imperatives, Truth, Falsity and Lies," *Philosophy of Science* (1959), Vol. 26, pp. 172–186.

mate topic of concern. My object today has been to lay out somewhat more analytically the groundwork for such a theory. Formula (16) analytically represents the purpose that the sounds candidly uttered by *a* signify. The primitive concern is to make something true; the primitive topic of concern is the proposition f(*b*,*r*), which *a* would make true. The ultimate concern, expressed by the sounds or marks that *a* creates, is to make it true that *b* stands in the relation f to some proposition; the ultimate topic of concern, indicated by the sounds or marks that *a* creates, is the proposition *r* to which *a* would put *b* in the relation f.

Papers of the Michigan Academy of Science, Arts, and Letters
Vol. XLV, 1960 (1959 Meeting)

LITERATURE

A85-88: CHAUCER'S SQUIRE AND THE GLORIOUS CAMPAIGN

ALAN GAYLORD

The University of Michigan

Lines 85–88 of Chaucer's *General Prologue*, in Group A of the *Canterbury Tales*, describe the military experience of the Squire:

> And he hadde been somtyme in chyvachie
> In Flaundres, in Artoys, and Pycardie,
> And born hym weel, as of so litel space,
> In hope to stonden in his lady grace.[1]

There has been no debate over what the allusion to the Squire's "chyvachie" amounts to. The word comes from the Old French *chevauch(i)é* (current at the time in France and widely used by Froissart), and meant (*a*) a troop of mounted knights and their retainers, a cavalcade; or (*b*) a cavalry expedition or raid.[2] Recent scholars have agreed in identifying this particular expedition with the so-called "Crusade of 1383,"[3] and their source may be traced back to the pioneer investigations of John S. P. Tatlock who, in his *The Development and Chronology of Chaucer's Works*, first suggested the identification and presented the proofs for it.[4] The simplest way to arrive at such an identification is to take the accepted date of the *General Prologue*, 1387, and the general age of the Squire, twenty, and work back to find campaigns in which he could have served. An age of about sixteen was a perfectly ac-

[1] In *The Works of Geoffrey Chaucer*, ed. F. N. Robinson (Boston, 2d ed., 1957), p. 18. Subsequent Chaucer citations are from this edition.

[2] "Chiuachye," in *Middle English Dictionary*, ed. Hans Kurath and Sherman M. Kuhn (Ann Arbor, Michigan, 1959).

[3] Cf. Robinson's notes to the passage, p. 653; and Muriel Bowden, *A Commentary on the General Prologue to the Canterbury Tales* (New York, 1957), p. 83f.

[4] Published for the Chaucer Society, Second Series, No. 37 (London, 1907), pp. 147–150.

ceptable one for a Squire first to taste battle; and the campaign of 1383 began in Flanders, passed through Artois, and struck briefly into Picardy.

The reference, needless to say, is not established beyond a doubt. But the circumstances lean heavily in that direction. In the first place, the Knight's campaigns are all identifiable; and since the rest of his description is contrasted in parallel manner by that of the Squire, it would seem likely that the Squire's "chyvachie" was other than an imaginative allusion. In the second place, there is little else that would include the three references in campaigns in a reasonable proximity to the assumed date of the *Prologue*. Chaucer had been abroad in the campaign of 1359–60 and had been taken prisoner then. But although that campaign may have contributed to Chaucer's particular impressions of continental warfare, it certainly would not have been called to mind by the *Prologue's* lines. For those lines would have been written some twenty-five years after the fact, and mention, furthermore, one region—Flanders—that Chaucer in his campaign had never even approached. It might be argued that "Flaundres, . . . Artoys, and Picardie" were picked out simply as names and for rhyming purposes; yet it does not appear that either "Picardie" or "chyvachie" are forced enough in either line to cause a frantic hunt for any sort of geographical or military rhyme. No, the rhymes seem natural and, in fact, paired appropriately; and the whole collection of items and references seems to coalesce to form a strong suggestion of an allusion that would be immediately recognized.

But there, for fifty-two years, the matter has rested. Apparently the allusion has been accepted in parallel with those of the Knight, treated with neither alarm nor particular curiosity, and regarded as nothing out of the ordinary. Although the historical facts concerning the Crusade are available if widely scattered, few literary historians have investigated them. Muriel Bowden apparently dipped into the account given by Charles Oman in his volume in *The Political History of England*;[5] she devotes a paragraph to the Crusade at one place, referring to its reputation, and in another place mentions the Squire's probable participation.[6] In her view, the

[5] Vol. IV, *The History of England from the Accession of Richard II to the Death of Richard III. (1377–1485)* (London, 1906), pp. 81–86.

[6] *Commentary,* pp. 10f., 83f.

Crusade is mentioned simply because it is "right" for the Squire: it is the only real campaign he could have accompanied. But that the acknowledged reputation of the Crusade may play some part, however small, in his total characterization, Bowden does not discuss; nor do any other scholars, past or present. It is precisely at this point that a new investigation of old sources may be proposed. The facts are there. What do they add up to and what kind of soldier do they make the Squire to be? The proposition of this paper is this: that the allusion to the "chyvachie" is uncomplimentary and ultimately satirical in regard to the Squire. To understand why this is so, we must know what kind of reaction Chaucer's audience was likely to show upon recognizing the allusion. And to understand this, we must return to the available information concerning the Crusade.[7]

The Crusade of 1383 was an episode in the Hundred Years War, connected with the revolt of Flanders. As such, it was bound up in a confusing complex of political and economic issues involving the rivalry between the textile centers of Ghent and Bruges, the economic relations of the Low Countries and England, the maneuvering behind the weak Richard II of parties and factions which included his uncle—the powerful but unpopular John of Gaunt, and the checks and mates involved in the chess of power politics played by the two popes, Urban VI of Rome and Clement VII of Avignon. As a Crusade, it was a disgrace to a Christian nation and, as a military "chyvachie," a miserable failure. The public revulsion that followed it was intense enough to cause it to remain vivid in memory for some time. As the most outstanding national involvement in re-

[7] See Figure 1 for a map of the Low Countries. Two monographs have been devoted to this Crusade: George M. Wrong, *The Crusade of MCCCLXXXIII., Known as That of the Bishop of Norwich* (London, 1892); and Gerhard Skalweit, *Der Kreuzzug des Bischofs Heinrich von Norwich im Jahre 1383* (Königsberg Inaugural-Dissertation, Königsberg, 1898). The latter draws upon a wider range of sources; but both have been dubbed "médiocres" by Édouard Perroy, in his treatment of the Crusade included in *L'Angleterre et le Grand Schisme d'Occident*, Vol. I, *Étude sur la Politique Religieuse de l'Angleterre sous Richard II (1378–1399)* (Paris, 1933), pp. 166–209. See also the summary by A. P. R. Coulborn of his unpublished London University Ph.D. thesis, "The Economic and Political Preliminaries of the Crusade of 1383" (London, 1932), in *Bulletin of the Institute of Historical Research,* X (June, 1932), 40–44. I wish to thank Mr. Zane J. Anderson, of the University of Michigan College of Architecture and Design, for making the map at Figure 1.

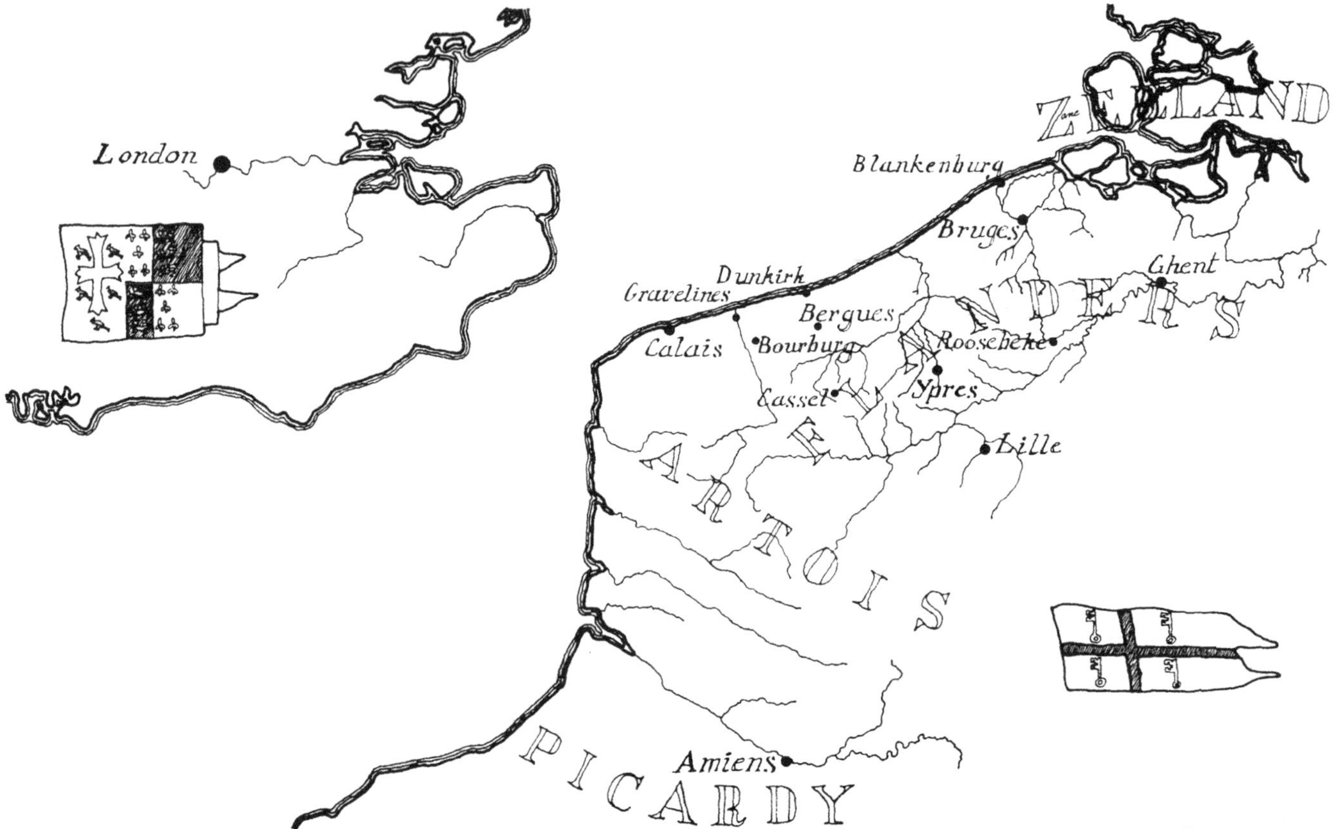

Fig. 1. The Area of the Crusade of 1383. Banners are those of Richard II and (typical) a crusading Pope.

cent years and the closest to home, it stood as an example of all that seemed to be going wrong with England. A brief account of its career may provide a rather startling picture of the Squire's first apprenticeship in arms.

When the Crusade was finally ratified by the English Parliament, it was expected to solve a number of problems affecting the nation at that time. One was economic, and concerned the political stability and loyalties of those cities in the Low Countries which dealt with English wool; namely, Ghent and Ypres, the manufacturing centers, and Bruges, the commercial and financial center of Flanders. English prosperity depended in part on the steady flow of wool to the staple at Calais, then on to the highway to Bruges, and then to Ghent. A closing of this road meant an obstruction in trade which could provoke a domestic crisis. And the road did close, in 1379, because of Bruges' intense rivalry with Ghent. When the citizens of Ghent, under Philip of Artevelde, took arms in May 1382, they were eventually encouraged by the English, and the way looked clear again. But by November, the French armies had entered Flanders and resubjugated all of the area which had recognized England, climaxing their decisive action in the battle of Roosebeke, 27 November 1382, in which Philip of Artevelde was killed. This precipitated a severe economic contraction back at the ports of England. In the crucial period that followed, the citizens of Ghent, not subdued by France but cut off from the rest of Flanders, were at London to do their own lobbying.

Another problem was religious and concerned the rival popes. Urban VI had been trying to enlist the support of sympathetic rulers for a number of years. The English had formally confirmed their loyalty to the Roman See in the Parliament of 1378. Urban's main desire was to bring the French back to the proper obedience, and he sought to do this in some manner through England. On 17 September 1382, bulls from him were published in England, urging a double crusade against the schismatics. One was to proceed against the Clementists in France, and the other against those in Castile. Parliament debated at some length which of these alternatives to choose. John of Gaunt was ready to crusade in Castile for the crown he considered rightfully his; for Constance, his wife, was the daughter of Pedro the Cruel (d. 1369), quondam ruler of Castile, and the Spanish were hated as both Clementists and allies of France.

But there were drawbacks to this and, as Parliament considered them, the battle of Roosebeke made clear which "way" was the most imperative. The duke would have to continue to wait. The combination of events and pressures were irresistible in the other direction.

Attention now must focus on the fiery Bishop of Norwich, Henry Despenser,[8] whom the Pope had appointed as special legate, a *Nuncius*, with extraordinary powers to organize a crusade against the Clementists in France and to cultivate in his nation zeal and substantial support for the project. He was relatively young (b. 1344), of noble blood, had been bishop for twelve years, was a warrior-knight, and an intensely aggressive and ambitious man. He had already been prominently involved in a ruthless and efficient crushing of the Peasants' Revolt in his county, and now burned to be the head of this Crusade. The machinery of announcement and exhortation that he controlled had been in operation for some time even before Parliament authorized him to lead the expedition against the enemy schismatics.

Roosebeke was indeed the turning point. Flanders had never been Clementist; in fact, the Flemish clergy, meeting in 1379 in fully representative synod, had decided upon loyalty to Urban VI, after extensive investigation. But with the coming of the French, the Low Countries were forced to recognize the sovereignty of a Clementist nation. This still by no means converted the Flemish to Avignon, nor did it mean that Ghent's claim of being the only remaining faithful in Flanders was anything but technically correct. But the road to the enemy lay through the corridor of Calais, Flanders, and Artois, and the Crusaders could set things right there as they proceeded toward the heart of France. The exigencies of an outlet for the staple combined with crown politics and religious conviction. Once approved and organized, the Crusade had all England in a fever of excitement.

Parliament, it is true, had ratified Despenser with great reluctance. Financially, the project seemed feasible because much of the expenses would be levied through the machinery of the Church,

[8] For summaries of the man and his career, see R. L. P[oole], "Despenser or Spencer, Henry le," *Dictionary of National Biography*, XIV (London, 1888), 410–412; and Herbert B. Workman, *John Wyclif a Study of the English Medieval Church*, II (Oxford, 1926), 64–70.

even now at work among the people. But there was much concern, even when available nobles could not be obtained, to granting full command to one who was a soldier of the Church, an "inexperienced Bishop."[9] The *Continuatio* of the *Eulogium Historiarum* provides us with a partial record of Parliament's doubts and Despenser's assurances.[10] And the final question: what would the Bishop do if the schismatic King of France renounced Clement and proclaimed his allegiance to Rome? Despenser: he would, in such a case, continue the war against France as a soldier of the English host.[11] The "Crusade," then, was to be a highly equivocal one.

Once organized and under way, the expedition had a number of encouraging successes as soon as it reached the Low Countries. But at first, those simple souls who had envisioned a holy crusade against the King of France were shocked to be ordered to the attack in Flanders, whose citizens, although rivals of Ghent, were only nominal Clementists. Froissart offers an account of the debate of an early council of war, in which Despenser's best captain, Hugh of Calverley, voiced his objections and was overruled. He had two reasons for not entering Flanders and two for proceeding to France: (*a*) the Flemish were truly as good Urbanists as the English and (*b*) the Flemish were no easy foe to conquer, "having had no other concern but war for 40 years"; (*c*) the French were open political enemies of the English and (*d*) were entirely Clementists.[12] The Bishop, reportedly "boiling hot" to reply, accused Hugh of inflexibility and opposed him with the reports of the advisers from Ghent who assured him of the profit of entering the "rich frontier" along the ocean. Calverley, who was ranked socially by the Bishop, at first held silence and then said, "*Par Dieu*, sir, if you ride out, *messire* Hugh of Calverley will ride out with you, nor shall you ever take a way he will not dare to enter." "I well believe it," concluded the Bishop, who greatly desired to embark.[13]

Thus the innocents were persuaded, and the battle was carried first to Gravelines, a town about fifteen miles from Calais. It was

[9] Thomas Walsingham, *Historia Anglicana*, ed. Henry Thomas Riley, Rolls Series, II (London, 1864), 84.

[10] Ed. Frank Scott Haydon, Rolls Series, III (London, 1863), 356.

[11] *Rotuli Parliamentorum*, III, 148.

[12] *Chroniques*, ed. Gaston Raynaud, XI (Paris, 1899), 93.

[13] *Ibid.*, p. 94f. It is not even necessary to believe this conversation took place in order to appreciate its dramatization of points of view.

taken after a day and a night of fierce fighting, during which time the Crusaders were exhorted to press on and annihilate the foe in hopes of earning as great a reward as if they were slaughtering Jews or Saracens.[14] The plundering and the carnage that ensued provide an interesting commentary on this holy army, with its large complement of parish priests, friars, and monks. The Crusaders, pulling the defenders from their hiding places, spared none but labored to obliterate the defenders to the last man.[15] Rich stores were captured, including ships and barges in the bay, and a large supply of horses to mount much of the previously unmounted company. While the army was fortifying themselves at Gravelines, the overlord of the area, the Count of Flanders at Lille, sent messengers to Despenser. Why was the Bishop warring against the Count's peace, when he was no enemy of England? The country had been enfeoffed to him by the King of France, at Tournai. The Bishop, after consultation, replied that they were soldiers of the Pope, not of England, and if the neighboring towns would pledge allegiance to Urban VI and assist in his Holy War, all would be well.[16]

The emissaries took this slippery reply back to Lille, and the Bishop's army marched next toward Dunkirk, thus driving deeper into Flemish territory while turning their backs to France itself.[17] On the road they encountered a spontaneous but motley opposition headed by the Bastard of Flanders, an illegitimate son of the Count. This was put to rout in great part through the deadly toll that the English longbowsmen took, and partly through the inexperience of the Flemish artisans who, not trusted overmuch by the Bastard, had been placed in the front of his lines. The enemy retreated and was followed into Dunkirk, which was taken with little loss to the English and great slaughter among the defeated.[18] The victory occurred

[14] Walsingham, *Historia*, II, 89.

[15] *Ibid.* The women, according to the Monk of Evesham (*Historia Vitae Et Regni Ricardi II*, ed. Thomas Hearne [Oxford, 1729], p. 45), were "reserved." Wrong interprets this as suggesting dire results (*Crusade*, p. 61).

[16] Froissart, *Chroniques*, XI, 98–101.

[17] Perroy reminds us, however, that the army was on the lands of the Dame of Cassel, whose sympathies had always lain with the Clementists; *Schisme*, I, 191.

[18] Estimates of enemy dead run from 6,000 (*Istore et Croniques de Flandres*, ed. Kervyn de Lettenhove, II [Bruxelles, 1880], 286) to 12,000 (Walsingham, *Historia*, II, 93).

on 25 May 1383. It was the Feast of St. Urban, which was taken as a favorable portent of victories to come. The coast, from Calais to Blankenburg, came easily into the control of the English, who were now at the very apogee of Fortune's wheel. But the successes of the campaign were over. The King of France was coming to the aid of the Count of Flanders, and the wheel was soon to sweep downwards.

The English nation at this time suspected nothing of the sort. For a while service in the Crusade, with its many fringe benefits, appeared in a most attractive light. The indulgences that were preached were overwhelming: to those participating in or assisting materially the Crusade, full absolution from punishments and guilt was granted both for the living and the dead (if sufficient were contributed). Knighton adds that it was reported that some of the Bishop's commissioners (who would have been the friars) claimed that at their command angels would come down from heaven, deliver the souls which had been put in purgatory to be punished, and lead them to heaven without delay.[19] Froissart mentions a current saying in England: Happy indeed, all those who could die in that season for such a noble absolution.[20] The appeal now was practically irresistible. Walsingham describes the effect of the victories in what is probably the true light: when the joyful news was brought to England by those who had massacred the inhabitants of Gravelines and Dunkirk and collected such booty of horses, cattle, and domestic goods, the whole country, lured by the hope of gain, was thrown into an uproar to depart. Without a second thought, many apprentices and servants left their masters and lords willy-nilly, and most of the rest of England followed their example; not only the layfolk but the religious also, who departed with or without permission from their superiors—not, Walsingham observes wryly, to serve Christ, but to see the world. The host that swarmed down to port, including mechanics and rustics, had little arms of their own, but sailed off nevertheless to join the Bishop.[21]

[19] Henry Knighton, *Chronicon*, ed. Joseph Rawson Lumby, Rolls Series, II (London, 1895), 198f. For a report of a Papal Bull, the privileges granted to the Bishop of Norwich, Despenser's Mandate for the Bull's publication, and the Form of Absolution, see Walsingham, *Historia*, II, 71–80.

[20] *Chroniques*, XI, 88.

[21] *Historia*, II, 95. The Monk of Evesham underscores the straggling and undisciplined quality of the mob; *Historia Vitae*, p. 46.

The increased and booty-swollen army and the inflated Despenser, who called himself now "The Conqueror of West Flanders," were convinced that the next move was to attack Ypres, on the highway to Bruges; for Bruges would surely fall if Ypres were overcome. The representatives from Ghent, again, had a good deal to do with this decision and the reasoning behind it. The direction was, indeed, correctly toward Artois. Some, however, had urged that it was time to pass directly into the interior of France. But Ghent had persuaded the Bishop that Ypres and Bruges must fall before its hegemony could be securely established and Flanders be made a fief of the English crown. Thus the army moved on and settled down for a siege at one of the greatest cities of the world at that time, Ypres. This choice of battleground demonstrated conclusively the ambiguities of the so-called Crusade. Ypres protested that it, too, recognized Pope Urban and demanded to be left alone. If the English would leave the siege and march toward the real schismatics, the citizens would even assist the army from their own resources. But the warnings and rebuttals of the captains from Ghent finally prevailed over such an appeal to the Bishop. He solemnly excommunicated the city of Ypres in the name of the Pope, whose representatives replied that they rejected the sentence and appealed to the same Pope. The solemnity of the Bishop's pronouncement was answered by the citizens with a ceremony and a religious procession of their own, with all the city bells ringing, "in honor of God."[22]

The siege, which lasted two months in the summer, turned out to be a fiasco. Help expected from England was delayed; the troops were restless under Despenser, the churchman; the rabble from England caused so much anguish that they were finally dismissed and left to forage and plunder the countryside, friends now to neither side. The city was heartened by the news that a great French force was to arrive about 15 August, but the invaders were made desperate. Careful preparations, religious as well as military, were made for an assault upon which all hope was staked. When the attack came, the fighting was fierce, but the city held; and the forces of Ghent and England fell back, for once and for all. The suburbs of Ypres were razed, the interior city was practically exhausted, the surrounding countryside was enervated, but the invaders had been repulsed.

[22] In *Istore et Croniques*, II, 301. Cf. also Froissart, *Chroniques*, XI, xxviii (Raynaud's notes), and 115f.

From that point on, their only progress was retreat, except for one sortie into Picardy. This had been the Bishop's idea, against the wishes of most of his officers, and was to be a thrust to surprise the French army which was to gather at Amiens. With only a part of his army, followed by the still faithful (if misgiving) Calverley, he rode through Artois into Picardy; but he was too early, the French were not yet gathered. The Bishop, having met no foe, reaped no victory, and established no line, returned to the rest of his troops with all possible speed, to continue fortification of the territory already captured. But when the French army finally entered Flanders, the English fell back to Cassel, and then to Bergues. Despenser had since retired to Calais, where he hoped for reinforcements.

Calverley was astounded when the full strength of the French force was revealed to him. Bergues was no place for defense. The English fell back to Bourburg and dug itself in. Froissart has Calverley say to his troops before entering: "We have made this season a most shameful campaign: a poorer or a more blameworthy one never went out from England. You have followed this Bishop of Norwich, who would have flown if he had wings. Now you see the honorable conclusion you gain.... There is Bourburg; retreat to it if you will, but I am going on to Gravelines and Calais. We are no people to fight the King of France."[23]

The campaign, the Crusade, was through. Its "chyvachies" had at first spread over Flanders like a dark stain, then wavered, shrunk, and finally diminished to the relative dot of the beleaguered port of Calais. Bourburg and Gravelines were also deserted, burnt by the English, and then reoccupied by the French. Now John of Gaunt, never an enthusiastic supporter of this Crusade, to say the least (even though for a while he had been in charge of part of it), was sent over to treat with the French King and come to terms. Both sides were eager for peace, and neither was in a position to bargain from strength. All Gaunt could accomplish at the time was a truce, to last from Christmas of that year to 1 October 1384. Despenser returned in disgrace to England. Parliament had been convoked since 26 October 1383, and on 13 November, Chancellor Michael de la Pole presented a list of grave charges to the Bishop. In the process of recrimination and defense that followed, Despenser and his officers (Calverley excepted) became the focus of the intensity of English reaction to

[23] *Chroniques,* XI, 132.

their recent humiliation. Despenser, of course, who "had carried his sword out before him wherever he went like a temporal lord" (in the words of de la Pole),[24] had prepared for his downfall through the limitations of his own independence and arrogance, which alienated the entire Court. His temporalities were removed by Richard (although they were regranted two years later), and some of his officers were imprisoned for their alleged duplicity and collusion with the enemy. Public opinion in all quarters had of course reversed itself as the English were routed; now ensued a general time of reaction, criticism, and disgust.

There was no glory attached to the memory of such a Crusade, only shame and recriminations. If it had once seemed a glorious heaven-blessed game, and a true campaign for Christ's vicar, the Crusade's reverses and the treachery of its officers soon changed all this. "They returned, dripping with blood and staining their country," writes one chronicler, "Blessed be God who confounds the insolent."[25] In the words of George M. Wrong, "The Crusade settled nothing; no burdens were lightened by it, and many were made heavier. Called into action to rescue the true Pope from the Schismatics, it injured chiefly the Flemish people who obeyed the same Pope and professed similar devotion."[26]

There is, of course, some question as to how deeply the nation examined its own conscience during this time, whether its fury was aimed at the immorality of failure or of original motive. The censure of Despenser lay not in that he had tried to do a bad thing, but, rather, that he had done a bad job. The chroniclers, blessed with the gift of hindsight, for the most part vituperated not the causes but the protagonists. And yet there were some who connected this failure to larger issues, as a symbol of the English nation's decline in power, and a symptom of the sickness within the ecclesiastical body. Perhaps the most outspoken critic of the whole undertaking, whose polemics and sermons rolled out as a thunder of prophetic judgment during the last years of his life, was John Wyclif. Even though full listing of the extent of his attack and the issues he treated would be

[24] In the Continuation of *Polychronicon Ranulphi Higden,* ed. Joseph Rawson Lumby, Rolls Series, IX (London, 1886), 25. For a full record, see *Rot. Parl.,* III, 151–158.

[25] *Continuatio Eulogii,* III, 357.

[26] *Crusade,* p. 90.

beyond the scope or purpose of this paper, a brief glance at his criticism will be useful in determining the variety of the English reaction to the Crusade.

The fullest and most formal work from him on this topic is his *Cruciata*, concerning the wars of the clergy, written probably in the summer of 1383, when the Crusade was in full progress.[27] To Wyclif, the affair in Flanders was absolute proof that the rival popes were the two-headed Antichrist, and that the "pseudofratres," the false friars, the mendicant orders in England, were the chief disseminators of the evil. For the friars had preached the Crusade; it was to them that Despenser entrusted the organization of public appeal. Through their office of confession, they had collected money—keeping 2½ per cent of the proceeds as their commission[28]—and had preached long and loud the amazing indulgences Urban VI was granting. To Wyclif such bait was false and such motives improper for a Church that was forgetting its mission of charity, humility, and poverty. All those that went to fight were thus involving themselves in a war which was actually against Christ. "Lord!" cried Wyclif in another tract, "where it be Goddis wille to lette to preche þe gospel, and ȝev leve and counseil to prestis to slee men? Hit semeþ þat suche dedis schewiþ whos beþ þes prelatis. . . . Ffor þes þat traveyllen wel for pees beþ children on Goddis half, and þei þat traveille by werris to have richesse and worldly glorie beþ children of þe fend and enemyes to Crist."[29]

Wyclif, it is true, cannot be cited as a spokesman for the orthodox; yet in his day he was not so much of a heretic that his attacks may not indicate to us the general area wherein dissent might arise. Evidence for this is supplied by John Gower, whose orthodoxy is above suspicion, and yet who likewise attacks the propensity of the clergy to deal in war instead of peace. His fullest treatment of the general

[27] In *John Wiclif's Polemical Works in Latin*, ed. Rudolf Buddensieg, II (London, 1883), 579–632.

[28] See the "Ordinationes Episcopi pro Cruciata publicanda," in Knighton, *Chronicon*, II, 201–203.

[29] *De Pontificum Romanorum Schismate*, in *Select English Works of John Wyclif*, ed. Thomas Arnold, III (Oxford, 1871), 249. This was evidently an English tract circulated in 1382 when the campaign was being organized; it was the vernacular counterpart to the Latin *De Dissensione Paparum*, which was aimed at "the cultivated classes of England" (*Polemical Works*, II, 567–576).

subject is in his *Vox Clamantis,*[30] written largely before 1381, where he may be found inveighing against the spiritual Estate's unleashing of Satan through its dedication to the wrong kind of sword:

> The piety in days of old used to implant faith, but our vindictive anger now uproots the proud deeds of our forefathers. As David bore witness, "The sword does not save, and he who shall put his trust in the bow is not saved thereby." But we are at variance with David's teaching as long as the hand of the priest is armed with a sword. The ark was Moses' strength in days of old, but our strength is the drawn bow, which strikes down the people of the world. . . .
>
> Behold, O Christ, Thou Who hast bound up Satan with Thy mighty power, we of the clergy have released him again. Now that he is released he has released all bondsmen, so that no one happens to render his devotions to God. . . . Meanwhile the priest, ready for arms, holds on to a prostitute with his right hand and a sword with his left. . . . But if this is the cause of the priest taking up arms, then war will last forever, far indeed from peace.
>
> It is not the function of a knight to offer sacrifice at the altar, nor that of a priest to carry on wars of state.[31]

The followers of Bishop Despenser would have qualified well for such an attack; and, indeed, a marginal note to this passage, perhaps added later by Gower himself to the manuscripts, refers to the Flanders campaign when both regular and secular clergy plundered just as much as the laity. It is in Gower's *Confessio Amantis,* completed by 1393, that an indirect allusion is made to the Crusade itself; it is in the "Prologus," where the three Estates are reviewed and excoriated and where Gower laments the decline of the Church from its former position:

> God wot the cause to the laste
> Of every right and wrong also;
> But whil the lawe is reuled so
> That clerkes to the werre entende,
> I not how that thei scholde amende
> The woful world in othre thinges,
> To make pes betwen the kynges
> After the lawe of charite,
> Which is the propre duete
> Belongende unto the presthode.

---

[30] In *The Complete Works of John Gower,* ed. G. C. Macaulay, IV, *The Latin Works* (Oxford, 1902), 1–313.

[31] III. vi. 361–368, 375–378, 387–390; p. 117. For this translation and information concerning dates and manuscripts of Gower's work, I am indebted to the courtesy of Professor Eric Stockton, of The University of Michigan, who is preparing a translation of the Latin works of Gower.

Bot as it thenkth to the manhode,
The hevene is ferr, the world is nyh,
And veine gloire is ek so slyh,
Which coveitise hath now withholde,
That thei non other thing beholde,
Bot only that thei myhten winne.
And thus the werres thei beginne,
Wherof the holi cherche is taxed,
That in the point as it is axed
The disme goth to the bataille,
As thogh Crist myhte noght availe
To don hem riht be other weie.
In to the swerd the cherche keie
Is torned, and the holy bede
Into cursinge, and every stede
Which scholde stonde upon the feith
And to this cause an Ere leyth,
Astoned is of the querele.[32]

With this brief survey of the attitudes of two prominent voices of Chaucer's England, it is time to return to the heart of the question and ask what Chaucer's own reactions were likely to have been to the Crusade. If both Wyclif and Gower share a general disapproval, it seems reasonable to expect that Chaucer would disapprove also and that he, notably no less sensitive than Gower to the present condition of society, could recognize the larger issues behind the late fiasco. It is not in Chaucer's nature to write a Tract for the Times, it is true; but the poet who wrote the *Friar's Tale*, the *Merchant's Tale*, and who translated the *Tale of Melibee* at an appropriate moment, was not one unaware of the present degradation of men and manners. Furthermore, certain characteristics of the Crusade involved issues to which Chaucer proved himself particularly sensitive. The estate of the clergy and the health of the medieval English Church were one of his continuing concerns, and there is no reason to believe his eyes were closed to the realities of Church politics and intrigues, in the Crusade that Wyclif was attacking. Perhaps what would have been surest to provoke him, however, was the stamp of the mendicant orders that was so prominently displayed on the Crusade's organization. All about him the friars would have been preaching, confessing, collecting funds, pocketing their commissions,

[32] In *The Complete Works*, II, *The English Works* (Oxford, 1901), 11f. It is probable that the "Prologus" was written several years earlier than 1393.

and absolving. The author of the *Summoner's Tale* would not have been likely to regard such activity in a favorable light.[33] Even when the military action of the Crusade had ceased, the friars continued their own campaigning. The official date for them to stop their public appeal was set for Easter, 1385, but it would appear that some continued even into 1386.[34] Here was fresh material for indignation.

It would seem safe to conclude from everything we have seen thus far that one would not likely have boasted of his part in the "glorious campaign" of the Bishop of Norwich. The courtly audience that Chaucer addressed would have added its own reasons—undoubtedly more of disgust than shame—to the ones Chaucer brought with him. And yet we find Chaucer suggesting that it was precisely in the campaign to "Flaundres, . . . Artoys, and Pycardie" that the Squire had served. How does such a compromising association fit with one who was "as fressh as is the month of May"? There seem to be two possible answers: either we find Chaucer here approving of such a campaign as a near-to-home testing of arms and chivalry,[35] or we find Chaucer deliberately adding a sour note. If we investigate the latter possibility, we are faced with the whole matter of Chaucer's portrait of the Squire and his attitude toward him; for such an underhanded or uncomplimentary allusion immediately suggests the possibility of irony—prominent, certainly, in other portraits in the *General Prologue*. But why should Chaucer have wished to satirize in any way such an attractive young man? A tentative answer to this larger problem may be suggested here—tentative because it concerns other issues not discussed in this paper.

Let us, first of all, not overemphasize the weight of the allusion, but rather take it in context, as only one part in a network of de-

[33] The friars in fourteenth-century England were far from being universally popular. The literature about them is available but scattered. Chaucer's view of them is probably apparent: cf. the portrait of the Friar, *Gen. Prol.*, vv. 208–269 (p. 19); *The Wife of Bath's Tale*, vv. 857–881 (p. 84); *The Summoner's Prologue* and *Tale* (pp. 93–100). For the friars as they were involved in satire and the theory of literature, see Charles R. Dahlberg, "The Secular Tradition in Chaucer and Jean de Meun," unpublished Princeton University doctoral dissertation (Princeton, 1953). And for more concerning Chaucer and the attack on the friars, see Arnold Williams, "Chaucer and the Friars," *Speculum*, XXVIII (July, 1953), 499–513.

[34] See Perroy, *Schisme*, I, 200, n. 3.

[35] Which is in harmony with Tatlock's original view; see *Development*, p. 148.

scriptive detail. It would surely be unwarranted to claim that Chaucer thereby condemns the Squire, destroys his reputation, or throws his pleasing personality into a sinister or morally suspect light. When Chaucer is really "agin" someone, his irony is constant and devastating. But here the touch is much lighter, not unlike the brilliant and graceful hints of satire discovered in his portrait of the Prioress. Let us look at the whole description of the Squire again. He seems to be everything a bright young squire of twenty years should be. His physique is strong, his hair curly, his clothes à la mode, his accomplishments various and courtly, his carving proper, and his loving hot and sleepless. With his father, he sums up the cultivated ideals of "chivalrie" and "curteisie." He is an ideal, moving slim-waisted and broad-shouldered into his picture frame much like those bright and elegant courtiers in the miniatures of the time (Fig. 2). But perhaps we should better say, he is a stereotype. His description is an echo of the past, a recapitulation of the idea of youth which the fourteenth-century courtly society made its wistful dream. For the world of expanding trade, complex international politics, and crumbling order in the social and religious realms was a far cry from the culture of the decentralized feudal courts of twelfth and thirteenth century France, where such images of the noble life were first created. It is easy, in fact, to locate the source of the Squire's behavior. We may say that he is like one who has listened well to the God of Love in the *Roman de la Rose*; and is his ardent disciple: "Whoso with Love wole goon or ride,/ He mot be curteis, and voide of pride,/ Mery, and full of jolite. . . ."[36] He would seem to have listened with all attention to that god's translation of Ovid into courtly terms, in his lecture on true courtesy and against villainy.[37] And we have seen his fashion before, too; for he emulates the elaborate embroidery which was such a prominent feature of the God of Love's garb in the *Roman*: "His garnement was everydell/ Yportreied and ywrought with floures,/ By dyvers medlyng of coloures."[38] Chaucer's Squire was "embrouded . . . as it were a meede/ Al ful of fresshe floures. . . ." By thus linking the Squire with that great satiric textbook of the courtly life, Chaucer makes him a virtual archetype of the fashionable young lover.

And what, then, do we make of the reference to the Crusade?

[36] The *Romaunt of the Rose*, ed. Robinson, vv. 2351–2353, p. 587.
[37] *Ibid.*, vv. 2175–2720, pp. 585–590.
[38] *Ibid.*, vv. 896–898, p. 573.

The lovers in the *Roman de la Rose* do not go off to fight battles for church and country. Precisely so. The allusion is the one harsh reality that disturbs an otherwise placid idealization of a character. It controls the portrait to the extent that it keeps it related to the harsh realities of contemporary English history. It is the nexus between the garden of courtly lovers and the battlefields of the Continent. It is aimed at an aristocracy which still liked to pretend it could get back to the good old days of chivalry and gallantry and color and the uncomplicated life. For the Squire is a dandy, a bit of a dude. True, his type was expected to be like that. Chaucer had been trained as a squire in Prince Lionel's household and knew the rules of the game as well as anyone. But the real world which squires had to enter as warriors, not as lovers, was no game. The citizens of Ypres and Dunkirk knew how much gallantry the late campaign had produced. The plains of Flanders kept their burned cities and their dead and offered no matter for new epics or romances.

All this is not to implicate the Squire necessarily in the deepest corruption of the Crusade. We do not know when he left, nor for precisely what motives. But we may realize that the nature of his achievement is highly qualified in the mirror of history: "And born hym weel, as of so litel space. . . ." If we accept the allusion, the line becomes somewhat ironic. The Squire served for only a short while, because his "chyvachie" was ignominiously shortened by the events that followed the break of the siege at Ypres. And we have the final conventional association: the Squire bore himself well "In hope to stonden in his lady grace." We know that the friars were able to get a great deal of their support from the noble ladies of the land, and that their early support may have tipped the scales in favor of the Crusade when national opinion was still in the balance. Knighton tells us that the friars collected an innumerable and incredible sum of gold and silver coin, jewels, necklaces, rings, bracelets, spoons, and other ornaments, especially from the ladies and other women of the land. One woman alone was said to have contributed £100. And thus, he observes, the private treasury of the kingdom that was in the hands of the women was hazarded.[39] But although the ladies gave their riches to the cause urged upon them by the friars, it would seem at any rate that there was little chance for elevated and courtly *frauendienst* in the course of the campaign, once under way.

[39] *Chronicon,* II, 198.

The conjunction of the familiar romantic terms with the very unromantic Crusade produces a subtle but nevertheless jarring discord. The golden glitter of the bright surface covers a somewhat tarnished interior. It is as if Chaucer said implicitly to his audience: yes, here is your dream of fond youth and pleasant fancies; here is your courtly ideal. But remember what that ideal involves today, and the kind of glory and campaigns it gathers for itself.

If there is irony in the portrait of the Squire, it completes the contrast between the son and his father. For the appeal of the Knight is not concerned at all with external things. His clothes are stained, his company of the smallest possible number, and his manner subdued and unostentatious. He neither sings nor dances, and we hear nothing about his qualities as a lover, except as they are attached to "chivalrie,/ Trouthe and honour, fredom and curteisie." But he tells a brilliant tale—one in which two attractive if somewhat foolish young squires come under the magisterial correction of a noble knight and lord who had "conquered al the regne of Femenye."

Another problem of interpretation arises here. How are we to be sure that father and son are contrasted in every point? Why should we suppose that the Squire did not go with his father and that the father did not go with his son? For if we were to deny these suppositions, it would be foolish to judge father by the crusades against the pagans and son by the crusades against the Christians, if both were present at both types of war. Roger S. Loomis suggested that "Chaucer's unrestrained laudation of the Knight might be taken as an expression of sympathy with a suspect class,"[40] because all his wars were crusades against the heathens and none against the Christians. But Gardiner Stillwell and Henry J. Webb[41] maintained to the contrary that since "Chaucer evidently wanted us to think of the Knight and the Squire in their medieval relationship of warrior and attendant," he evidently "did what an accomplished poet might be expected to do; he indicated that the Knight had seen action/ As wel in cristendom as in hethenesse,/ enumerating the heathen battles at once and reserving the expeditions against the French until describing his squire."[42] There, apparently, the matter has rested, and F. N.

[40] "Was Chaucer a Laodicean?" in *Essays and Studies in Honor of Carleton Brown* (New York, 1940), p. 140.

[41] "Chaucer's Knight and the Hundred Years' War," *MLN*, LIX (January, 1944), 45–47.

[42] *Ibid.*, p. 46.

Robinson cites the two articles together in his notes to the Knight's description in the *General Prologue* (p. 652). But the matter is more complex than that. Stillwell and Webb were correct in underscoring the reference to the Knight's service in both Christian and heathen countries (*Gen. Prol.*, v. 49), and it would seem preposterous to claim flatly that Chaucer made his Knight an ideal crusader in order to criticize covertly the English wars against the French and the Spanish. But the Knight is not a young man; he is an old campaigner, and could well have assisted in the earlier English glories of the Hundred Years' War. Not so the Squire. He is young when the English nation has lost much of its enthusiasm for such wars. He can witness no smashing and clear-cut victories when one could with little qualm insist that right had triumphed over wrong. He has, it is true, been present during one expedition—not a war, but a "chyvachie." But if we are able to identify that one, we still will not be able to allow its single instance to represent "the expeditions against the French," as Stillwell and Webb implied. Nor need we insist that the Knight accompanied his son there. Custom did not demand this. And although Stillwell and Webb imply that the Squire served his father everywhere, Chaucer only makes it clear that it is for the Canterbury pilgrimage in particular that the Squire and the Yeoman attend the Knight. One could argue, in fact, that there is a suggestion that the Squire and his father did not come together to the pilgrimage; for while the Knight is "late ycome from his viage" and wears a tunic stained from the iron war-hauberk recently over it, the Squire is "as fressh as is the month of May," and beautifully dressed. This would hint that the son was not with his father at least during his most recent "viage."

Of course father and son are not in competition. The Squire is still young and may someday be a knight like his father. But he is the symbol of a new generation, and his physical years are in this sense of little importance. It is a generation whose holy wars are political contests, Holy Church a festering place of intrigue, nepotism, and corruption. And although its young men, eager for the excitement of knighthood, may put on the French manner, and dance and sing and "floyt," and spend their nights in restless yearning, the chivalry for which they prepare themselves is one enforced by the will, a matter of the letter more than the spirit. The Squire is courtly to the letter; but the glorious campaigns for which he trains so elegantly turn out to be in reality "chyvachies," and not Crusades.

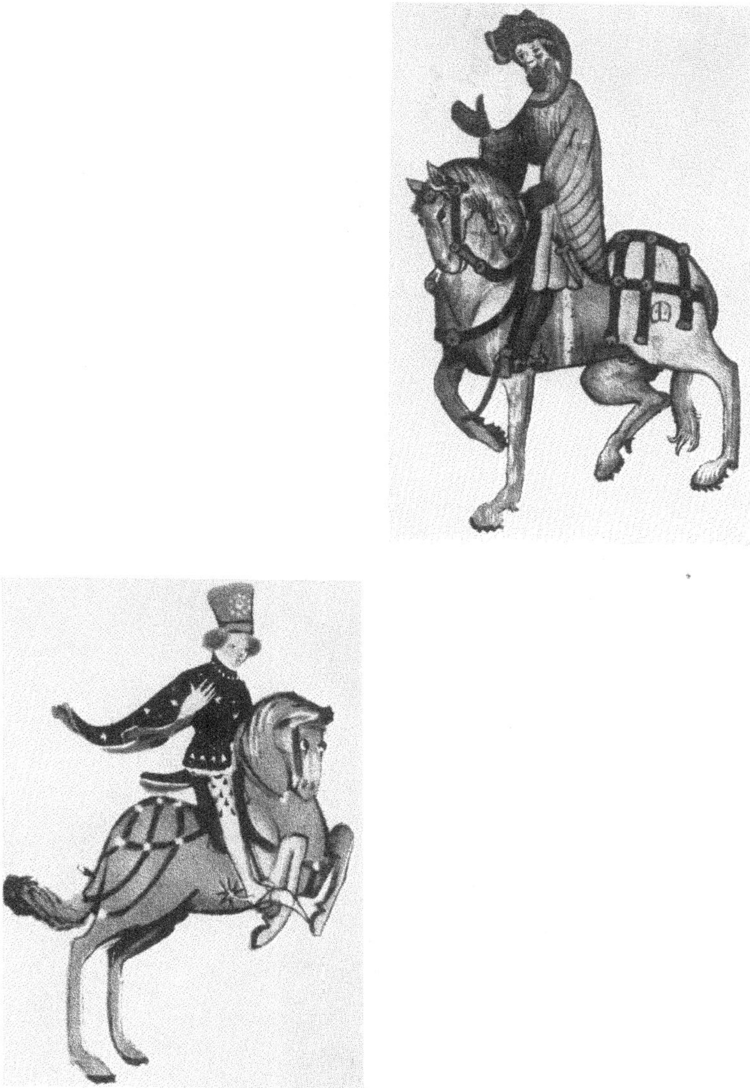

Fig. 2. The Son and His Father. Chaucer's Squire and Knight; from the Ellesmere Manuscript.

Papers of the Michigan Academy of Science, Arts, and Letters
Vol. XLV, 1960 (1959 Meeting)

PRIMITIVISTIC MOTIFS IN SIDNEY'S *ARCADIA*

ERNEST S. GOHN
Eastern Michigan University

Since Greenlaw's early essay on the *Arcadia*,[1] scholars have been busy reinterpreting this rambling historical romance. The direction of this recent criticism has been away from that of the eighteenth-century essayists, who thought of Sir Philip's 'trifle . . ., triflinglie handled' as some sort of *bergerie* in which we find "bleating lambs on pretty pastures, piping shepherds, beloved shepherdesses, kings and queens, ladies and gentlemen, all loving and hating and singing and mourning in rhythmic prose of considerable beauty, interspersed with odes and sonnets."[2] And as this criticism moves away from old misconceptions, it tends to move toward interpretations that stress either Sidney's political and ethical philosophy or else his conception of the *Arcadia* as a type of heroic poem.[3] Recent scholarship generally has emphasized Sidney's deliberate idealization of the principle of action. Indeed, it is hard to see how the eighteenth-century and Victorian readers could avoid Greenlaw's conclusion that "Sidney does not hold up the pastoral life of Basilius as a model; he does not find in it an admirable withdrawal from the cares of life; it is no idyllic existence in the forest of Arden, but a criminal evading of re-

[1] Edwin A. Greenlaw, "Sidney's Arcadia as an Example of Elizabethan Allegory," *Kittredge Anniversary Papers* (Boston, 1913), pp. 327–337.

[2] Grant C. Knight, *The Novel in English* (New York, 1931), p. 13. It is surprising to note that such an eighteenth-century view was published so recently. For a review of eighteenth- and nineteenth-century criticism of the *Arcadia*, see Marcus S. Goldman, *Sir Philip Sidney and the Arcadia*, Illinois Studies in Language and Literature, XVII (1934), 122–143.

[3] For Sidney's political and philosophical ideas, see especially William Dinsmore Briggs, "Political Ideas in Sidney's *Arcadia*," *SP*, XXVIII (1931), 137–161 and "Sidney's Political Ideas," *SP*, XXIX (1932), 534–542. See also Marcus Goldman, *op. cit.*, pp. 168–185; R. W. Zandvoort, *Sidney's Arcadia: A Comparison Between the Two Versions* (Amsterdam, 1929), pp. 149–159; E. A. Greenlaw, *op. cit.* and his later "The Captivity Episode in Sidney's Arcadia," *The Manly Anniversary Studies in Language and Literature* (Chicago, 1923), pp. 54–63. The best treatment of the *Arcadia* as a type of heroic poem is that by K. O. Myrick, *Sir Philip Sidney as a Literary Craftsman* (Cambridge, Mass., 1935).

sponsibility that will bring ruin to any state."[4] On the other hand, the emphasis on the political philosophy of the *Arcadia* has led, I believe, to an oversimplification of the problem, an oversimplification expressed by Lois Whitney: "We need scarcely be surprised, in view of the nature of the . . . doctrine of the function and importance of disciplined reason, that Sidney was not an enthusiast for the 'natural man,' nor a primitivist in any sense."[5]

But if Professor Whitney is right, as she surely is, in arguing that the explicit doctrine taught in the *Arcadia* is anti-primitivistic, this does not mean that there are no primitivistic ideas implicit in the text—ideas which do much to shape the work as a whole. It must be remembered that while Sidney has a lofty conception of the philosophical function of the poet, and while he may be trying to come to this conception in the *Arcadia*, he is still not writing a philosophical treatise but a narrative. And Professor Whitney ignores one question when she says that Sidney is not a primitivist in any sense—why did Sidney place his action in Arcadia at all? Moreover, is the ideal which Sidney posits for his heroes necessarily anti-primitivistic because it is not anti-intellectual? For certainly an author may be "pro-intellectual" and still be a primitivist in certain respects.

As Professors Lovejoy and Boas have defined it, chronological primitivism is a "kind of philosophy of history, a theory, or a customary assumption, as to the time—past or present or future—at which the most excellent condition of human life, or the best state of the world in general, must be supposed to occur."[6] In any discussion such as this, the first question is, then, what time in history does Sidney assume to have been the best? The answer is fairly obvious, though confused. While Sidney never so much as mentions a golden age, it is clear that the age of chivalry was for him the highest point in history. Although the action ostensibly takes place in ancient Greece, this fiction is not very deceptive. Sidney's persistent attention to the medieval code of love embraces the chivalric ideal. The descriptions of single combat and of war make them clearly

[4] Greenlaw, "Sidney's Arcadia as an Example of Elizabethan Allegory," p. 337.

[5] Lois Whitney, "Concerning Nature in 'The Countesse of Pembrokes Arcadia,'" *SP*, XXIV (1927), 217.

[6] Arthur O. Lovejoy and George Boas, *Primitivism and Related Ideas in Antiquity* (Baltimore, 1935), p. 1.

chivalric.[7] When Musidorus upbraids Pyrocles for falling in love, the clinching point of his argument is that Pyrocles, before he fell in love, had studied virtue and had put his virtue into practice with "worthie enterprises."[8] None of the action in the *Arcadia* would seem out of place in Malory.[9] The emphasis on friendship, while not necessarily a medieval concept, is presented by Sidney in the truly chivalric tone. After rebuking Musidorus for scolding him, Pyrocles ends pathetically:

> I am sicke, and sick to the deathe, I am a Prisoner, neyther ys there any Redress, but by her, to whome I am a slave: Now, yf yow list, Leave him, that loves yow in ye highest degree, but remember ever, to carry this with yow, that yow abandon youre Frende in his greatest neede. And herewith the deepe woounde of his Love beeyng rubbed a fressh, with this newe unkyndnes, began as yt were to bleede ageane, in suche sorte that hee was unable to beare yt any longer: But, gusshing oute abundance of teares, and Crossing his armes over his wofull harte hee sancke downe to the grounde.[10]

One aspect of chronological primitivism is found in Kalander's description of the degeneration of Arcadia itself from the condition of the country during his youth:

> Then spared he not to remember how much *Arcadia* was changed since his youth: activitie & good felowship being nothing in the price it was then held in, but according to the nature of the old growing world, still worse & worse.[11]

The fact that Arcadia has declined from its pristine beauty suggests, at least, the relationship between chronological and cultural primitivism as found in the *Arcadia*. The great confusion of time in the work adds to this impression. All that we can gather is that at some time in the past—presumably in the Middle Ages—there existed the ideal life. And, as the state of Arcadia has degenerated from the ideal according to the nature of the 'old growing world,' the state itself as presented in the *Arcadia* is not at its highest level. But here again the confusion of time in the work leads one to think of Arcadia at the

[7] Albert Feuillerat, ed., *The Countesse of Pembrokes Arcadia*, in *The Complete Works of Sir Philip Sidney*, Cambridge English Classics (Cambridge, 1912–26), I, 283, 386ff., and 413. The whole of the captivity episode is filled with chivalric echoes, as is much of the rest of the work.

[8] *Ibid.*, I, 55–58.

[9] See Marcus S. Goldman, *Sir Philip Sidney and The Arcadia*, pp. 186–210.

[10] *Works*, IV, 21.

[11] *Ibid.*, I, 60.

time of the narrative as being near to the ideal—except that Basilius has neglected his duty, temporarily. The ideal concept of society is not only something that existed in the past but, like the Platonic *idea*, is an eternal goal. Throughout the romance there is a constant emphasis on the idea that we must impose on nature some ideal plan.[12] Ostensibly this is anti-primitivistic. At the same time, however, Sidney constantly assumes that this ideal did exist in the past; the *Arcadia*, then, is at least chronologically primitivistic.[13]

Although the *Arcadia* holds up a chivalric ideal, thus embodying the backward look of chronological primitivism, the most important kind of primitivism in the work is cultural. Although the actions of the characters who parade through the country are not always admirable, Arcadia itself is an ideal place. In his opening description of the country Sidney speaks of it in glowing terms:

> Arcadia amonge all the Provinces of Grece was ever had in singuler reputation, partly for the sweetnes of ye Aire and other naturall benefittes: But principally, for the moderate & well tempered myndes of the people, who, (fynding how true a Contentation ys gotten by following the Course of Nature, And howe the shyning Title of glory somuche affected by other Nacions, dothe in deede help litle to the happines of lyfe) were the onely people, wch as by theire Justice and providence, gave neyther Cause nor hope to theyre Neighboures to annoy them, so were they not stirred with false prayse, to truble others quyett. Thincking yt a smalle Rewarde for ye wasting of theire owne lyves in ravening, that their posterity shoulde long after saye, they had done so: Eeven the *Muses* seemed to approove theire good determinacōn, by chosing that Contrie as theire cheefest reparing place, and by bestowing theire perfections so largely there, that the very *Shepeardes* them selves had theire fancyes opened to so highe Conceiptes (as the moste learned of other nations have bene longe tyme since content) bothe to borrow theyre names, and imitate theire Conning.[14]

---

[12] The perfect expression of this ideal is found in the description of the Knights' retreat from battle: "The Knightes as soone as they heard the retraite (though they were eagerly set, knowing that courage without discipline is nearer beastlinesse than manhood) drew backe their swords, though hungrie of more blood" (*Works*, I, 444).

[13] An interesting, and somewhat amusing, aspect of chronological primitivism is found in the praise of the arts of surgery 'in that age': "For bothe those greate Ladyes were excellently seene in that parte of surgery, an Arte in that Age greatly esteemed, bycause yt served as a Minister to vertuous Corage: which in those worthy dayes was even by Ladyes more beloved, then any owteward bewty" (*Works*, I, 121; IV, 45–46). Evidently ladies no longer love virtuous courage.

[14] *Works*, IV, 1.

The virtues here are not essentially primitivistic. They are the old classic virtues of temperance and moderation. This does not, however, diminish the importance of their being found in Arcadia. The important thing here is that Sidney seems to imply that the shepherds who follow the course of nature live in serenity. It is clear that Sidney means that they follow nature in that they have well-tempered minds. The net result, though, is a simple life—and therefore desirable.[15] The impression of the desirability of such a simple life is strengthened by the fact that Sidney always speaks of Arcadia in such idyllic terms.[16] It is a "Worlde not so farr gon into paynted vanityes,"[17] a world where the shepherds are happy, "wanting little, because they desire not much."[18] Although Musidorus' amazement at the wit of the shepherd's learning as found in Claius and Strephon is explained away,[19] there is no doubt that the shepherds in Arcadia are adept at versifying:

> But certainely, all the people of this countrie from high to lowe, is given to those sportes of the witte, so as you would wonder to heare how soone even children will beginne to versifie. Once, ordinary it is amonge the meanest sorte, to make Songes and Dialogues in meeter, either love whetting their braine, or long

[15] Cultural primitivism, as defined by Lovejoy and Boas, *op. cit.*, p. 7, is "the discontent of the civilized with civilization, or with some conspicuous and characteristic feature of it. It is the belief of men living in a relatively highly evolved and complex cultural condition that a life far simpler and less sophisticated in some or in all respects is a more desirable life."

[16] For other passages see *Works*, I, 54, 13–14, 57, 118–119, and all of the general nature descriptions. Musidorus' rebuttal of Pyrocles' enthusiasm is not found elsewhere: "But I marvell at the excessive praises you give to this countrie; in trueth it is not unpleasant: but yet if you would returne into *Macedon*, you should see either many heavens, or find this no more then earthly. And evẽ *Tempe* in my *Thessalia*, (where you & I to my great happinesse were brought up together) is nothing inferious unto it" (*Works*, I, 58). Here it seems that Sidney is recalling his major aim in his work: that it contradicts other implicit and explicit judgments of Arcadia is only another example of the dichotomy which this paper points out.

[17] *Works*, IV, 3.

[18] *Ibid.*, I, 14. It should also be noted that even when Basilius is condemned for misjudging Dametas, simplicity is implicitly praised. That the Duke interpreted rudeness as plainness—and Sidney adds in a parenthesis, "allthoughe there bee great difference betwixt them"—does not mean that simplicity is condemned (*Works*, IV, 27–28).

[19] *Ibid.*, I, 27.

peace having begun it, example and emulation emending it. Not so much, but the clowne *Dametas* will stumble sometimes upon some Songs that might become a better brayne: but no sorte of people so excellent in that kinde as the pastors; for their living standing but upon the looking to their beastes, they have ease, the Nurse of Poetrie.[20]

While Sidney emphasizes the principle of reasoned action throughout the *Arcadia,* he also makes it clear that Arcadia is a place where people, harassed by the world, may retire for peace. In the introduction to the "First Eglogues," after describing how the shepherds relegated the duty of polishing the verses of their un-thought-on song to one man, Sidney says:

> But the peace wherein they did so notably flourish (and specially the sweete enjoyng of theyre peace to so pleasant uses) drew dyvers straungers aswell of greate, as meane howses; Especially suche, whome inwarde Melancholyes, made weary of the Worldes eyes, to come and live amonge them, applying them selves to theyre trade, which likewyse was many tymes occasyon to beutify more, (then otherwyse yt woulde have bene) this pastorall exercyse.[21]

All of this may be interpreted as belonging to the Arcadian pastoral tradition in which Sidney is working. The simple shepherds traditionally were adept at versifying. The fact, however, that a poet does work in a certain tradition indicates at least some sort of intellectual affinity for the assumptions of that tradition. And the assumptions of the pastoral tradition point toward soft primitivism. When Sidney puts his heroic action in Arcadia, when the Arcadia he describes is a place to which people can go who are weary of the world's eyes, and when the normal course of affairs in Arcadia is void of worldly vanities, he must also admit a certain amount of this soft primitivism into his world.

One might argue that Sidney's ideal of a rationally ordered society, even one entailing simplicity, is merely an echo of the classic concept of the golden mean and is not to be construed as demonstrating a bias for cultural primitivism, were it not that this element of classic rationality in the *Arcadia* is tinged with Protestant zeal—and there is a strong inclination toward natural religion. In Dorus's song in the "Second Eglogues," Sidney expresses, in the first stanza, the typical desire for the delights of solitariness—in solitariness man's senses can behold the order of creation, and from this learn what

[20] *Ibid.*, I, 27–28.
[21] *Ibid.*, IV, 51.

the Creator is. Sidney evidently feels some discontent with a civilization that prevents such study of the Deity. The second stanza of the same song is more direct in condemning the evils of civilization:

> O sweete woodes the Delighte of Solitarynes?
> O howe muche do I like youre solitarynes,
> Here, no treason ys hidd vailed in Innocency,
> Nor Envyes snaky eye fyndes any harboure here,
> Nor Flaterers venymous Insinuations,
> Nor Cunning Humoristes pudled opinyons.
> Nor Courteous Ruyning of proffered usury,
> Nor tyme prattled away, Cradle of Ignorance,
> Nor tryffling Tytle of vanity daseleth us.
> Nor golden Manackles stande for a Paradyse,
> Here wronges name ys unhearde, Sclaunder a Monster ys.
> Keepe thy Spirite from abuse, here no abuse dothe haunte,
> What Man graftes in a Tree dissimulation.[22]

Sidney is obviously weary of the vanities of this world, especially the court; in his ideal Arcadia he finds a simplicity and a solitariness where it is possible to view the works of God without interruption.[23] The shepherds inhabiting Arcadia are mainly good—although the most important ones in the plot are not. Their lives are in pointed contrast to ordinary civilized life:

> But among the shepheards was al honest libertie, no feare of daungerous tel-tales, who hunt greater prayes, nor indeede mindes in them to give tell-tales any occasion; but one questioning with another of the manuring his ground, and governing his flock, the highest pointe they reached to was to talke of the holines of mariage.[24]

---

[22] *Ibid.*, IV, 157–158. See also Sidney's explicit attack on courtly life, "Dispraise of a Courtly life," *Works*, II, 326–338; Amphialus' song to Philoclea, *Works*, I, 395; and the "Fourth Eglogues," *Works*, IV, 313.

[23] When Sidney is presenting his reasoned arguments, in the person of Musidorus, he condemns solitariness: "But, I thincke yow will make mee see that the vigor of youre witt can shewe yt self in any subject, or else yow feede sometymes youre solitarynes, with the Conceyptes of the *Poetts*: whose liberall pennes can as easily traveyll over mounteynes as Mole hilles, and so like well disposed men, sett upp every thinge to the highest Noate, especially, when they putt suche wordes in the mouthe of one of these Fantasticall mynde infected people, yt Children and Musicians calle Lovers" (*Works*, IV, 14). In this conflict of attitude one can again see how Sidney's doctrinal, rational self is at odds with his occasional lapses into primitivistic tenderness.

[24] *Works*, II, 62–63.

And when Basilius dies,

> The Shepeheardes fynding no place for them in these garboyles to which theyre quyet hartes (whose hyest ambition was in keeping them selves up on goodnes) had at all no aptenes, retyered themselves from among the Clamorous multitude.... And there ... the pore men sate pensive of theyre present miseryes, as yf they founde a werysomnes of theyre wofull wordes; Till, at last good oulde *Geron* ... began in this sorte to Complayne. Alas pore sheepe sayde hee wch hetherto have enjoyed youre fruitfull pasture in suche quyetness, as youre wolle (amongst other thinges) hathe made this Country famous?[25]

Thus Sidney praises the quiet life. And for him the quiet life is in Arcadia. In another place Sidney, although speaking of events that took place outside of the country of Arcadia, condemns the "new-fanglenes both of manners, apparel, and each thing els,"[26] which was rampant in Macedon before Euarchus ascended the throne. In one respect, such objection to new fashions is a kind of chronological primitivism, but to the Elizabethan, especially the Elizabethan of Puritan inclinations, "new-fanglenes" in dress and manners connoted all the overrefinements of the new Italianate Englishman. Such civilized refinements attract Sidney's scorn, and he can, in another place, even show himself something of a technological primitivist:

> But when his seate so Rooted hee [Man] had founde,
> That they now skilde not, howe from hym to wende,
> Then gan in giltles Earthe full many a wounde,
> Iron to seeke, which ageanst yt self shoulde bende,
> To teare the Bowells that good Corne should sende,
> But yet the Comōn Damm̃ bemone,
> Bycause (thoughe hurt) they never heard her grone.[27]

Such primitivistic motifs are admittedly not too prominent in the *Arcadia*. What is important, as I have indicated before, is that the state of Arcadia, the place where most of the action takes place, is a relatively simple country. It must also be remembered that the simple life of the shepherds attracted people from high estate as well as those from low. Sidney is no democrat, but he does not make the distinction which would limit honorable simplicity to peasants. And Arcadia itself has degenerated since and because of Basilius' retirement. The difficulty is that Sidney has two opposing ethical

[25] *Ibid.*, IV, 306.
[26] *Ibid.*, I, 186.
[27] *Ibid.*, IV, 240.

standards which do not coalesce. Setting out to lead men on the paths of virtuous action, he yet had a wistful hankering for a past when life was easier and issues were simpler. Such chronological and cultural primitivistic yearnings form only a minor theme of Sidney's romance. It so happens that the eighteenth- and nineteenth-century readers hit upon this minor theme and discussed it as if it were major. But what they recognized is there. It may be that the opposition of minor to major theme is a cause of the singularly incoherent impression one has of Sir Philip's work. In Crocean terms, Sidney's intuition was not complete. His artistic vision was ultimately discordant.

PAPERS OF THE MICHIGAN ACADEMY OF SCIENCE, ARTS, AND LETTERS
VOL. XLV, 1960 (1959 MEETING)

PSYCHOLOGY AND THE COHERENT FORM OF SHAKESPEARE'S *OTHELLO*

JOHN V. HAGOPIAN
The University of Michigan

PSYCHOLOGY as a formal discipline is not yet a century old—which means that, as intellectual disciplines go, it is still in its infancy. How many centuries it will take to reach the height of its powers and final decline—as metaphysics, for instance, has done—no one can say. It should not be surprising, therefore, that much of what psychologists have to say about literature should strike us as ridiculous and absurd. Ella Freeman Sharpe, for example, cites four quotations from *Hamlet* about "fatting," "slave's offal," drinking "hot blood," and the ocean eating "flats," and concludes: "That the fundamental problem [in *Hamlet*] is the *oral sadism attendant upon oral frustration* is clear enough in the text of the play." She says further, in a marvelous example of Freudian gobbledygook: "The super-ego sadism turns upon the ego and destroys it when that ego is identified with the frustrating love-objects. Yet the sadistic super-ego, as we know, being unconscious, has its alliance with the destructive, hostile, aggressive id-impulses that in the oral stage manifest themselves in eating fantasies directed against the parents."[1]

I hope that what I am about to say about *Othello* will not amount to this sort of thing. But I find this reduction of Shakespeare's drama to psychoanalytic allegory, ridiculous as it is, no less repulsive than the reduction to *moral* allegory so popular among traditional Shakespearean scholars. Henry James once said of Hawthorne: "[he] is nothing if not allegorical, and allegory, to my sense, is quite one of the lighter exercises of the imagination . . . it has never seemed to me a first-rate literary form."[2] I find Shakespearean drama, and especially *Othello*, to be a first-rate literary form, and I do not think justice is done to it by those who equate Desdemona with absolute

[1] Ella Freeman Sharpe, *Collected Papers on Psycho-Analysis* (Hogarth Press, 1950), p. 211.

[2] Henry James, as quoted by Joan Bennet, "The Art of Henry James," *Chicago Review*, IX (Winter, 1956), p. 17.

innocence and purity, Othello with great and simple virtue flawed only by the belief that men are what they seem, and Iago with "motiveless malignancy," the principle of evil incarnate—shake well, and draw a didactic significance. S. L. Bethell, for example, says that

> Iago is so strongly associated with the diabolic that we are justified in interpreting his character in terms of demonology. All his stated motives may be genuine, but the deepest, as Bradley saw, is the desire to plume up his will. He is mastered by the sins which caused the angels to fall, Pride and Envy.[3]

The most recent and most elaborate study of this play, Robert Heilman's *Magic in the Web,* does a marvelous job with the poetry of *Othello* but concludes with a view of its dramatic pattern in just about such simple terms of moral allegory as I have described.

Good drama—Shakespearean drama—is far less abstract than that. It presents us with a single, unique, whole human action involving the interrelationship of complex characters acting upon each other with clear (if not always conscious) motivation. It is an analogue of experience and not a commentary on it. There are two branches of psychology that can help us to come to terms with such an art form: one is Gestalt psychology (the study of perception involving the integrated structures or patterns of experience)[4] and the other is Freudian or psychoanalytic psychology (the study of the subconscious motives for human behavior). Surely, there is some significance in the fact that the only reviewer to note the monumental importance of Freud's first publication (*Studies in Hysteria,* 1895) was Alfred von Berger, not a psychiatrist but director of the Imperial Theater in Vienna.[5]

The form of the action that *Othello* presents seems to have certain curious excesses, not satisfactorily accounted for as a total complex by any interpretations I have seen. Briefly these excesses seem to be: (1) Iago's villainy, and the enormous consequences of his originally vague (but apparently limited) desire for revenge on

[3] S. L. Bethell, "Shakespeare's Imagery: The Diabolic Images in Othello," *Shakespeare Survey 5* (1952), 75. This view of the play is at least superior to G. B. Shaw's "tested by the brain it is ridiculous; tested by the ear it is sublime," or G. Wilson Knight's "a thing of meaningless beauty in the Shakespearean universe."

[4] Cf. Ian Rawlins, *Aesthetics and the Gestalt* (Nelson, 1955).

[5] Ernst Kris, *Psychoanalytic Explorations in Art* (International Universities Press, 1952), p. 265.

Othello; (2) Desdemona's attraction to Othello, her elopement with him—and then her extraordinary interest in Cassio's reinstatement, a crucial fact that convinces Othello of Iago's charges; and (3) Othello's readiness to believe the absurd, impossible insinuations of Iago on the basis of inadequate evidence. If this play does have a smooth, cause-effect relationship of all the parts in presenting a unified whole, we must find some satisfactory account of these apparent excesses that will square with human behavior. And in this respect I believe psychoanalytic theory can be used with good effect. It is an axiom of Freudian thought that whenever in human behavior response to a stimulus is far more powerful than the nature and intensity of the stimulus seems to warrant, we are justified in searching for reinforcements to that stimulus from the subconscious. It is precisely in search for such subconscious reinforcement that we come to an understanding of the motives that explain the interrelation of Othello, Iago, and Desdemona.

I believe that the text provides sufficient evidence for seeing Iago's initial drive for revenge as activated by his tremendous disillusionment in failing to be promoted. On this point there need no longer be much controversy—even so eminent a traditional scholar as G. B. Harrison recognizes that:

> Iago's motives for the whole tragedy are laid bare in the first bitter outburst to Roderigo (I, i, 8–33).... Iago is a professional soldier. He has climbed up from the lowest rank until he is now within reach of the top. He naturally expects that Othello will select him as second in command.... When he is rejected he suffers [a] bitter blow.... At any cost he will ease the writhing anguish of his hurt vanity by destroying those who have wronged him and all that they value most.[6]

The fact that years of loyal and valorous service have gone unrewarded explains much more than "motiveless malignancy" the despairing cynicism of Iago's injunctions to Roderigo: "Virtue! a fig! ... Put money in thy purse" (I, iii, 319, 338). It is a striking fact that Iago has with everyone in the play a solid reputation for honesty. And when his honest service is ignored at the very moment he expected it to be rewarded and when the reward goes to a mere courtier who had served as go-between in Othello's courtship of Desdemona,

[6] G. B. Harrison, ed., *Shakespeare: The Complete Works* (Harcourt, Brace, 1952), p. 1058.

should we be surprised that the moral basis of his universe collapses and he turns to villainy?[7]

But Iago is not merely a simple honest soldier who has been disillusioned—he is a man with an uncanny, shrewd insight into the characters about him. His revenge succeeds, nay *exceeds*, because he works on Othello where he is weakest—in his relationship with women. To assert, as Mark Eccles does, that "Iago could have deceived *anyone*"[8] is to be blind to Othello's peculiar vulnerability—though, to be sure, it would be going too far to turn the tables and assert that *anyone* could have deceived Othello.

Evidence for Othello's sexual inadequacy is plentiful. "Vouch with me, Heaven," says Othello in his speech to the Senators upon Desdemona's appeal that she be permitted to accompany him to Cyprus,

> I therefore beg it not
> To please the palate of my appetite,
> Nor to comply with heat—*the young affects*
> *In me defunct*[9]—and proper satisfaction,
> But to be free and bounteous to her mind. (I, iii, 262–267)

And when the duke agrees but calls for haste ("You must hence tonight")

> Des: Tonight my Lord?
> Duke: This night.
> Othello: With all my heart. (I, iii, 277–278)

I will speak of Desdemona's "Tonight my Lord?" in a moment, but here it is Othello's "With all my heart" that interests me. I find it significant that this bridegroom of only a few hours should so eagerly welcome a postponement of the consummation of his mar-

[7] There may be something to the view that "Iago's destructive hatred can be traced to his struggle against his feminine identification with Desdemona; repressed homosexuality drives Iago to destroy Desdemona, his rival." M. Wangh, "Othello: The Tragedy of Iago," *Psychoanalytic Quarterly*, XIX (1950), 202. I do not believe it is necessary to accept this (or the manifold variations on it that one finds in the psychoanalytic periodicals) in order to see the play as a successful gestalt of dramatic interpersonal relationships.

[8] Mark Eccles, ed., *Othello* (Crofts Classics, 1946), p. ix.

[9] Italics mine. There is no point in going into the textual problem here. We are safe in accepting G. B. Harrison's reading and his gloss, "in me the passion of youth is dead" (p. 1066).

riage.[10] Of course, this may be merely the language of diplomacy, but it explains so much and is so thoroughly confirmed by subsequent speeches and actions that it must be taken at face value. Othello himself, considering in Act III the possibility of Desdemona's unfaithfulness, lends support to it:

> Haply for I am black
> And have not those soft parts of conversation
> That chamberers have, *or for I am declined*
> *Into the vale of years*—yet that's not much—
> She's gone. (III, iii, 263–267)

There is a clown in this play who dismisses the musicians hired by Cassio to play beneath Othello's bedroom: "the General so likes your music that he desires you, of all loves, to make no more noise with it" (III, i, 11–13). But as we know from *Twelfth Night* "music is the food of love," and we are instructed in *The Merchant of Venice* that

> That man that hath no music in himself,
> Nor is not moved with concord of sweet sounds,
> Is fit for treasons, stratagems, and spoils.
> The motions of his spirit are dull as night,
> And his affections dark as Erebus.
> Let no such man be trusted. Mark the music. (V, i, 83–88)

Desdemona hath music in herself, as we know from the "Willow Song," but it would seem that the motions of Othello's "spirit are dull as night, and his affections dark as Erebus."

We must consider, too, that Iago *succeeds* by basing his entire plot on the sexual weakness in Othello:

> It cannot be that Desdemona should long continue her love to the Moor.... She must change for youth. (I, iii, 339, 346)

> When the blood is made dull with the act of sport, there should be, again to inflame it and to give satiety a fresh appetite, loveliness in favor, sympathy in years, manners, and beauties, all which the Moor is defective in.... Very

---

[10] That the marriage was not consummated at the Sagittary inn before Othello was interrupted by Iago would seem clear from Othello's speech to Desdemona in Cyprus:

> Come, my dear love,
> The purchase made, the fruits are to ensue;
> That profit's yet to come 'tween me and you." (II, iii. 8–10)

nature will instruct her . . . and compel her to some second choice. (II, i, 225–231)

> "Her appetite shall play the god
> With his weak function" (II, iii, 325–326)

It requires no psychoanalytic gobbledygook to support the observation that a man who is sensitively aware of his sexual inadequacy and who is married to a passionate young woman would be quick to fear the worst—particularly when he knows his wife is capable of lying to him (as Desdemona does concerning the handkerchief) and when she so vehemently pleads on behalf of a handsome young soldier (Cassio is the only male in the play, apart from Othello, actively engaged in heterosexual relations) who has been legitimately deprived of his post for malfeasance of duty in time of war.

We come now to Desdemona. Let us examine this complex "imitation of an action" from her perspective. This

> . . . maid so tender, fair, and happy
> So opposite to marriage that she shunned
> The wealthy curled darlings of [Venice] (I, ii, 66–68)

was so stirred by the masculinity of the big, black Moor that she was more than half the wooer in their courtship, eloped with him in violation of all the social mores of her day, and then boldly stood up in the Senate not only to defend her action but to plead for an opportunity to consummate the marriage:

> If I be left behind,
> A moth of peace, and he go to the war,
> The rites for which I love him are bereft me
> And I a heavy interim shall support
> By his dear absence.

When told that they "must hence tonight" it is *she* who is shocked and disappointed: "Tonight my Lord?" And when Othello is so obviously eager for postponement, is there any wonder that she should begin to doubt herself. While waiting for Othello in Cyprus, she turns to Iago with a strange request in such a time of stress: "What wouldst thou write of me, if thou shouldst praise me?" (II, i, 118). And when Iago demurs she insists: "Come on, assay. . . . Come, how wouldst thou praise me?" (II, i, 121, 125). Iago assays:

> She that in wisdom never was so frail
> To change the cod's head for the salmon's tail,

> She that could think and ne'er disclose her mind,
> See suitors following and not look behind,
> She was a wight, if ever such wight were—
> Des: To do what?
> Iago: To suckle fools and chronicle small beer.
> Des: O most lame and impotent conclusion! (II, i, 154–161)

This woman whose sexual expectations have been so passionately aroused and who has been disappointed by another "lame and impotent conclusion," naturally seeks, ever so tentatively and no doubt subconsciously, a relationship with someone else—as Iago clearly predicted. And the most eligible alternative is, of course, Cassio. "I give thee warrant of thy place," she tells him, after he has been dismissed for drunkenness on guard duty:

> Assure thee,
> If I do vow a friendship, I'll perform it
> To the last article. My lord shall never rest;
> I'll watch him tame and talk him out of patience;
> *His bed shall be a school,*[11] his board a shrift;
> I'll intermingle everything he does
> With Cassio's suit. Therefore be merry, Cassio,
> For thy solicitor shall rather die
> Than give thy cause away. (III, iii, 19–28)

This is certainly an excess interest—the affection for Cassio is powerfully expressed and so, too, is the aggression against Othello. Desdemona is true to her word. She unmercifully goads him to receive Cassio until he bursts out: "Prithee no more! Let him come when he will. I will deny thee nothing" (III, iii, 75–76). And later, after she lies to the incensed Othello about the precious lost handkerchief, we have that powerfully intense dialogue that mounts to the point where they are practically shouting "Cassio! The handkerchief! Cassio! The handkerchief!" at each other, until Othello can no longer bear it and rushes away in an anguished panic (III, iv, 83–96).

Othello's agonized awareness of his own failure in satisfying Desdemona makes him prepared to believe the worst about her passion on behalf of Cassio. Iago has set off a psychological cause-effect chain reaction that inevitably destroys them all in the grand, bloody conclusion.

[11] Italics mine.

There remain, of course, problems—the time scheme, Iago's references to the possibility of his having been cuckolded, the curious line about Cassio being "almost damned in a fair wife," and others. But the main lines of action shape up a marvelous dramatic gestalt; the apparent excesses disappear as we adopt a proper psychological perspective.

Papers of the Michigan Academy of Science, Arts, and Letters
Vol. XLV, 1960 (1959 Meeting)

MORAL AND TRAGIC RECOGNITION: THE UNIQUENESS OF *OTHELLO, MACBETH,* AND *KING LEAR*

EDMUND H. CREETH

The University of Michigan

At the dawn of Elizabeth's reign, the obscure playwright W. Wager conceives for the hero of his moral play *Enough is as Good as a Feast* a death unprecedented in the play's tradition but characteristic of Elizabethan tragedy. Wager's protagonist dies under the spiritual care of a chaplain—whose name is Ignorance. Wordly Man perishes unenlightened and therefore unredeemed, his mind to the last "corrupt and il."[1] Having seen a tempter with a disguise and false name draw out his worser nature by making greed look legitimate and moderation vicious, we are left unsatisfied when Wordly Man falls deluded. In drama, deception asketh an end. There is properly no dénouement in Wager's play, for the knot of Wordly Man's folly remains tied.

Contrary to what we may believe from our familiarity with Shakespeare, the typical Elizabethan tragic hero, acting in ignorance, does not himself perceive at death the meaning of his experience, though the moral lesson of his fall may be patent to us. The earliest English tragedies moralize by the method of the fearful object lesson. The title figure in the germinal *Tragedy of Gorboduc* (1562) is annihilated by his rebellious subjects without revealing the least understanding that his death, the deaths of his sons, and the chaos in Britain stem from his initial folly in harkening to bad counsel and dividing the rule of the state. Both the first edition of the *Mirror for Magistrates* and the first Englishing of a Senecan tragedy were published within a year of Elizabeth's accession—a political event to which the birth and early character of English tragedy owe much—and neither lays groundwork for the dramatization of spiritual regeneration through knowledge. As Elizabethan tragedy unfolds,

[1] Henry E. Huntington Facsimile Reprints, II, introductory note by Seymour De Ricci. Wager's other extant play, *The Longer thou Livest the more Fool thou art,* ends similarly.

playwrights remain insensitive to the possibilities for emotional catharsis in the spectacle of tragic recognition, as, for example, in the last scene of *Romeo and Juliet*.[2] They do not moralize by the positive method of depicting spiritual gain through rejection of error, except that criminals in the domestic tragedies ordinarily repent before punishment.[3] In general, any struggling through to clearer vision remains equally foreign to the Jacobean tragic stage. "In the world of Marston, Chapman, Middleton, Tourneur, Webster, Beaumont, Fletcher, Ford," Una Ellis-Fermor has generalized, "there is crime and suffering, often of Aeschylean depth, but no hint of the Aeschylean resolution of evil through the education that suffering brings."[4]

Against this background, the three great tragedies with which Shakespeare inaugurates the Jacobean period stand as a group in sharp relief. Each guilty of tragic misjudgment, Othello, Macbeth, and Lear each endures a saving realization of his error. Knowledge enables them to assert their best qualities in the face of catastrophe generated by their folly. Their experience is in this respect unlike that of any other tragic heroes in the work of Shakespeare or of his contemporaries, and it is noteworthy that the plays which they distinguish should have been composed consecutively in the brief period 1604–1606. The uniqueness of *Othello, Macbeth,* and *King Lear* lies not simply in their employment of tragic recognition, however, but in its employment within traditional dramatic structure. I wish to show in this paper that the broad outlines for the spiritual careers of these protagonists belong, surprisingly, to pre-Elizabethan dramatic convention. The establishment of this relationship will be a step toward understanding the brief emergence of tragic recognition in Shakespearean tragedy.

A recognition scene, leading to contrition and spiritual restoration, had formed a regular part of Tudor moral interludes structurally related to Wager's *Enough is as Good.* Like Wager, their authors lay initial stress upon the virtue, however flawed, of the protagonists. They introduce a tempter or tempters whose true nature will be dis-

[2] See Bertrand Evans, "The Brevity of Friar Lawrence," *PMLA,* LXV (1950), 841–865, for an argument that fate works through the ignorance of the characters.

[3] As a rule, the repentances come from the moral tracts on which the plays are based.

[4] *The Jacobean Drama* (London, 1936), p. 18.

guised from him and who will lead him, in a scene of temptation, to suppose that vice is not vicious and hence to embrace it. John Skelton's hero Magnificence, for example, exhibits kingly virtue in the form of Aristotelian moderation. Fancy, representing allegorically an aspect of the hero's mind, presents himself under the name of Largesse. Through trickery and false logic, Fancy causes Magnificence to put himself in the hands of corrupt ministers. Ignorant of their true characters, the King squanders his wealth, boasts De Casibus fashion that he rules Fortune, shows his spiritual debasement by delight in the babbling of Folly. But in Skelton's plan, which is not original with him, delusion is the material for moral awakening. A significant stage direction marking the point at which the King's mind begins to grasp the truth reads, "Here goth Foly away." "Why," exclaims Magnificence to Fancy in nascent realization, "who wolde haue thought in you such gyle!" He learns the real name of Largesse and the extent of his own misconduct. "Alasse my Foly! alasse my wanton Wyll," he weeps.[5] After adversity and despair, he regains integrity, and solvency, through penance and the grace of God. Redress clothes him appropriately.

Recognition of moral truth occurs regularly in the work of other Tudor moral dramatists between 1498 and 1558, and always at the same point in the plot—that is, after depiction of initial innocence, temptation, and sinful folly. It is stirred in a variety of ways. Wit, in the play by John Redford, peers at himself in his glass of Reason and cries, "gogs sowle a foole/a foole by the mas!"[6] He has acquired through idleness the fool's cap and smudged face of Ignorance. In Henry Medwall's *Interlude of Nature* (1498), Reason himself admonishes and awakes the hero, Man.[7] Good Counsel, revealing the trickery of the Devil, that author of lies, performs the same service for Lusty Juventus, who falls to the ground and cries, "Alas, alas, what haue I wrought and done? . . ./Thus in my first age to worke myne own destruction."[8] Among protagonists who are concrete individuals, Melebea, persuaded by the bawd Celestina that surren-

[5] Ed. R. L. Ramsay, E. E. T. S. Extra Series 98, p. 58, p. 64.

[6] *Wit and Science*, ed. A. Brown, W. W. Greg, F. P. Wilson, Malone Society Reprints (1951), 11. 828 ff.

[7] Ed. A. Brandl, *Weltlichen Dramas in England Vor Shakespeare* (Strassburg, 1898), p. 146.

[8] *Lusty Juventus*, Tudor Facsimile Texts, E_1 *v*.-E_3 *v*.

dering herself to Calisto would be Christian charity rather than sin, is stirred to moral awakening by her good father Danio. "All pensyfe and sore abasshyd" after hearing his words, she declares, "Alas dere fader, alas, what haue I done?/Offendyd god as a wrech unworthy."[9] (This recognition, a contribution of the English morality tradition, is of course wholly independent of *La Celestina*, the playwright's source.) To the heroine of Lewis Wager's Edwardian morality *Marie Magdalene* comes a personified abstraction called Knowledge of Sin, representing something opposite to W. Wager's Ignorance.[10] King Assuerus, to cite finally a moral play based on the *Book of Esther*, learns the true merit of the Jews, whom he had been led to persecute, and the true character of their calumniator, Aman. To this tempter he exclaims, much as Magnificence had to Fancy,

> O kaytiffe most crafty o false dissembler,
> With thy flattering tonge thou haste deceyued me,
> All noble princes by me may be ware
> Whom they shall truste and put in auctorite.[11]

Such recognition of folly enables each of these protagonists to regain the particular state of virtue proper to him, as a rule defined upon his first appearance.

The structure of these Tudor interludes, including the scene of recognition, stems from fifteenth-century drama. Its basis is theological. Instead of a noble prince, the original central figure is universally representative, Mankind or Man. His career is an aberration between two states of grace, the first owing usually to the sacrament of baptism and the last to the sacrament of penance. Awakened by Christian admonition to the fact that he has been deluded by infernal counsel, he fights off despair and proceeds, through regular steps of contrition, confession, and satisfaction, back to grace. The phases of innocence, temptation, sin, recognition, and restoration dramatize the universal conflict between good and evil for the soul of a flawed protagonist representing all men. The plot, which may be

[9] *Calisto and Melebea*, ed. W. W. Greg, Malone Society (1908), ll. 930 ff.

[10] Ed. F. I. Carpenter (Chicago, 1904). Knowledge of Sin follows a character representing *Old Testament* law. Christ banishes the tempter Infidelity and raises the contrite sinner. See ll. 1027–1454. The remainder of the play follows Luke.

[11] *Godly Queen Hester*, ed. W. W. Greg in Bang's *Materialien*, V (1904), ll. 971–974.

called for convenience the Struggle with Evil, is the ritual of the individual life. The Lancastrian plays embodying it are those of the Macro collection of manuscripts: *The Castle of Perseverance, The Morality of Wisdom,* and *Mankind.*[12] In the first of these, the plot is worked out twice, once for the period from birth to maturity and once again, with appropriate variations, for old age.

The design of the Struggle with Evil becomes a convention of Tudor drama, outlasting in the theater the theological matrix and sometimes the allegorical techniques with the aid of which it took form. Inherently didactic, it serves in a variety of new contexts: religious polemics, the upbringing of youth, political satire, the conduct of magistrates. But when it enters the world of Elizabethan tragedy, as we have seen, it suffers distortion. Benign in its implications, this plot instructed the audience by presenting finally a pattern of behavior which the viewer might emulate in order to purge himself of his own sins. Elizabethan tragedy moralizes by the opposite method—by an appeal to fear. How could recognition of moral error, originally intended to fit Mankind for divine mercy, be retained in company with a just reward for the misdeeds of an Elizabethan heir to Mankind's role, a Worldly Man, a Cambises, an Apius, or a Philologus? Playwrights in the time of W. Wager and Thomas Preston had not the tragic eyes to see that salutary recognition of folly is fit accompaniment for the retribution earned in folly. The old element of delusion, traditionally and logically bound up in the structure, survives but remains unresolved.

It is easy to show, however, that the spiritual careers of Othello and Macbeth follow in full the general design for a protagonist's experience which we find in the Macro plays and their Tudor parallels —perhaps the most common plot in pre-Shakespearean English drama.

The protagonist of *Wisdom,* represented analytically, begins in perfect spiritual union with Wisdom, who is Christ. Over her white cloth-of-gold, signifying reason, Anima, the soul of man, wears a mantle of black, signifying the potential for sin in the passionate side of man's nature: "Thus a sowle ys bothe fowlle & fayer." Sensuality, "þis dyrke schadow I bere of humanyte," takes over, we are told, when suggestion deludes the mind. In the next scene Lucifer, who

[12] *Macro Plays,* ed. F. J. Furnival and A. W. Pollard, E. E. T. S. Extra Series 91.

hates man out of envy, introduces himself in soliloquy and outlines his plan to deceive the protagonist and lead him to damn himself. In the third scene Lucifer, disguised, tempts man's Mind, Will, and Understanding to consent to mortal sin by persuading them that their holy life is hateful to God. But in the last phase of the action Wisdom confronts the mind of man with his corrupted soul:

> Se howe ye haue dysvyguryde yowur soule!
> Be-holde yowur selff; loke veryly in mynde![13]

Taken aback, Mind trembles to realize that by violating the mutual love of man and Christ he has put himself in danger of hell, but dread dissolves into contrition. Sorrowful in heart, the composite protagonist makes confession and satisfaction, and receives absolution. This scene represents the theological prototype of the scenes of moral awakening of interlude protagonists like Magnificence, Wit, and Assuerus. Man presently returns dressed as at first, that is, symbolically restored like Magnificence to his original state. "Fyrst ye were reformyde by baptyme of ygnorans,/and clensyde from the synnys orygynall," Wisdom explains, "Ande now ye be reformyde by þe sakyrment of penaunce,/Ande clensyde from the synnys actuall." The exemplary return to initial virtue was originally motivated by what Wisdom calls "þe recognycion ye haue clere."

Compare the formal context of recognition in *Othello.* Married at the outset to perfect virtue incarnate in Desdemona, Othello himself accurately describes his initial condition:

> I must be found.
> My parts, my Title, and my perfect Soule
> Shall manifest me rightly. (I, ii, 30)

But his baptismal innocence of Venetian intrigue renders him susceptible. Iago, by suggestion, draws out the Moor's primitive passion and gives it sway over his reason. Othello's blood begins his "safer Guides to rule," and at the close of the temptation scene (III, iii) he decides to do violence upon his good wife as if in the cause of justice. He mistakes evil for good, his spiritual degeneration manifest in his uncontrolled behavior and in his speech—"(pish) Noses, Eares, and Lippes." The final scene restores Othello's old command

[13] *Macro Plays,* p. 65.

of the situation and his old eloquence, as when he coolly met Brabantio's wrath and defended his marriage before the Duke.

Othello is chided to moral awakening by Emilia: "Oh Gull, oh dolt,/As ignorant as durt. . . ." "Moore, she was chaste: She lou'd thee, cruel Moore. . . ." Knowledge frees him from Iago's domination. It leads him through despair to contrition, confession, and satisfaction. Wisdom urged, "But very contrycyon. . . . A tere of the ey, with sorow veray, . . . Go, seke þis medsyne, soull." Othello's eye drops tears as fast as the Arabian trees their "med'cinable gum." He confesses that, perplexed in the extreme, he threw a pearl away. He makes satisfaction appropriate to the Renaissance tragic theater.

We can readily imagine the play which Thomas Preston or such another might have made from Cinthio's tale of the Moor, the Ensign, and Disdemona. He would have assigned to the Ensign the morality role of tempter and a deceptive name, perhaps Honesty. He would presumably have magnified the Moor's initial merits according to dramatic convention. But we may be certain that the play would have ended with a flourish of mundane justice for the Moor and with no tragic recognition of his error. The Moor in the *Hecatommithi* is banished for his cruel treatment of his Venetian wife and later dies deservedly at the hands of her relatives. He never discovers, so far as one knows from Cinthio, that his consent to the murder of his beloved was folly. Our Elizabethan would hardly have enlightened and restored him as prelude to punishment for his misdeed. The Moor would die like Gorboduc, Apius, Cambises, or Worldly Man, an ignorant example to others, *com'egli meritava.*

One consequence of approaching *Othello* through the morality tradition is to see Othello's maltreatment of Desdemona as truly vicious. The crimes of Macbeth are of course more consciously so, and his recognition is thus different in quality. Infernal suggestion works upon his flawed soul to foster the delusion that he has metaphysical sanction and that crime will be to his advantage within the bounds of this life.[14] He is as thoroughly disabused as Othello.

After a moral struggle with Largitas equivalent to Macbeth's soliloquy of soul-struggle, Avaricia, in the second part of *The Castle of Perseverance*, lures Mankind down from the scaffold of virtues with

[14] For the view that Macbeth is the victim of a fatal hallucination see Sir Arthur Quiller-Couch, *Shakespeare's Workmanship* (New York, 1918), p. 18 ff.

the argument that an increase in worldly status is appropriate to old age. Generosity observes that those who seek worldly good sacrifice their souls, but Mankind is prepared, or thinks he is, to jump the life to come:

> If I myth al-wey dwellyn in prosperyte,
> Lord God, þane wel were me!
> I wolde, þe medys, forsake þee,
> & neuere to comyn in heuene.[15]

Though virtuous abstractions lament in choric fashion that Mankind is fallen in error, he did so of his own free will. His cry is always more and more and never enough.

Here I should like to return parenthetically to W. Wager. Up to this point in the plot, Worldly Man's experience parallels that of both Mankind and of Macbeth. Covetous tricks him into forsaking Contentation, alienates him from his companion Enough (a kind of Banquo), and causes him to pursue a course of pitiless acquisition. Tenant, Servant and Hireling revile the tyrant. But Wager, unlike both the *Castle* playwright and Shakespeare, withholds moral recognition. Worldly Man can only wonder at the soul-sickness brought him by God's Plague, a sickness about which there is nothing salutary. His comic Physician cannot minister to a mind diseased, and his spiritual guide is Ignorance. Believing to the last in the Devil's agent Covetous (alias Policy), Worldly Man falls dead in the futile effort to will his property to his heirs—a final gesture of worldliness. Presently Satan fetches him off to hell.

Struck by the dart of Death, dramatic ancestor of God's Plague, Mankind in contrast learns the bitter truths of his situation. The World, who had hired Covetise and in whom the protagonist had put his trust, laughs sarcastically at his dying appeal for aid. Bitterest of all is the sure knowledge that what he has acquired at such cost will go, not to his kin, but to strangers, to I-wot-neuere-whoo. At death, Mankind is able to address to the congregation a moral injunction that they profit by his experience. It is echoed in many moral plays and in Macbeth's penultimate speech. "I was ful wod," Mankind (or Anima, his soul) subsequently acknowledges,

> þat I forsoke myn Aungyl good
> & with Coueytyse stod
> tyl þat day þat I schuld dey.

[15] *Macro Plays*, p. 159.

In an Aeschylean finale, Mercy and Peace argue successfully that Mankind deserves pardon because of his purgatorial suffering, so that his three enemies, the World, the Flesh, and the Devil, do not after all win his spirit.

Exposed to successive evidence of the treachery of the instruments of darkness, Macbeth in Act V reveals concurrently a regenerative agony of spirit. Beginning to "doubt th'Equiuocation of the Fiend,/ That lies like truth" (V, v, 42), he pulls in resolution. By no means does his mind remain "corrupt and il," for the corrupting ignorance gives way to knowledge. Macbeth already knows that strangers, the seed of Banquo, rather than his own kin shall inherit the crown. He perceives that the sacrifice of his eternal jewel has cost him "Honor, Loue, Obedience, Troopes of Friends," values he well knows to be proper to old age. He has earned the very opposite of those golden opinions from all sorts of people which Macbeth the soldier properly required of life. Assuming armor, though "Tis not needed yet" (V, iii), Macbeth symbolically restores his outward appearance ("lapt in proofe"), as the protagonists of moral drama commonly do in the last phase of the action, and sets aside the ill-fitting robes of kingship, the state which for him is sinful. Macduff, in a sense not of woman born, reveals to him the full duplicity of the equivocating Weird Sisters:

> Dispaire thy Charme,
> And let the Angell whom thou still hast seru'd
> Tell thee, Macduffe was from his Mothers womb
> Vntimely ript. (V, viii, 13)

Before he raises his warlike shield, Macbeth, cowed momentarily, utters in reply a general moral imperative like that assigned by the *Castle* dramatist to Humanum Genus:

> And be these Iugling Fiends no more beleeu'd
> That palter with vs in a double sence,
> That keepe the word of promise to our eare,
> And breake it to our hope. (V, viii, 19)

The "vs" is not kingly; it implies the tragic hero's bond with the common experience of humanity.

Closely before or after *Macbeth* in time of composition belongs *King Lear*, its main plot involving Shakespeare's major study in the effects of error and moral realization. Obviously, the pattern is different from that of *Othello* and *Macbeth*. Lear's good qualities are not asserted at the outset. There are neither tempters nor temptation.

The outline of Lear's experience—initial vanity, *agon* of recognition, spiritual exaltation in humility—follows, however, the alternate plan lent by theology to the English moral drama of the fifteenth century. Outside the Macro group, the two anonymous moral plays definitely of that century are *The Pride of Life*[16] and *Everyman*.[17] Despite important differences in detail and quality, provenance, and date, these dramatizations of the Coming of Death may be classed together. Their message that we should be spiritually ready for death is no different from that of the more discursive Struggle with Evil. The point of attack is simply later. Although several Tudor editions attest the popularity of *Everyman*, no extant plays from the sixteenth century are patterned after the Coming of Death.

Each morality hero initially exhibits proud folly. The King of Life trusts his knights, Strength and Health. He reacts with something like Lear's unreasonable violence to his wife's suggestion that he remember his mortality, and he doubts her love. Everyman, a far meeker figure, naïvely supposes that Kindred, Fellowship, and Goods will sustain him in his sudden need. The text of *The Pride of Life* breaks off, but a Prologue tells us that the King endured a stormy encounter with Death, who "dredit no thing his kniȝtis," and that his soul was saved by intercession of Our Lady. "As for the agony of Everyman, it lies in large part in a growing spiritual recognition that the things of this world which he has possessed and depended upon are now failing him in his great need and are not worth possessing."[18] Knowledge takes Everyman to Confession and gives him a garment of sorrow. After communion and extreme unction, Everyman steps into the grave with Good Deeds, whom he had once slighted.

Lear's gradual discovery of the falseness of his evil daughters and the sustaining goodness of Cordelia, his growing vision of community

[16] Ed. O. Waterhouse, *Non-Cycle Mystery Plays and The Pride of Life*, E. E. T. S. Extra Series 104 and Brandl, *op. cit.*

[17] Ed. W. W. Greg in Bang's *Materialien*, IV and XXIV (1904 and 1909). The two editions by John Scot.

[18] W. Farnham, *The Medieval Heritage of Elizabethan Tragedy* (Berkeley, 1936), p. 205.

with loop'd and window'd raggedness form a clear parallel to the action of *Everyman.*[19] Moral recognition, then, foreign to tragic drama in Shakespeare's time except for *Othello* and *Macbeth,* recurs in *King Lear* within a received pattern of moral drama.

In three consecutive plays, Shakespeare sets within a tragic frame the designs for the spiritual life supplied originally to moral drama by Christian theology. On one level the action represents the typical experience of men. The result is the unique emergence of the tragic recognition of error. Our pent-up awareness of the protagonist's folly finds emotional outlet in his own realization. This and its attendant regeneration of his spirit offer as of old a pattern morally worthy of emulation. Disabused, the Renaissance descendant of typical man exhibits the most noble aspect of his humanity. But this demonstration now renders as intense as possible our sense of tragic loss. His material fortunes obey Elizabethan laws of a great man's fall and wretched end. His greatness of spirit, emancipated and exemplary, feeds death.

It is, in Iago's phrase, a pregnant and unforced position from which one maintains structural survival of the moral drama in *Othello, Macbeth,* and *King Lear.* Having demonstrated in many Tudor plays its ability to inform and give new meaning to literary materials, the morality tradition intersects Shakespeare's course of development as a tragic dramatist at a logical point. It does not participate significantly in his earliest plays. *Julius Caesar* and *Hamlet,* first tragedies of Shakespeare's maturity, however, prefigure a conception of tragedy in which the structure of moral drama might operate. Cassius is not quite a diabolic tempter of flawed Brutus, and Hamlet wrongly suspects that the spirit that he has seen may be the devil, who "hathe power/T'assume a pleasing shape" and who perhaps out of Hamlet's weakness and his melancholy abuses him to damn him. But the flawed nobility of Brutus and Hamlet is preparation for the dual character of the central figures in the next tragedies, who will share the stage with subordinate characters correspondingly divided into factions of good and evil. The Moor is the first of Shakespeare's tragic heroes susceptible and reasonable enough to play Mankind's role. The dramatist follows an interlude of perplexing dark comedy with *The Tragedy of Othello,* a play of classic economy and clarity, with Iago in the role of Vice. He reworks the career of the legendary Macbeth to make of it another version of the archetypal pattern of moral struggle, sin, and contrition, with three Weird Sisters from Holinshed as instruments of darkness. He brings to foolish Lear a final, beatific vision through the alternate design for English moral plays, making of Goneril and Regan, not tempters, but those upon whom the central figure initially depends, only to find them utterly false. In the Roman heroes of his last, more typically Jacobean tragedies, defects of character mingle inseparably with virtues. Shakespeare relaxes his preoccupation with the problem of evil and sets aside the traditional formulas with which he had stated and investigated it.

The recrudescence of traditional structure corresponds precisely to the Elizabethan twilight. Alfred Harbage argues persuasively that around the close of the Elizabethan period the impulse to justify the popular dramatic tradition is a motive in Shakespeare's creative life because of competition from the rival tradition of coterie playwrights.[20] Paul Siegel, similarly, finds Shakespeare engaged in a defense of Christian humanist values in the day of their peril.[21] For a dialectic reassertion of Christian humanism the structure of moral drama, with destructive, Machiavellian forces embodied in the villains, offers a natural vehicle. The early Tudor combination of theological structure with literary material was itself a manifestation of Christian humanism on the level of vernacular drama.

Published studies have already related the experience of Lear and Macbeth to that of morality heroes.[22] As for Othello, Bernard Spivack's recent *Shakespeare and the Allegory of Evil*[23] documents beyond doubt the rise of Iago from the conventional tempter, or Vice, and it is not a difficult corollary that Othello himself re-enacts in specific and concrete terms the career devised to stage allegorically the life of typical man and transmitted as a dramatic convention,

[19] See O. J. Campbell, "The Salvation of Lear," *ELH*, XV (1948), 93–109.

[20] *Shakespeare and the Rival Traditions* (New York, 1952), p. 295.

[21] *Shakespearean Tragedy and the Elizabethan Compromise* (New York, 1957), p. 87 *et passim*.

[22] Campbell ("The Salvation of Lear") describes *Lear* as "a sublime morality play." E. R. Hunter, "*Macbeth* as a Morality," Shakespeare Association *Bulletin*, Oct., 1937, pp. 217–235. Cf. Farnham, *Shakespeare's Tragic Frontier* (Berkeley, 1950), p. 79 and H. Craig, "Morality Plays and Elizabethan Drama," *Shakespeare Quarterly*, I (1950), 64–72; *English Religious Drama* (Oxford, 1955), pp. 377–389.

[23] New York, 1958.

along with the Vice, to Tudor playwrights. Each of the three tragedies has also been studied in terms of tragic recognition.[24] But it has not hitherto been seen that the morality structure and tragic recognition are bound up together and that both occur uniquely in the one group of consecutive tragedies from Shakespeare's pen.

It does not follow that the plays are allegories. Like Chaucer's *Troilus and Criseyde* they stem in part from literary transformation of allegory. The old, universal fable, not the need to understand each participant as abstract or supernatural, survives. Like certain Greek tragedies, notably *Oedipus Rex,* they fuse and mutually transform ritual and legend. Nor does it follow that we should imaginatively supply the Christian afterlife, though Othello and Macbeth, being Christians, speak of damnation and Lear looks beyond mortality. The Shakespearean tragic heroes are saved only in the sense that they perceive their folly and reject the evil forces in their lives before the terminal experience of death. They are in no sense damned; the dramatic ancestors of Iago and the Weird Sisters had indeed sought to damn the central figure, but they failed when he recognized his moral error.[25]

The discovery of morality structure in Shakespeare, though it by no means illuminates every aspect of these plays or revolutionizes our judgment of them, gives historical support to many contributions of criticism to their understanding and solves certain problems connected with Shakespeare's treatment of their sources. It demonstrates in a new way that he is a more truly native dramatist than his Jacobean fellows. Perhaps most important, the discovery enables us to re-evaluate the final, tragic moments in the plays as they might have been understood in Shakespeare's day. We shall not, having done so, agree that Othello is cheering himself up or that Macbeth, Faustus-like, spiritually disintegrates. At death, they have overcome the passions of jealousy and ambition, as Lear overcomes van-

[24] E. E. Stoll, "Oedipus and Othello: Corneille, Rymer, and Voltaire," *Revue Anglo-américaine,* XII (1935), 385–400, *Shakespeare and Other Masters* (Cambridge, 1940), p. 28. F. L. Lucas, *Tragedy in Relation to Aristotle's Poetics* (London, 1928), p. 93. A. H. Gilbert, "Scenes of Discovery in *Othello,*" *PQ,* V (1925), 119–130.

[25] Cf. Campbell, "The Salvation of Lear," and Siegel, "The Damnation of Othello," *PMLA,* LXVIII (1953), 1068–1078, revised as Ch. VI of *Shakespearean Tragedy,* cited above.

ity, through an *agon* of realization that is tragic because unavailing but also moral because its inherited structural context makes it exemplary. Othello brings back to the English stage the common awakening of a host of earlier, lesser protagonists. "O Foole," he cries in the voice of Mankind, of Magnificence, and of Wit staring at his begrimed image in Reason's glass, "foole, foole!"

Papers of the Michigan Academy of Science, Arts, and Letters
Vol. XLV, 1960 (1959 Meeting)

CATHOLIC TRADITION, PURITAN LITERATURE, AND EDWARD TAYLOR

NORMAN S. GRABO

Michigan State University

TWENTY years ago the poems of Edward Taylor were wrested from two centuries of obscurity, shaped into print,[1] and accorded an almost immediate status as the best poetry produced in colonial America during its first two hundred years.[2] Their protection from public judgment was easily, and teasingly, explained: the dying minister forbade their publication. That his wish was carried out is a matter of historical record. The teasing part of this final injunction is its motive: why did he insist that the poems not be published? To this history has as yet provided no answer.

But the absence of evidence has proved a spur to speculation, and in the twenty-odd years since the publication of the poetry, there has grown about Taylor's last prohibition a sense of mystery, a suggestion that Taylor set pen to paper only "secretly,"[3] and that therefore he must have had something to hide. The secret has not been hard to find: there must be in Taylor some belief or attitude that orthodox New England Puritans could neither have sanctioned nor permitted. On the one hand his theology has been attacked, accused of being a fundamentally pagan and Platonic conception flimsily covered with Calvinistic trappings;[4] on the other hand his temperament is, in its

[1] Thomas H. Johnson, *The Poetical Works of Edward Taylor* (New York, 1939); "Some Edward Taylor Gleanings," *NEQ*, XVI (1943), 280–296; "The Topical Verses of Edward Taylor," *Pub. Col. Soc. Mass.*, XXXIV (1937), 513–554.

[2] This, of course, is largely due to the influence of Perry Miller; see for example *The New England Mind* (Cambridge, 1939), pp. 288, 361.

[3] See for example Kenneth B. Murdock, *Literature and Theology in Colonial New England* (Cambridge, Mass., 1949), p. 167; *The Literature of the American People*, Quinn, Murdock, *et al.* (New York, 1951), p. 57; Perry Miller, *The American Puritans: Their Prose and Poetry* (New York, 1956), p. 301, and *The New England Mind: From Colony to Province* (Cambridge, Mass., 1953), p. 31.

[4] This challenge has been developed by Willie T. Weathers in two articles: "Edward Taylor and the Cambridge Platonists," *AL*, XXVI (March 1954), 1–31; "Edward Taylor; Hellenistic Puritan," *AL*, XVIII (March 1946), 18–26.

attraction for sensual and sensory images and in its kinship with the inclinations of Donne, Herbert, and the other poets called "metaphysical," too close to high church or Roman Catholic pleasures to fit our concept of "Puritan." One critic at least completes the inference, pointing out that Taylor's use of Biblical images "at times becomes as richly sensuous and personal as that of the Catholic and Anglo-Catholic poets who seem to be his inspiration."[5] One remembers that Crashaw's steps to the temple led directly to Rome and suspects that Taylor is tracing the same path.

The orthodoxy of Taylor's theology has been ably demonstrated elsewhere.[6] I wish to address myself to the question of Taylor's inherent Catholicism, not to deny it, because it is there, but to suggest that it is thoroughly possible to create a Puritan literature within a Catholic tradition, that Puritans in both Old and New England knew that it was possible, and that Edward Taylor knowingly undertook the task. The formal meditation, defined and popularized by such books as Lorenzo Scupoli's *The Spiritual Combat*, Saint Francis de Sales' *Introduction to a Devout Life*, and Saint Ignatius Loyola's *Spiritual Exercises*, has such a tradition, influencing Puritan expression without a doctrinal compromise on the part of either faith.[7]

The meditation, regarded during the first half of the seventeenth

---

The latter article contends that Taylor, taking his theology from the ancients, has merely converted Zeus to God, Aphrodite to Divine Love, Hercules to Divine Grace, Eros to the Soul of the Elect, and so forth.

[5] Mindele Black, "Edward Taylor: Heaven's Sugar Cake," *NEQ*, XXIX (June 1956), 172. The tendency to link Taylor with the Anglican metaphysicals is widespread, following the early comments of Johnson in *The Poetical Works*; see Austin Warren, "Edward Taylor's Poetry: Colonial Baroque," *Kenyon Rev.*, III (Summer 1941), 355–371; and W. C. Brown, "Edward Taylor: an American 'Metaphysical,'" *AL*, XVI (Nov. 1944), 186–197.

[6] See particularly Donald E. Stanford, "An Edition of the Complete Poetical Works of Edward Taylor" (unpublished Ph. D. dissertation, Stanford University, 1953), and "Edward Taylor and the Lord's Supper," *AL*, XXVII (May 1955), 172–178. The remarks of Samuel Eliot Morison are also pertinent to this consideration: *The Intellectual Life of Colonial New England* (New York, 1956), p. 240.

[7] This tradition and its importance to English metaphysical poetry has been fully discussed by Louis L. Martz, *The Poetry of Meditation* (New Haven, 1957). John Malcolm Wallace has demonstrated its influence in Traherne: "Thomas Traherne and the Structure of Meditation," *ELH*, XXV (June 1958), 79–89.

century "as an exercise essential for the ordinary conduct of 'good life' and almost indispensable as preparation for the achievement of the highest mystical experience,"[8] was a means of directing devout men to "thinke of heavenly things, not to learne but to love them. . . ."[9] The form the meditation assumed was not utterly inflexible but was generally marked by a series of ordered and predictable stages. First the "exercitant," as he is called in Loyola's *Spiritual Exercises*, attempts to fix his attention on some "heavenly thing," say some Scriptural incident, to secure as lively an apprehension of it as he can by an effort of memory and imagination. Then he submits the object of his contemplation to a lengthy, rigorous, and thorough intellectual examination until his understanding of it is complete. The next step is to judge or evaluate the object, after which the exercitant submits his evaluation to his will and affections, which are moved according to the attractive excellence or repulsive sinfulness of the object. The final stage is some expression of these moved affections—usually joy, love, praise, gratitude, or sorrow—in a colloquy with the Lord. The efficacy of such formal meditation in preparing for the symbolic union with the Lord in the sacrament of the Holy Eucharist is commonly urged by Catholic writers, and since the Lord's Supper is one of the two sacraments retained by the Puritans, it should not be surprising that they also retained the practice of meditation.

In 1648 the popular *Shorter Catechism* of the Westminster Assembly—the statement of doctrine on which Puritans both of Old and New England agreed—makes the practice of meditation a clear duty in approaching the Lord's Supper,[10] and a year later Puritans had their own handbook of meditation in Richard Baxter's *The Saints*

[8] Martz, p. 16.

[9] Cited in Martz, p. 15.

[10] *The Confession of Faith* (Boston, 1713), drawn up by the Westminster Assembly of Divines, states the duty as follows:

Q. 114. How are they who receive the Sacrament of the Lord's Supper to prepare themselves before they come unto it?

A. They that receive the Sacrament of the Lord's Supper, are, before they come, to prepare themselves thereunto, by examining themselves, of their being in Christ, of their Sins and Wants, of the Truth and Measure of their Knowledge, Faith, Repentance, Love to God and the Brethren, Charity to all Men, forgiving those that have done them wrong, of their desires after Christ, and of their new Obedience, and by renewing the Exercise of these Graces, by serious meditation, and fervent Prayers.

Everlasting Rest.[11] One of the most popular companions to the Bible that England has ever seen, *The Saints Everlasting Rest* enjoyed the careful attention of devout persons of all classes,[12] a fact due largely to the obvious catholicity of Baxter's spirit. Plato, Pico della Mirandola, and a string of Catholic Neoplatonists frequent the marginal glosses of the book, and once, citing Gerson's *De Monte Contemplationes*, Baxter exclaims: "Read this you Libertines, and learn better the way [of] Devotion from a Papist."[13] If Baxter differs from his Catholic predecessors, it may be in the greater emphasis he places upon the second stage of the meditation, the intellectual analysis, which he calls "Ratiocination, reasoning the case with your selves, Discourse of minde, Cogitation or Thinking, or, if you will, call it Consideration."[14] Like Loyola and De Sales, Baxter insists that the most important part of the meditation is the very end: love, not knowledge, is the way to eternal rest, but it must be a rational love. "Must not every thing," he asks, "first enter your judgment and consideration before it can delight your heart and affection?"[15] For Baxter such consideration or "reasoning the case with your selves" becomes "a preaching to ones self,"[16] a "Soliloquy" that takes the shape of a formal sermon, perhaps the most common literary form in colonial New England.

New Englanders knew Baxter. Three of them—prominent Boston divines and closely involved in the activities of Harvard College, to which Baxter had proved a munificent benefactor[17]—reflect the continuing concern for the meditative tradition in the New World.

Increase Mather, for one, insists that two duties are basic to the

[11] *The Saints Everlasting Rest: or, a Treatise of the Blessed State of the Saints in their enjoyment of God in Glory. Wherein is showed its Excellency and Certainty; the Misery of those that lose it, the way to Attain it, and the assurance of it; and how to live in the continual delightful Foretastes of it, by the help of Meditation....* (London, 1649). The fourth part is devoted to the method of meditation.

[12] Helen C. White, *English Devotional Literature [Prose] 1600–1640*, Univ. of Wis. Stud. in Lang. and Lit., No. 29 (Madison, 1931), pp. 261–262.

[13] Baxter, p. 662.

[14] *Ibid.*, p. 719.

[15] *Ibid.*, p. 607.

[16] *Ibid.*, p. 750.

[17] See Clifford Blake Clapp, *The Gifts of Richard Baxter and Henry Ashurst to Harvard College*, reprinted from the *Pub. Col. Soc. Mass.*, XX (Cambridge, Mass., 1918).

good life: self-examination and secret prayer.[18] The latter he identifies with meditation, the aim of which he says is "intimacy of Communion with God"; its method is primarily a careful, "particular" examination of conscience, proceeding like Baxter's soliloquy, and as Baxter suggests, expressed as a formal sermon. Such a review of one's sins stirs the affections to a "*godly sorrow* for sin" and also to "a hungering after Christ. The Soul must stand affected towards Christ as a hungry man doth towards food."[19] Fully aware that there is only one feast at which this "hungering after Christ" may be satisfied, Mather too stresses the importance of meditation in preparing for the Lord's Supper.

Elaborating his views in the "Epistle Dedicatory" of his *Meditations on the Glory of the Lord Jesus Christ,* Increase Mather describes the end of the meditative exercise as a "Facial Vision" of God. "This *Meditation,*" he insists, "should make [exercitants] willing to be absent from the Body, that so they may be present with the Lord." To this end the meditation must "be attended with a *Transforming Power,* changing them into the same Image" as the Lord Himself.[20] From his style one could hardly guess that Increase himself ever enjoyed the rapturous vision he described as the fitting end to the meditative process.

But the prolix Cotton Mather, Increase's son, apparently achieved a higher rung of the *scala meditatoria* than his father did, and from there perceived the value of meditation to the Lord's Supper much more intensely. In 1690 he issued *A Companion for Communicants,* designed, he tells us, to correct that "Paucity of Books among us, which might be Companions and monitors" to the Saints preparing for the sacrament.[21] "My advice to you," he declaims, "is, that you immediately set apart a time for due and deep thoughts on your Condition: and that you give not over those thoughts, till you find

[18] Increase Mather, *Practical Truths Tending to Promote the Power of Godliness: wherein Several Important DUTIES, are Urged, and the Evil of divers common SINS, is Evinced; Delivered in Sundry SERMONS* (Boston, 1682), p. 80.

[19] *Ibid.*, p. 135.

[20] Increase Mather, *Meditations on the Glory of the Lord Jesus Christ: Delivered in several sermons* (Boston, 1705), pp. ii–iv.

[21] Cotton Mather, *A Companion for Communicants, Discourses upon the Nature, the Design, and the Subject of the Lord's Supper. With devout methods of preparing for and Approaching to this* BLESSED ORDINANCE (Boston, 1690), sig. A_1^v.

your Hearts moved, melted, broken, unto a compliance with all the Articles in such an Holy Covenant as this."[22] These "due and deep thoughts" are, as we might expect, to be exercised "not *Rashly*, but with clear *Understanding*, and much Deliberation. When this is done," Mather concludes, "then *come away* to the table of your Saviour." Like Baxter and his own father, Cotton Mather stresses a soliloquy-like self-examination. "Deal with your own hearts," he dexhorts communicants, "as you would by a Slippery Fellow, who declines to let you know, what he is,"[23] and then reflect upon the actions and elements of the sacrament itself. It is at this last stage that Mather's style becomes, if not ecstatic or mystical, intensely emotional, even sensual. Creating a vivid image of Christ crucified, Mather demands that the communicant "Hear the echoes of his *Groans*, and look again into every one of His Grievous, Hideous *Wounds* and Imagine that you have like Joseph of Arimathea, His Dead Body all Ghastly, and Bloody in your arms."[24]

Perhaps the best example of such reflections upon the actions and elements of the Lord's Supper is Samuel Willard's *Brief Meditations*.[25] Ebenezer Pemberton, who introduces this posthumous publication, echoes Willard's high esteem for the exercise, pointing out that "the very Power and Capacity for meditation argues the Dignity of Humane Nature; and the right Exercise of it will advance the Soul to a divine and Angelick Perfection."[26] Willard finds this exercise a strenuous one. "Oh that I could get up my stupid Heart into some lively Apprehensions," he complains,[27] and when he succeeds, his utterances of love acquire an intensity more akin to Catholic mysticism than any we have yet encountered: "Is not this *Sacrament* appointed to be a Medium of Communion between Christ and his Spouse? Is it not for the expressing of their mutual Conjugal Love to each other? . . . Have I fixed my Choice upon him, and rejected all other Lovers for him? and shall I not be able to give a Reason of this Choice of mine? Can I indeed love him, and fix the Eyes of my Mind in Contemplation of his transcendent Excellen-

[22] *Ibid.*, p. 78.

[23] *Ibid.*, p. 107.

[24] *Ibid.*, p. 153.

[25] Samuel Willard, *Some Brief Sacramental Meditations Preparatory for Communion at the Great Ordinance of the Supper* (Boston, 1711).

[26] *Ibid.*, p. i.

[27] *Ibid.*, p. 133.

cies . . . ? If I truly believe on him, he is Preciousness it self unto me. Surely then I must be able to tell what it is that makes him so." Faith is not a Fancy, divine Love is not a brutish Affection, but most highly rational.[28]

Willard's rationalism time and again leads him in his *Brief Meditations* to an expression that is intensely personal, often frankly sexual, and unashamedly sensual in its appeal.[29]

Here, then, we have three arch-Puritans, the sturdiest supports of New England Calvinism, men whose faith has never been called Catholic, all aware of the Roman lineage of their devotional practice, yet completely agreeable to its adoption, regardless of its shady parentage. Edward Taylor belongs truly to this group; like Willard and the Mathers, Taylor knew Richard Baxter well, at least from the time he entered Harvard in 1668, and we know that at his death his library contained several Baxter titles.[30] When Taylor shipped for America he set down in his diary a record of his "exercising" from Scripture, a spiritual exercise that accords, so far as one may judge, exactly with Baxter's soliloquy—"a preaching to ones self."

This pattern of devotion recorded in the diary is more obvious in Taylor's later work. As minister to the frontier community of Westfield, Taylor prepared sacrament-day sermons to be delivered every six weeks or so on the occasion of administering the sacrament.[31] After composing the sermon for such an occasion, Taylor turned to write a preparatory meditation, usually based, he tells us, on the doctrine preached in that sermon. These meditations are what we know as Taylor's "Sacramental Meditations"; to study them as they were written—with their sermons—is to reveal Taylor's own modification of the Catholic tradition in which he worked. The sermons are relatively emotionless, intellectual discourses, moving inexorably from Text to Doctrine to Proof to Application, an order that parallels the very meditative procedure the Jesuits had propagated under the instruction of Loyola himself. Only one part of Loyola's formal meditation is lacking in Taylor's sermons—the final colloquy with the

[28] *Ibid.*, p. 143.

[29] *Ibid.*, pp. 26–27, 74–75, and 205.

[30] Johnson, *The Poetical Works*, provides an index to Taylor's library (pp. 201–220); see items 38 and 83.

[31] Edited as "Sacramental Meditations" by Johnson and others, the poems in the Yale manuscript are titled as follows: "Preparatory Meditations before my Approach to the Lords Supper. Chiefly vpon the Doctrin preached vpon the Day of administration."

Lord. This we find supplied in the poetic meditations themselves. Taylor ends his sermons with the exhortation to express one's love, joy, and gratitude to God in "Sparkling Songs" of praise.[32] His "Sacramental Meditations" are really his own response to the exhortation of his sermons. They do not record enthusiastic emotional outpourings, but reasoned and just reactions to rigorous intellectual analyses of "heavenly things, not to learne but to love them."

Increase Mather can talk of a transforming power that yields a mystic vision of the Lord, Cotton Mather of holding the dead body of Christ "all Ghastly, and Bloody in your arms," and Samuel Willard of the love of the Saint for Christ in language no less erotic than we find in Francis de Sales: in no case do we react with alarm. We see them not attracted toward Rome, but following the dictates of the Westminster *Catechism* and the encouraging example of prominent English Puritans like Richard Baxter. So should we see Edward Taylor. His poetry was not secret; private and personal, yes, but he had no more to hide than had his Boston colleagues, all of them probably sharing the sentiment of William Nichols, who grudgingly admitted in his 1701 English edition of De Sales' *Introduction to a Devout Life* that "notwithstanding the great and deserved aversion which this nation has to Popery, yet the Books of their Divines upon Devotional and Practical Subjects, have met with as favorable reception among Us, as if the Authors had been of a better Religion."[33] Undoubtedly Taylor and his fellow New Englanders, too, might have wished that their meditative tradition had been formulated by men of a better religion, but no matter. The meditation, compromising neither faith, drew the human soul out of the shadow of its elaborately constructed edifice of dogma and church polity, led it thence at least part way along the path that ascends the mystic mountain, and exposed it there naked in the radiant vision of its Divine Maker. The experience was intensely personal, its record most private, and the struggle to attain it exceedingly rigorous—an experience reserved for the strong in faith. Such a faith was Edward Taylor's; it found expression in a fine contribution to Puritan literature and inspiration in a rich Catholic tradition.

[32] I have discussed this at considerable length in "Edward Taylor's *Christographia* Sermons: Edited from the Manuscript, with a Discussion of their Relationship to his 'Sacramental Meditations'" (unpublished Ph. D. dissertation, UCLA, 1958).

[33] Cited in White, p. 115.

PAPERS OF THE MICHIGAN ACADEMY OF SCIENCE, ARTS, AND LETTERS
VOL. XLV, 1960 (1959 MEETING)

CANNONS NO CANON: POPE'S EPISTLE TO BURLINGTON

ROBERT G. SCHAFER
Flint College, The University of Michigan

ALEXANDER POPE was one of the more conspicuous English contributors to the lively eighteenth century discussion of architectural standards. Writing in elegantly proportioned verse, Pope made his main contribution through a vigorous poem which is now known as the fourth of the "Moral Essays," but which has also been variously titled "An Epistle to the ... Earl of Burlington," "On False Taste," and "Of the Use of Riches." Unfortunately, Pope's contemporaries regarded this poem as just another satire, thus failing to appreciate the aesthetic theory which the poet very evidently intended his verses to convey. Consequently the real contribution of one of Pope's most constructive works was very readily overlooked, nor has it been fully recognized at the present time. In this paper I should like to wade through the accumulated irrelevancies, coming up with what seems to me the clear meaning of the poem.

First, the irrelevancies. The "Epistle to Lord Burlington" created an immediate sensation when published in December of 1731. True to its author's reputation for waspish malice, however, it made its impact less on the grounds of its intrinsic merit than upon a misreading of its intent. For to its readers the poem appeared to be an outright attack upon the aesthetic sensibilities of one or more members of a group small in numbers but mighty in influence: the great magnates then conspicuously competing for the honor of creating the biggest, most impressive, and most magnificent private dwelling in England.

In support of this impression it will be recalled that by 1731 the building competition had gone so far that the great palaces of Blenheim, Seaton Delaval, Claremont, Stoneleigh, and Cannons had all recently been completed on the grand scale. Among others Holkham, Houghton, Castle Howard, and Wentworth Woodhouse were either being planned, or were being constructed, on a scale equally massive. In addition, throughout the country there was rising a host of lesser places in emulation of one or another of these models.

In these circumstances there would seem to be considerable justification for the public's ready assumption that the author of *The Dunciad,* in writing on false taste, only intended another satire, this time directing his spleen against a group of pompous builders. The general tone of the poem suggested some such purpose; the circumstances of the time favored it; the poet's reputation lent itself to such an interpretation; and, far from least, in the middle of the poem there appeared a detailed, realistic, and cutting description of a great mansion of the type being criticized.

It was the latter, the rollicking description of "Timon's Villa," which gave the poem its immediate notoriety. Presumably generic, this monument of tastelessness was an establishment where vast expense had created a "stupendous air" which, when analyzed, was seen to be composed of "huge heaps of littleness" scattered all around. Drawing close to give his readers a better look, Pope pictured extensive gardens wherein were squirting Cupids, ill-placed lakes, and an unpleasing symmetry of plots and groves. Especially ludicrous were these artificial groves, dotted as they were with "Trees cut to statues," all jumbled in with "statues thick as trees."

Pressing on to make his example more specific, Pope took his readers on a tour of the house. Bowed into its vast depths by the proud owner, the visitor would find in the study "books, not authors": here an Aldus imprint, there a Du Sueil cover. Neither Locke nor Milton—much less a modern author—profaned this room. Having learned so much the guests were summoned to the chapel. There, 'midst prayers, they heard the music jingling, saw the painted ceilings, and were soothed by "cushion and soft dean" 'til dinner chimes could end the gentle sermon. To table came a hundred persons, not to dine, but to witness "A solemn sacrifice perform'd in state." Past the regimented visitors flew numerous courses of unpalatable food, brought in small helpings by officious servants. Soon the guests, "Treated, caress'd and tired," took their leaves, cursing the lavish cost and lack of skill displayed equally throughout this great "Quarry above ground."

No sooner did the poem reach its public than London's gossips, ignoring the message it bore, set about guessing who might have inspired the portrait of Timon. With experienced acuity they soon reduced the prospects to one: James Brydges, the "princely" Duke of Chandos. Since the identification was a plausible one, it embarrassed both men exceedingly, and caused Pope to deny vehemently that Chandos, and his great new mansion of Cannons, were being described in the poem.[1] Despite his denials the damage was done and the identification, once having been established, persisted. Since that time, whatever Pope may have intended, Cannons, and "Timon's Villa," have been synonyms for architectural grandeur unrelieved by good sense in conception, or good taste in construction.

It is fruitless to hunt for the various small points in which the villa of Pope, and Cannons, do not coincide. The damaging fact was that certain larger similarities could not be blinked. This fact the public recognized, and refused to have invalidated. After all, it was only six years since Daniel Defoe had published his *Tour thro' the Whole Island of Great Britain*,[2] in which he had labeled Cannons England's most magnificent palace. In glowing terms he had praised the massive appearance, the outer beauty and the inner glory, of "the finest house in England."

Not only were Cannons' gardens unsurpassed, Defoe had written, but, within were marble halls which had been decorated by the finest Italian painters. The two chief beauties of the house, however, were the library, known to scholars for its wealth of printed books and manuscripts, and the chapel, for which the Duke maintained a set of chaplains and a full choir to perform "the best Musick, after the manner of the Chappel Royal, which is not done in any other Noble man's Chappel in Britain; no not the Prince of Wales', though Heir Apparent to the Crown." It was widely known that the great Handel had been choirmaster, organist, and composer to Chandos; in 1731 Dr. Johann Pepusch, only less well known than Handel himself, held these offices. Pepusch's trained orchestra of twenty-four—some said thirty—musicians furnished "excellent Musick" not only for His Grace's chapel, but also for his dining room; this too Defoe found a singular custom for an English household.

[1] The evidence indicating that Pope did not intend to attack Chandos has been exhaustively summarized by Professor George Sherburn in "'Timon's Villa' and Cannons," *Huntington Library Bulletin,* VIII (1935), 131–152; see also *The Poems of Alexander Pope,* Twickenham ed., III, ii (London and New Haven, 1951), xxvii–xxxiv, xli–xlii, 127 n. 2.

[2] Daniel Defoe, *A Tour Thro' the Whole Island of Great Britain,* II (1725). Defoe's account, from which the selections in the text are taken, is based upon an earlier description of Cannons published in John Macky, *A Journey Through England in Familiar Letters,* II (London, 1722).

Thus far in his description Pope seems, consciously or otherwise, to have followed Defoe. He could have gone further with the celebrated journalist, who had been particularly impressed by Cannons' guard of half a dozen Chelsea Pensioners. Clad in red livery these watchmen strolled the grounds at night, and regularly in their rounds gave "the Hour to the Family at set appointed Places and Times." An unfriendly critic could hardly have made Chandos' establishment appear more ludicrous. But "in a Word," the admiring Defoe had concluded, "no nobleman in England and very few in Europe, lives in greater Splendor, or maintains a Grandeur and Magnificence, equal to the Duke of Chandos." It is no wonder that, with this description of Cannons current in the public mind, Pope was unable to deny convincingly that he had been describing it under the pseudonym of "Timon's Villa."

Nonetheless, he tried. As the clamor continued, Pope in letters to private persons and to the newspapers, disclaimed any intention of picturing Chandos. "Why, in God's Name, must a *Portrait*, apparently collected from 20 different Men, be apply'd to one only?" Nor could Cannons be seen in Timon's Villa. Were Chandos' "Trees cut into *Figures of Men*?" Pope asked. Did he exclude "all *Moderns* from his *Library*?" "Is the *Musick* of his Chapel bad, or *whimsical*, or *jiggish*?"[3] Whatever the answers to these questions, the public regarded Pope's dwelling on such minor differences between the two mansions as irrelevant, as efforts to save the Duke's face or Pope's reputation, or, perhaps, as an attempt to avoid a suit for libel. In any case minor variations could be dismissed as products of poetic license, or even as thinly veiled efforts to dissemble. It was no use for Pope to point out the fact that, for example, Cannon's ceilings had been painted by Belluci and Zeeman, not Verrio and Laguerre, the artists of the poem. Useless it was, also, to say that Cannon's domestic staff numbered fewer than the poem's 100 persons: in praising Cannons, Defoe had peopled it with 120 servants, and no one had corrected him.

If Chandos had been regarded in 1731 as other than a benevolent and kindly person Pope's apparent satire might have been overlooked. But an ungracious and uncalled-for attack on a man reputed to have been a patron to the poet himself, and known to have been generous

[3] Sherburn, p. 134: the letter is probably, though not certainly, Pope's.

to other artists of the day, seemed particularly reprehensible. Hence, preferring to believe the worst of Pope—if not of his victim—the public refused to be undeceived by his protestations. As a result his poem was widely misinterpreted as a personal attack on yet another "Dunce." Thus it was that, having been made to appear a frivolous and ill-natured effusion, the poem failed at first in its immediate object, which was to call attention to the pressing need for raising the architectural and aesthetic standards of the time and to suggest specific means for doing this.

That this was Pope's primary purpose cannot now be doubted. Although he took the opportunity to lash out at these inviting new targets, the wealthy builders and collectors, he was mainly interested in establishing a new, and better, basis on which future builders could rest their architectural conceptions. Thus, rather than attacking specific persons, as his contemporaries thought he was doing, the poet actually had in mind three other objectives: to call attention to the need for improvement in standards of taste generally; to criticize some current aberrations in the application of the existing standards; and to recommend a different, and better, standard as the new model. Seen in this light his poem appears to be, not negative and destructive as generally assumed, but positive and constructive, although it is perhaps lacking in specific detail and it is certainly uneven in stress and argument.

Let us consider the structure of the poem. It falls, obviously, into three major parts. The first section, lines 1–98, contains a very general discussion—diffuse and rambling, it can be said—of taste, false taste, and no taste. In this part of the poem Pope lays about with gusto, attacking, praising, and blaming various contemporaries, some by name, others by implication, for their achievements as art patrons. Also, and confusingly, in this section he presents two of the three criteria on which he would like to rest the new aesthetic standard he is proposing.

In the poem's second section, lines 99–176, there falls the vivid description of Timon's Villa, the horrible example of all that was bad in the current practice. It was apparently only this part of the poem which the busy public found time to read, thereby missing the main point of the whole essay in verse, which came almost immediately after, in the third and final part, lines 177–204.

The fact that the first half of the poem is a rather confused mélange of negative and positive conceptions partly excuses an undiscerning public's failure to grasp the development of the poet's thought. Despite the lack of clarity, however, the introductory couplets, which are seemingly an attack upon the follies of misers, prodigals, and their agents, do establish a necessary foundation for the new aesthetic. They do this by putting taste and sense in proper relationship. According to Pope the basis for any aesthetic appreciation must surely lie in the "certain truth" that

> Something there is more needful than expense.
> And something previous e'en to taste—'tis sense.

This platitude, with which it is difficult not to agree, and whose significance therefore is easy to miss, is followed almost immediately by the second criterion for establishing a proper canon of building, namely, reference to Nature as the only suitable arbiter of taste. For his lovely and forceful expression of this point Pope has been given great credit, for it is in the paean to Nature, lines 47–98, that he helped to publicize the contemporary naturalistic style being developed by Lord Burlington, to whom, further to emphasize the point, the work was dedicated.

Still, in spite of certain felicities, if the poem had ended at line 98 it would very likely have made little impact on its public, and thus possibly have contributed only slightly to its author's reputation. The skilled satirist who wrote it was easily able to avoid this kind of failure. In order to point the contrast between the present, inferior standard and that ideal harmony he wished to advocate, he moved, at line 99, from the general to the particular. "At Timon's Villa let us pass a day . . ." he invited, thus, as things turned out, proceeding to distract his readers from his main argument.

Despite its diversionary effect the description of Timon's Villa serves a very useful purpose, and is therefore an integral part of the poem. That purpose was to enable Pope to make explicit, and comprehensive, his objections to all those lapses of taste he had previously alluded to generally, and piecemeal. Thus the Villa was made truly generic so that its description could be in the nature of a catalogue of the follies of various kinds committed by all sorts of people. The Villa was intended, in short, to be put forward as a canon, not of virtues, but of errors. It is only by chance that Pope's negative yardstick, so forcefully described, came to be closely identified with Cannons, for it could, with equal accuracy, have been equated with any one of half a dozen other great homes of the time.

If meaningless grandeur, represented by Timon's Villa, is to be no suitable canon for builders and architects, what then is a better one? This is the question used by Pope to mark the transition from the middle to the final section of the poem. Pope states it more gracefully:

> "Who then shall grace, or who improve the soil?"

And he immediately gives this answer:

> "Who plants like Bathurst, or who builds like Boyle."

These men were named in order to illustrate the central point of the essay, now introduced, that

> " 'Tis use alone that sanctifies expense."

Here then in line 179 appears Pope's perfectly serious suggestion that use—the principle of utility to be rediscovered by Jeremy Bentham nearly half a century later—was to be applied as an acceptable standard for improving the quality of contemporary building.

The use Pope had in mind, as he went on to make clear, was a practical one. He praised that lord who introduced new methods of farming, who pastured cattle in his deer park, who made plantations not for show but for the sake of future buildings and ships. These purposes, although designed for private benefit, contributed to the public good. They thus contributed equally to Pope's vision of building, landscaping, and improvement undertaken in consonance with the site.

In this last section of the poem it was also made clear that Pope's version of Nature meant more than pretty scenery, a stately prospect, or an unkempt wilderness. Earlier he had urged his readers to do what was necessary and desirable, but in the doing, he had begged, "Consult the genius of the place in all." The aesthetic quality of improving done in this spirit would then be based upon the harmony between building and surroundings, and in the balance struck between man's needs and nature's setting.

Pope brought his vivid lecture on the use of riches to a pleasant ending with a notable vision of a happier, busier, safer Britain whose energies would be devoted to works "imperial" and "worthy kings."

He enumerated such tasks as improving harbors, building better roads, worthier churches, and sturdier moles. Perhaps struck by the appalling thought that many of these necessary public structures might be conceived in the spirit of Timon's Villa, the far-seeing poet had continually repeated his basic theme; "In all, let Nature never be forgot."

To his distress, however, it was neither his new aesthetic nor his optimistic conclusion that made the strongest impression on his contemporaries. Nonetheless, in a few years, when the tumult caused by the satire had died away, it was seen that the canon of a more natural taste which Pope had so vigorously championed had been accepted to a remarkable degree, even though the principle of utility which underlay it had not been widely noticed.

PAPERS OF THE MICHIGAN ACADEMY OF SCIENCE, ARTS, AND LETTERS
VOL. XLV, 1960 (1959 MEETING)

HENRY FIELDING'S COMIC ROMANCES

SHERIDAN BAKER

The University of Michigan

HENRY FIELDING and "comic epic in prose" have become almost synonymous. But, though the words are undeniably Fielding's, is this really his own taxonomic term? As Ian Watt has recently pointed out, Fielding's remarks about the "comic prose epic," and perhaps his belief in his novels as epics, were actually rather incidental and certainly brief.[1] Fielding's own name for his new genre is "comic romance," and, though he obviously liked the term no better than "comic epic," a reading of *Joseph Andrews* and *Tom Jones* as comic romances, I think, will somewhat correct our views of these two books.

Our critical error arises from Fielding's own uncertainty[2] and from his suspicion of the word "romance." Fielding opens his Preface to *Joseph Andrews* by stating: ". . . it is possible the mere English reader may have a different idea of romance from the author of these little volumes. . . ." Let me repeat, it is with an "idea of romance" that he begins, not of a comic epic, setting the two little volumes of his kind of romance against the conventional one: the vast French romances in big gilt-edged folios. His models, which he thinks unfamiliar to "the mere English reader," are the Spanish and French works he enumerates in Book III: Cervantes, *Gil Blas*, "Scarron, the Arabian Nights, the History of Marianne and Le Paisan Parvenu." Fielding's Preface then follows a practice common even to the French romances he scorns: he dignifies his work by claiming that it belongs under the "Epic" heading.[3] He then employs a rather common ar-

[1] *The Rise of the Novel* (Berkeley and Los Angeles, 1957), pp. 248–251.

[2] Ethel M. Thornbury, *Henry Fielding's Theory of the Comic Prose Epic* (Madison, 1931), p. 105: "He does not make a distinction between the comic romance and the comic epic." William R. Irwin adopts this idea and phrase from Thornbury, though he cites her page 50 instead: *The Making of Jonathan Wild* (New York, 1941), p. 96. F. Homes Dudden then plagiarizes from Irwin (*Henry Fielding: His Life, Works, and Times*, Oxford, 1952, p. 330, n. 4).

[3] Thornbury, pp. 38–45, 66–67; Arthur L. Cooke, "Henry Fielding and the Writers of Heroic Romance," *PMLA*, LXII (1947), 984–994; G.-E. Parfitt,

gument that poetry doesn't make the epic, that prose will do, and goes on to speak contemptuously of "those voluminous works, *commonly* called Romances"—I italicize the antithesis, the one with which he began and which now continues on through his famous definition: "Now, a comic romance is a comic epic poem in prose. . . . It differs from the serious romance in its fable and action . . . it differs in its characters by introducing persons of inferior rank . . . whereas the grave romance sets the highest before us. . . ."

I have deliberately omitted the ruinous phrase: "differing from comedy, as the serious epic from tragedy," in which we find "serious epic" where the basic antithesis should lead us to expect "serious romance." Fielding's own uncertainty, in other words, has allowed his predicate nominative momentarily to send his critics, from Dobson on, down the wrong road. But the subject of his defining paragraph is, nevertheless, "comic romance."

Unlike history with its unreal people, Fielding argues, in Book III, his fictive models are "true histories." This is the term Fielding liked best and perpetuated in text and title—he even calls Hogarth a "comic history painter" (Preface). And this is the phrase made much of by Cervantes,[4] the phrase also prominent in Scarron's *Comical Romance*, to use Tom Brown's translation, a work on Fielding's list that has been, I think, rather consistently underrated. I find it highly significant that Fielding begins *The Opposition* with a dream prompted by "the comical romance of Scarron."[5] *The Opposition* was published just two months before *Joseph Andrews*—while *Joseph Andrews*, in other words, was still in process. We find Scarron's "comical romance," like *Joseph Andrews*, talking about actual history and upholding *Don Quixote* against "imaginary Heroes," against "fabulous Stories, which have no Foundation in History," against "*Cassandra's*, *Cleopatra's* and *Cyrus's*."[6] Indeed, Tom Brown's

---

L'Influence française dans les Oeuvres de Fielding et dans le Théâtre anglais contemporain de ses Comédies (Paris, 1928), p. 115.

[4] Bruce W. Wardropper, "The Pertinence of *El Curioso Impertinente*," *PMLA*, LXXII (1957), 592–593; Sidney E. Glenn, "Some French Influences on Henry Fielding," *Abstracts of Theses . . . University of Illinois, 1931* (Urbana, 1932), p. 11.

[5] *The Complete Works of Henry Fielding, Esq.*, ed. William E. Henley (New York, 1902), XIV, 323; Wilbur L. Cross, *The History of Henry Fielding* (New Haven, 1918), I, 229; Parfitt, p. 114.

[6] *The Whole Comical Works of Mons. Scarron*, 4th ed. (London, 1727), I, 123.

Scarron seems almost to speak from Fielding's own pages: "which I have told you somewhere before, in this true History"—"as you may have read in the second Part of this true History"—"you might have perceived throughout the whole Comical Adventures of this Famous History."[7]

Cross credits Cervantes as the source for Fielding's book-and-chapter divisions, facetious headings and transitions, "drollery of style," the mock heroic diction, the ridiculous detail.[8] Granting this, granting Scarron's own debt to Cervantes, I am convinced that Fielding filtered some of his Cervantes through Brown's Scarron, whose first chapter begins:

> Bright *Phoebus* had already perform'd above half his Career; and his Chariot having past the Meridian.... To speak more like a Man, and in plainer Terms; it was betwixt five and six of the Clock....

And ends:

> And whilst the hungry Beasts were feeding, the Author rested a while, and bethought himself what he should say in the next Chapter.[9]

Indeed, Fielding's narrative mannerisms are evident on a great many of Scarron's pages,[10] and he certainly seems to have suggested Fielding's term "comic romance."

[7] *Scarron,* I, 244, 293, 317.

[8] Cross, I, 322. An anonymous reviewer of *Tom Jones* in *The Gentleman's Magazine,* XX (March, 1750), 117–118, notes that Fielding is "imitating the manner of *Cervantes, Scarron,* and *le Sage,* in the titles of his chapters," quoted by Frederick T. Blanchard, *Fielding the Novelist* (New Haven, 1927), p. 44.

[9] Tom Brown has slightly heightened Scarron's mock-heroic passage, which begins simply "Le soleil...." Cf. the chapter beginnings of *Joseph Andrews,* I, viii; *Tom Jones,* VIII, ix; IX, ii; X, ii; XI, ix. Wayne C. Booth points out that Scarron's claim here of not knowing what is coming next, later underpinned by the admission that perhaps he has "a fix'd Design" (I, p. 48), is his unique contribution to self-conscious storytelling: "The Self-conscious Narrator in Comic Fiction before *Tristram Shandy,*" *PMLA,* LXVII (1952), 169–170. Fielding always takes the opposite direction, the one more common to Scarron, the one initiated by Cervantes, which simply directs the reader on to "what will be found in the next chapter." But Scarron uses the formula oftener and elaborates it further than Cervantes' simple statement (used 9 times in 126 chapters as compared with Scarron's 14 in 60).

[10] Booth, attending to those passages in which the author intrudes "to comment on himself as writer" (p. 165), overlooks the frequent brief phrases and asides which sound almost like Fielding himself. Sir Walter Scott believed that

Of Fielding's critics, only G.-E. Parfitt has done much more than to mention Scarron. Parfitt points to two battles royal at inns and their parallels in Fielding and asserts, correctly I think, that the climactic scene in *Joseph Andrews* (IV, xiv), in which Adams takes a wrong turn into Fanny's bed, is the most Scarronesque of all.[11] But Parfitt gives less than a page to Scarron, and, ignoring the strong stylistic similarities, insists more on differences than on resemblances. Indeed, Parfitt and Cross, and more recently Alan McKillop, in naming among Fielding's sources "the realistic fiction of France" (i.e., the works of Scarron, Le Sage, and Marivaux) have managed to conceal altogether the distinctly romantic cast of Fielding's two great novels.[12]

Scarron's central characters are a young man and woman of mysterious but undoubtedly noble background, disguised as actors, who are traveling chastely as brother and sister until the impediment to their marriage and to their rightful fortunes—they are, of course, hopelessly in love—is removed at last. This is hardly realistic, and it is suspiciously like the traveling companionship of Joseph and Fanny, perhaps even the source of their momentary brother-and-sister-hood. Within a boisterous context of picaresque adventures and prose fabliaux, in other words, Scarron accepts a great deal of pure romance. His main story follows romance from mysterious origin to noble end, and all his inserted tales are straight cloak-and-dagger.

Almost precisely the same may be said of Fielding.[13] It is highly

---

Fielding's mock-heroic passages imitated Scarron rather than Cervantes (Blanchard, p. 327). Sidney E. Glenn thinks Fielding indebted to Scarron for no more than his chapter headings (p. 17).

[11] Parfitt, p. 114. See also Cross, I, 314–315, 322, 330. Parfitt does not give analogues in Scarron for Fielding's climactic scene; there are two, Part I, Chapter xii, and Part III, Chapter iv (*Scarron*, I, 47–48, 301–302). But Scarron does not give this ancient theme from fabliaux—the mistaken turn into the wrong bed (see Chaucer's "Reeve's Tale")—sexual complications. Fielding pretty clearly also has in mind Don Quixote's encounter with Maritornes (III, ii) and perhaps something of Chapter xlviii in Part II. For the thematic importance of this scene in Fielding, see Mark Spilka, "Comic Resolution in Fielding's 'Joseph Andrews,'" *CE*, XV (1953), 11–19.

[12] Cross, I, 321; Parfitt, p. 111; Alan McKillop, *The Early Masters of English Fiction* (University of Kansas, 1956), p. 110. E. M. W. Tillyard has recently redressed the balance (*The Epic Strain in the English Novel*, London, 1958).

[13] In 1928 Oliver Elton pointed out the major flaw in the "comic-epic" view

significant that Fielding claims as universal truth, as the highest fictive realism, exactly those parts of Cervantes that would strike the modern reader as most romantic. Of "the shepherd Chrysostom, who . . . died for love of the fair Marcella," Fielding asks: ". . . will any one doubt but that such a silly fellow hath really existed? Is there in the world such a sceptic as to disbelieve the madness of Cardenia [sic], the perfidy of Ferdinand, the impertinent curiosity of Anselmo, the weakness of Camilla, the irresolute friendship of Lothario. . . ."[14]

The romantic framework of Fielding's two principal novels is indeed stronger than we suspect and is by no means all burlesque. Andrews, like Havelock the Dane, begins his high life belowstairs; both he and Jones begin obscurely and mysteriously. Both must adventure for extended periods of time away from their unobtainable ladies, and their constancy is in no way basically comic like Quixote's. Quixote is a man of fifty who fancies himself a young knight dedicated to a beautiful young lady so unobtainable as to be, really, nonexistent. Andrews and Jones are young modern Englishmen of low degree, with no illusions, whose ideal young love is quite real, and both reveal unconscious traces of knighthood. When wounded in "battle," for instance, both disclose the knightly clue—a remarkably white skin.[15] Andrews, for no functional purpose, is even an accomplished rider, able to bewitch money into the worldly pockets of Sir Thomas Booby with noble horsemanship. The framework of romance is more evident in *Joseph Andrews* than in *Tom Jones*: babies exchanged in the cradle, lovers taken as brother and sister, the gold amulet tied to

---

of Fielding, noting that *Tom Jones* "is nearer drama than to epic, for an epic has no secret": *A Survey of English Literature 1730–1780* (New York, 1928), I, 195. But Elton overlooked the fact that the drama had derived the "secret" from the general stock of fairy tale and romance, and the epic theorists have overlooked Elton. W. Dibelius recognized the basic romance pattern in Fielding: "Englische Romankunst," *Palaestra*, XCII (1911), p. 92.

[14] *Joseph Andrews*, III, i. In 1752 Fielding objects to these same passages (after the epic *Amelia*, published December, 1751). Writing of Charlotte Lennox's *Female Quixote* in his *Covent-Garden Journal* (No. 24, March 24, 1752), Fielding remarks that Cervantes "in many Instances, approaches very near to the Romances which he ridicules. Such are the stories of Cardenio and Dorothea, Ferdinand and Lucinda, &c." Quoted in Thornbury, p. 126.

[15] *Joseph Andrews*, I, xiv; *Tom Jones*, IV, xiv.

Joseph's arm, the strawberry mark, the gentle lineage. And the burlesquing of romance is more evident, too.

But Joseph must play Sancho to Adams' quixotic chivalry, and consequently Jones is much more the knightly hero, half comically, half ideally. He is the Quixote of his adventure, the benevolent idealist to Partridge's Sancho, though at times the roles reverse.[16] It is Jones who rescues the damsel in distress (Mrs. Waters, as Adams rescues Fanny and calls her, in fact, "damsel"), and who defends the aged Man of the Hill, with his own sword "stained with the blood of his enemies" (VIII, x). Jones's knightly magnanimity is not quite the wonderful quixotic thing that blesses Adams. Professor Crane has already noticed that Fielding "repeatedly depicts Tom, especially when he is talking to Sophia or thinking about her, in terms of the clichés of heroic romance."[17] And Fielding's intent seems serious. Jones's love is more romantic than Joseph's; his heroine is a "Sophia" not a "Fanny." Nevertheless, even Andrews, at least once, can speak the sublime diction of love: "O that I could but command my hands to tear my eyes out, and my flesh off!" (IV, xi). Cervantes' elaborate and delicate mockery of courtly love becomes in Fielding almost completely serious—comically counterpointed, to be sure, as in the Adams-Andrews dialogue from which I have just quoted, but never considered ludicrous in and of itself. Fielding's lovers, indeed, are at a center pretty much inviolate to the rough comedy outside. Fielding's comic romances, in other words, seem to deal with romance in two opposite ways, often managing a synthesis: they accept courtly love and much of its romantic sublimity, and they mock heroic adventure with the picaresque scuffling of low life.

But it is the romantic structure that sustains the story. No matter how Fielding intended his offcast footman and his outcast foundling to mock the adventuring knight, they complete the romantic pattern in the end: their mysterious and natural nobility turns out to be sanctioned by birth, they step up into the elevated position so long deserved, they win the lady in the bower and live happily ever after. This is comic only in Dante's sense: the hero's travels end happy and

[16] Cross, II, 205; Gustav Becker, "Die Aufnahme der Don Quijote in die englische Literatur," *Palaestra*, XIII (1906), 152–153.

[17] R. S. Crane, "The Plot of *Tom Jones*," *Journal of General Education*, IV (1950), 129.

high. It is ludicrous, a burlesque of never-never land, only in a partial and almost cosmic sort of way—perhaps, one might say, only wistfully so.

In fact, throughout Fielding's work and career we can see a blend of the romantic attitude with the burlesque of romance, and we can see the serious wish deepening behind the conventional romantic ending. We can also observe the more interesting fact that Fielding is mixing his own self-portrait into that of his romantic hero, identifying himself in varying degrees with both his comic and romantic aspects. We will remember that Fielding at eighteen had attempted to carry from her uncle's clutches the fifteen-year-old heiress and mistress of his heart and that something of this seems to appear in the cloistered damsel of his first play.[18] In *The Author's Farce* (1730), written when Fielding was twenty-three, we find even more striking evidence. Cross pointed out some time ago (I, 81) that in "Harry Luckless," the protagonist, Harry Fielding was giving an obvious self-caricature of the playwright and his troubles. But no one has yet called attention to the fact that, in Harry Luckless, Fielding is also comically making himself the hero of a romance. In Luckless he indeed acts out a complete little burlesque of romance, one which includes the very details of romance that Fielding three years later was to omit from his translation of Molière's *L'Avare*, apparently as too romantic: boat wreck, an agate bracelet, rescue by an old servant.[19] Like the gold amulet which Joseph ties to his arm (I, xv) and which seems as if it might have once been intended to identify him to Adams (I, xiv), a jewel once tied to Luckless's arm uncovers his true identity as heir to the Kingdom of Bantam. Fielding even includes for himself as Harry Luckless a heroine of cognate name, using and burlesquing that siblingish amity and identity often found in the lovers in romance, as for instance in that flowery couple, Floris and Blancheflour: Harry's lady is Harriot, and when at the end her nobility is discovered along with his, the pair is proclaimed Henry and Henrietta, King and

[18] Cross, I, 50–51, and 54: ". . . Fielding, by a mere change in the conclusion, adjusted his own romance to the exigencies of the stage"; and III, 268: "Such was the wild and headlong dash of the boy when spurred on by the romance and the animal within him."

[19] I suspect that these romantic details in *The Author's Farce*—a jewel tied to the arm instead of the agate bracelet—indeed come from *L'Avare*, thus giving another clue to Fielding's reading. Cross (I, 63) sees suggestions from Molière in Fielding's first play, two years earlier.

Queen of Bantam,[20] and, of course, it is "Henry I." All this has been overlooked in the play's welter of social satire; all this is burlesque. But the point of highest interest is that Henry Fielding is identifying himself, in public, with the hero of his absurd little romance. He has, in other words, comically allowed himself to play the hero of romance and to fulfill the romantic wish: the play's penniless gentleman playwright has inherited his due of rank and riches and love. The self-pity of the temporarily unsuccessful young playwright with his perennial human daydream of success and recognition has been buried and transmuted into comedy—indeed, into comic romance.

Fielding's portraits of Andrews and Jones include the chestnut locks, the fair skin, the nobility, the sweetness and even the effeminacy of the conventional knight of romance,[21] and there is considerable evidence that in back of these conventional traits appear also the lineaments of Henry Fielding himself.[22] The boy who, we may infer from *Joseph Andrews* (I, i), had imagined himself Guy of Warwick and all the Champions of Christendom apparently was, in spite of himself, father of the man who railed against "immense" and "idle" romances "filled with monsters."[23] The evidence of *Amelia* only bears this out—though the structure of *Amelia* is consciously epic,[24] the devices of romance are more flagrantly evident than ever (Booth's entrance to his lady's house in a basket, for instance). I omit *Amelia* from my present discussion not because it lacks romance, but because Fielding's comic romance has changed, and the new kind of writing has grown prematurely old, has become interested in other things.

But the idea of comic romance pervades *Joseph Andrews* and *Tom Jones*, and the term, today more contradictory than when Field-

[20] It is tempting to read "Bantam" as a variant of Lilliput, but the text touches only on outlandishness (cut from the 1734 version). I believe Fielding borrowed "Bantam" from Congreve's *Love for Love* (II, i), which play he parallels closely in phrase and situation in his opening scenes (cf. *Love for Love*, I, i).

[21] Arthur J. Tieje, *The Theory of Characterization in Prose Fiction Prior to 1740*, University of Minnesota Studies in Language and Literature, No. 5 (Minneapolis, 1916), p. 5.

[22] Cross, I, 168–169, 330, 348–350; II, 170–171, 328.

[23] *Joseph Andrews*, III, i; *Tom Jones*, IV, i.

[24] George Sherburn, "Fielding's *Amelia*: An Interpretation," *ELH*, III (1936), 1–14; Lyall H. Powers, "The Influence of the *Aeneid* on Fielding's *Amelia*," *MLN*, LXXI (1956), 330–336.

ing used it, pretty well defines these two books. They are comic, and they are romantic. They contain Fielding's own romantic fancy along with his comic wisdom. The trueness and the durability come perhaps not so much from their realistic glimpses of eighteenth-century life or their realistic demonstrations of man's eternal self-deceptions and hypocrisies as from the even more essential realism that sees the comic impossibility of the ideal and romantic glories of life yet affirms their existence and their value. All of us, no less than Fielding, want to live happily ever after; we too like to fancy our innately noble selves rewarded at last by a climb from scullery to castle. And all of us—at least when we read Fielding and Cervantes and Scarron (who was actually shriveled and bent by pain when he wrote his comic work)—can appreciate the amusing impossibility and illusory nature of such a solution, and can see, for a moment, that the human condition itself, caught between fact and aspiration, is, after all, a kind of comic romance.

Papers of the Michigan Academy of Science, Arts, and Letters
Vol. XLV, 1960 (1959 Meeting)

JUBILATE AGNO: THE MIND OF CHRISTOPHER SMART

ARTHUR SHERBO

Michigan State University

Shortly after the middle of the eighteenth century, a century that has sometimes been maligned as an age of reason, a little man much given to drink—a forgivable failing—and to loud public prayer—after a time no longer forgivable—was confined for insanity. The man was Christopher Smart; the place was London; the exact date and place of his first confinement are not known. It is perhaps fortunate that Smart was living in London; had he been in a small town or village his eccentricity might have been a matter of interest to newcomers, but its novelty would soon have worn off for the townsfolk, and Smart would have continued unmolested. But Smart lived in London, and doubtless he made a public nuisance of himself with his too-urgent exhortations to passers-by to kneel and pray audibly with him. He had no sense of the fitness of things; even the pleasure parks of London witnessed his exploits when the spirit moved him: "For I blessed God in St. James's Park till I routed all the company," he writes. It must have been highly disconcerting, even annoying, and possibly physically repugnant for a sweet-smelling young gallant, intent on a conquest, to be approached by a complete stranger whose breath was heavy with drink and whose linen was rarely virginal—and whose unexpected purpose was to force him to his knees there before the amused glances of the others. Smart, one suspects, would not have been loath to practice the laying on of hands. Perhaps it was a complaint to authorities already familiar with Smart that finally brought about his confinement, either privately with some member of his family or in an institution. If so, we may be grateful, for without Smart's intermittent confinement for about seven years—not in one place, nor without possible temporary freedom—without the necessity to turn in on himself, it is probable that Smart would never have written that strange document *Jubilate Agno* and that magnificent poem *A Song to David.* "For stuff'd guts make no musick; strain them strong and you shall have sweet melody." Smart, who is a determined punster in *Jubilate Agno,* is probably punning here on "stuffed guts," which cannot make music (in stringed instruments) until "strained," and "stuffed guts" (a full stomach, repletion, stultification) which will not produce great music or poetry. "Sweet melody," Smart seems to say, can come only when the artist is under extreme pressure of some kind.

As a hack-writer Smart knew what it was to work under pressure. He liked his bottle and he had a family to feed; hence it became increasingly difficult to find the money to keep himself and his family provided with the respective necessities. He worked almost feverishly under these pressures, but only rarely did he manage, in a line or in a passage, to write the kind of poetry of which he was capable. He was a true poet, but the strains to which he was subjected were not such as to wring from him the sustained music that lay within him and broke out in parts of the *Jubilate Agno* and in *A Song to David.* There are enough suggestions of poetic strength in the verse written before 1756 or 1757 to prepare one to some extent for the poetry written during and shortly after the years of duress, but they cannot, it is obvious, fully account for the appearance, in 1763, of *A Song to David.* Something happened to Christopher Smart in those seven years about which we know so little, and it is in *Jubilate Agno,* written probably during the last four of these seven years, that the poet himself tells us what was happening to him.

A warning is in order at this point: a neurologist I met last year, a very learned student of the history of the treatment of the insane in eighteenth-century England, cautioned that I must not give any credence to what Smart was writing in *Jubilate Agno* while confined for insanity. This raises a problem with which I must wrestle—at least to a draw, I hope—at length in the future, but I must ask you to give me the benefit of your doubt. I must say, as one example, that the well-known passage on Smart's cat Jeoffrey in the *Jubilate* is, of course, drawn from the life. He had a cat, a real one, not the product of a diseased imagination, and he described his cat in words that show his powers of observation and description.

But the *Jubilate* tells us much more than the fact that Smart had a cat which he loved: it tells us that he was a man of wide range of learning; it tells us the names of many of his friends; it reveals his likes and dislikes; it serves as a kind of running account of the period of his confinement; it tells or hints at his treatment under confinement (and here the biographer must tread warily); it displays on almost every page his fascination with language; and it is full of statements and half-statements that raise seemingly unsolvable questions. And, of course, it confronts us with line after startling line of poetic beauty: "For the nets come down from the eyes of the Lord to fish up men to their salvation." Furthermore, if I am right in my belief that the *Song to David*, the translation of the Psalms, and probably the *Hymns and Spiritual Songs* were composed while Smart was in confinement, intermittently or otherwise, and while he was writing the *Jubilate*, we must revise existing notions of his progress as a poet—and the evidence for my belief is in the *Jubilate* itself.

There is much I should like to say about the *Jubilate*—about Smart's possible use of Freemasonic symbolism, about his interest in the affairs of the everyday world outside the walls of his place of confinement, and about the mechanical regularity with which he wrote one, two, or three pairs of lines each day of his imprisonment. But I have had my say about these aspects of the *Jubilate* elsewhere.[1] What I want to concentrate on now, however, is the question: what does a man like Smart, former student and teacher at Cambridge University, poet, and man-about-town in London, familiar with most of the famous men of his day, think about in those months and years when, except for a possible occasional visitor, he must look into himself—when he cannot take refuge from himself in drink? First, unlike Robinson Crusoe whose enforced solitude was of much greater duration, one would expect that he would think of love and of sex, of his wife and of other women. And he does, but not to the extent one would think. Smart married about the summer of 1753, and he was already the father of two children when he was confined. Moreover, his early poetry shows him to have been of an amorous nature; when thirteen, for example, he wrote a poem to "Ethelinda" on her doing his verses "the honor of wearing them in her bosom," concluding by asking her to throw the verses "from that downy bed,/ And take the poet in their stead." And, also as a youth, he had fallen in love with Lady Anne Vane, daughter of a nobleman and hence beyond his reach. Anne Vane, twice married by then, appears in the *Jubilate*, but her memory does not torment the poet as a later

[1] JEGP, LIV (1955), 664–669; *Harvard Library Bulletin*, X (1956), 201–207; *MLN*, LXXI (1956), 177–182.

poet, John Clare, was to be tormented by his memories. Smart writes:

> For I saw a blush in Staindrop Church, which was of
> God's own colouring.
> For it was the benevolence of a virgin shewn to me
> before the whole congregation.

He blesses her, later in the poem, using her married name, "God be gracious to Anne Hope," and he refers more than once to the home of the Vanes. He also recalls Harriotte Pratt, sister of a fellow Cambridge man, with whom he had fallen in love and with whom he could probably have been happy. And he refers, of course, to his wife: "For they have seperated me and my bosom, whereas the right comes by setting us together"; "For the angel Gratitude is my wife—God bring me to her or her to me." (This latter is a few lines after he declares that the sun and moon "are together in the spirit every night like man and wife"); and he congratulates himself that he, like Esdras and Sir Isaac Newton, now practices chastity. One editor of the *Jubilate* finds the theme of cuckoldry running through the poem;[2] I am not so sure it is there.

Smart, whose wife left London some time after his confinement and never, so far as our knowledge of the poet extends, saw him thereafter, missed his two little daughters. He alludes to them in general terms, "For the Fatherless Children and widows are never deserted of the Lord," and, a few lines later, writes, "For I bless God for my retreat at Canbury, as it was the place of the nativity of my children. For I pray God to give them food which I cannot earn for them any otherwise than by prayer." I do not know if Smart ever saw his children again after 1763 when he left confinement; I hope he did. The enforced separation from his family and friends and from the pleasures of London life (he congratulates himself, poor unfortunate, on his sobriety while in confinement) leads one to ask what he had to say about life in the asylum. At the beginning of the *Jubilate Agno* he is bitter, comparing himself to famous contemporary freaks ("For I have seen the White Raven and Thomas Hall of Willingham [a young giant] and am myself a greater curiosity than both"); complaining that he is on public show; protesting that he is molested ("For they work me with their harping-iron, which is a barbarous

[2] See the edition by William Bond, 1954, *passim*.

instrument, because I am more unguarded than others"), etc.; but as time passes he appears to be reconciled to his lot. But we know that from his release in 1763 to his death in 1771 he never forgave those who had him confined or whom *he* thought responsible for his confinement.

I can do no more here than to suggest something of the variety and complexity of the *Jubilate* by quoting some scattered lines, with only the briefest introductory comment.

Remembrance of Cambridge University: "For I pray God for the Professors of the University of Cambridge to attend and to mend."

Awareness of his state: "For I am under the same accusation with my Saviour—for they said, he is besides himself."

And: "For I pray the Lord Jesus that cured the LUNATICK to be merciful to all my brethren and sisters in these houses."

Interest in science: "For the PERPETUAL MOTION is in all the works of Almighty God. For it is not so in the engines of man, which are made of dead materials, neither indeed can be."

On his own poetry: "For my talent is to give an impression upon words by punching, that when the reader casts his eye upon 'em, he takes up the image from the mould which I have made."

On flowers: "For flowers are musical in ocular harmony."

On Gray's *Elegy*: "For many a genius being lost at the plough is a false thought, the divine providence is a better manager."

On the English language: "For every word has its marrow in the English tongue for order and for delight."

Fondness for puns: "For two creatures the Bull and the Dog prevail in the English. For all the words ending in ble are in the creature. Invisi-ble, Incomprehensi-ble, ineffa-ble, A-ble."

On Thomas Hobbes: "Let Crispus rejoice with Leviathan—God be gracious to the soul of HOBBES, who was no atheist, but a servant of Christ, and died in the Lord—I wronged him, God forgive me."

And, finally, these four lines:

For God the Father Almighty plays upon the HARP of stupendous magnitude and melody.
For innumerable Angels fly out at every touch and his tune is a work of creation.
For at that time malignity ceases and the devils themselves are at peace.
For this time is perceptible to man by a remarkable stillness and serenity of soul.

Christopher Smart must have had his moments of "remarkable stillness and serenity of soul" in the asylum—God knows he had little opportunity for them before or after his confinement.

Papers of the Michigan Academy of Science, Arts, and Letters
Vol. XLV, 1960 (1959 Meeting)

LITTLE NELL REVISITED

MARK SPILKA

The University of Michigan

In the history of Dickens criticism, the case of Little Nell remains curiously unsettled. Some scholars defend her portrait on the grounds of changing taste: for Victorians, sweet; for moderns, maudlin: each to his own age. Still others find a defect in the author, a form of self-deception or of pathological blocking, which somehow nullifies his art. In his excellent study, *Dickens and His Readers,* George Ford reviews these various theories and rejects them as inadequate.[1] The death of Little Nell, he argues, is "inexcusably dreadful"; but neither scholar nor historian, biographer nor psychologist, can say exactly why. The historical relativist, for instance, can establish changing modes of taste, in Dickens' readers, but cannot judge the writer's work. Or the biographer can reveal sincerity in the author, or experience, but without explaining his performance. In writing *David Copperfield,* Dickens took a shocking episode from childhood and handled it with skill and power; but in writing *The Old Curiosity Shop,* he explored an equally painful episode—and in all sincerity created Nell. Not even the psychologist, says Ford, can settle such discrepancies: like the biographer, he can only indicate the author's hidden motives, or explain his psychic satisfactions, but without accounting for aesthetic worth.

Here the case rests, if we are to accept Ford's views on "the limits of explanatory criticism"; and certainly his position on biography and history seems sound enough. But his treatment of psychology seems only half-correct. Psychology, as we know it, is largely mechanistic; though it cannot account for anything so vital as the creative process, it can, I think, account for psychic *mechanisms* which block that process or distort it. Granting Dickens' conscious sincerity, that is, in creating Nell, suppose he was too disturbed, unconsciously, to depict her with authentic fullness? This puts the question of sincerity on a level Ford ignores, in his discussion, though,

[1] "Little Nell: The Limits of Explanatory Criticism," *Dickens and His Readers* (Princeton, 1955), pp. 55–71.

in all fairness, the psychologists whom he cites ignore it too. Thus, if Aldous Huxley notes the "pathological" quality of Dickens' sentiment, he fails to define its psychic basis; or if Jack Lindsay illuminates the neurotic side of Dickens' life, he fails to connect it with the texture of his art.[2] Yet these same insights, if combined and expanded, might readily solve the perplexing case of Little Nell. My aim, in this essay, is to attempt that combination, so as to isolate a defect in the author's sensibility and to indicate, in passing, a similar defect in his nineteenth-century readers. This method has its dangers—the results are speculative, at best, when the premises themselves are held suspect: but since the stake is only an author's weakness, not his strength, and since the same applies to his early readers, we can safely risk the dangers.

To begin, then, with the novel itself. The "psychic clue" to *The Old Curiosity Shop* lies, I think, in the flight from sexuality which defines the central action. On the surface, that flight is social allegory. Little Nell, the touchstone of human worth, is surrounded on all sides by poverty and industrial blight, and exposed by greed to the illness which at last destroys her. Her grandfather, Mr. Trent, has lost his shop through gambling, which is Dickens' metaphor for business life: "Who were those with whom I played?" the old man asks. "Men who lived by plunder, profligacy, and riot; squandering their gold in doing ill, and propagating vice and evil. My winnings would have been bestowed to the last farthing on a young sinless child whose life they would have sweetened and made happy. What would they [the other players] have contracted? The means of corruption, wretchedness and misery. . . ."[3] But inevitably he has lent himself to such corruption; his shop is now possessed by Quilp, the

[2] *Ibid.*, pp. 62, 65. In the same vein, A. O. J. Cockshut sees "a mixing of irreconcilable types of affection" as the cause of sentiment in Dickens, and even connects this with unconscious conflict: but he defines such conflicts vaguely (*e.g.*, "passion for the macabre"), and fails to explore their sources in biography or their historical appeal. His major premise, that sentimentality is always the product of unconscious pretense, seems superficially like mine; but he twists the point around, in the course of exposition, and labels *any* psychic pretense as sentimental, regardless of its emotional texture. My own analysis is confined to a single mode of pretense in a single age; I also consider sentiment, in itself, an expression of sweet, asexual feeling, soft or tender in texture—not any feeling whatsoever. See Cockshut, "Sentimentality in Fiction," *Twentieth Century*, April 1957, 354–364.

[3] *The Old Curiosity Shop* (London, 1951), p. 74.

avaricious dwarf, who presides there like the spirit of inhuman commerce. When Nell and the old man flee that spirit, they search for a better world in the open countryside, and trust themselves to God: "It is far better to lie down at night beneath an open sky," they feel, "than to rest in close rooms which are always full of care and weary dreams."

This is the conscious plan which Dickens worked with, this flight from city to country, or from economic greed to "the place where God dwells"; but less consciously he invested the novel with feelings which he scarcely understood. For Quilp, his greatest comic villain, owes most of his vitality to sexual rather than economic drives. His wife considers him a second Richard Crookback, and tells her critics "that the best-looking woman here couldn't refuse him if I was dead, and she was free, and he chose to make love to her."[4] Her remark is open to question—Quilp lacks the sexual force of an adult seducer; but where sweet, saintly creatures like Nell and Mrs. Quilp are concerned, his powers are ample and appropriate. Nell's fate is sealed, at any rate, once Quilp proposes to marry her "when Mrs. Quilp the first is dead." For by Dickens' logic, this ugly lascivious dwarf, this boy's idea of avid lust (who savors girls as if they were candy), is Nell's inevitable mate unless she can elude him. It seems significant, then, that Quilp now sleeps in the young girl's bed; that the rusty door is locked, when Nell and her grandfather flee their shop; and that Nell must return to her room for the key, to stand "transfixed with terror at the sight of Mr. Quilp, who was hanging so far out of bed that he almost seemed to be standing on his head, and who, either from the uneasiness of this posture, or in one of his agreeable habits, was gasping and growling with his mouth wide open, and the whites (or rather the dirty yellows) of his eyes distinctly visible."[5] In symbolic terms, this repulsive, phallic image has replaced Nell's "light and sunny dreams"; she flees it to the chaste life of perpetual childhood—but since there is no perpetual childhood, she must die rather than risk the taint of sexual experience, as Dickens here conceives it.

Of course, this hidden allegory subverts the conscious plan of the novel, turning Nell into a frightened, sexless goodie and Quilp into a boy's idea of Richard Crookback, more appealing, in his re-

[4] *Ibid.*, p. 32.
[5] *Ibid.*, p. 96.

pulsiveness, than his victim in her goodness; but, in fact, it is Quilp who has survived the nineteenth century as the novel's great attraction, even as Nell became its greatest blot. Considering this odd reversal, we need to look more closely into the author's cherished dream, the lost realm of infancy, converted here into the open countryside. In that happy kingdom, Nell rests forever on the threshold of puberty, while her grandfather, now senile and decrepit, stands on the other side of active, conscious sexuality (his nocturnal restlessness and gambling mania tell us otherwise, but for Dickens he was sexually quiescent).[6] Even Quilp, the erotic villain, has barely crossed the threshold into adolescence; for his size, ugliness, and actions suggest the raw frenetic lust of an awkward boy, whose sex breaks out in comic nastiness and jiggling frenzies. Indeed, the little boy on his wharf, who sticks to him so doggedly, seems like an outright tag for his condition. But there are other signs of psychic adolescence. Quilp never sleeps with his wife in this novel; he spends the night in his chair, or sleeps alone at his wharf, called Bachelor Hall, supposedly to thwart and annoy his faithful mistress. Actually, it is Quilp who seems to be thwarted, as if his marriage were a sexual failure and his antics the result of plain frustration. To this extent, Quilp is a projection of the author's private life. As Jack Lindsay has shown, the dwarf's abuse of his wife and mother-in-law, and his desire to get rid of his wife and marry Nell, derive from Dickens' discontent with his own family, and his obsessive love for his wife's young sister, Mary Hogarth.[7]

[6] See Freud's analysis of compulsive gambling as a substitute for onanism, in "Dostoevsky and Parricide," *Stavrogin's Confession*, by Fyodor Dostoevsky (New York: Lear, 1947), pp. 108–114. See also Robert Lindner's excellent essay, "The Psychodynamics of Gambling," *Explorations in Psychoanalysis*, ed. Robert Lindner (New York, 1953), pp. 197–217, which confirms and extends Freud's theory. That Dickens sees himself as the grandfather, Nell's loving but furtively guilty companion, is the point here. On the conscious plane, of course, Trent is Dickens' father, whose improvidence, in real life, is equated here with gambling; but on the unconscious plane, where gambling represents lust, Dickens himself becomes the grandfather and indicts his own desire for the real Nell, Mary Hogarth. In this sense, the old man represents the passive side of Dickens' guilt, while Quilp is the active projection of desire. This kind of psychic split occurs, more consciously, in *A Tale of Two Cities* and *The Mystery of Edwin Drood*.

[7] *Charles Dickens: A Biographical and Critical Study* (London, 1950), p. 193. Lindsay's analysis of the novel occurs on pp. 190–196. For the Mary Hogarth

The story of this attraction is bizarre enough. Mary had lived with Dickens and his wife during the first year of their marriage. She was sixteen at the time, demure, pretty, and, in Dickens' eyes, angelic; her hero-worship intoxicated the young writer, who considered her "the chief solace of his labours," and clearly preferred her to his wife. When she died in his arms, after a sudden illness, he slipped a ring from her finger and wore it till his death. For years he longed for burial beside her; he also dreamed of her repeatedly, and called her memory an essential part of his being, as inseparable as his heartbeat. Some biographers have tried to soften these details, but their obsessive quality seems obvious. In the childlike Mary, the eternal sister, he had found a substitute (says Lindsay) for the lost love of his youth, his own sister Fanny. With Mary beside him, he could return to the sweetness of his childhood Eden, when he walked with Fanny in the churchyard near his home, before his parents sent him off to the blacking warehouse. In this way he could have his wife and sister too, without ever breaking the moral code; his physical and psychic drives were satisfied, until death destroyed the "idyllic" pattern. Even then, disguised as Grandfather Trent, he could still run off with Mary in perfect innocence, and possess her openly in death: "They laid him by the side of her whom he had loved so well; and in the church where they had often prayed, and mused, and lingered hand in hand, the child and the old man slept together."[8] In fiction, Dickens' wishes were fulfilled; he could lie beside Mary Hogarth, at the very site of his childhood walks with Fanny.

This is as far as Lindsay takes his shrewd analysis. For many sensitive readers, it will seem far enough; but our problem is to take it a step farther, and apply it to the aesthetic quality of the novel. For if Quilp's marvelous comic antics are, as Lindsay hints, an honest release of Dickens' sexual frustrations, the grandfather's love seems morbid, and the depiction of Little Nell is plainly faked. Dickens' feelings toward Mary Hogarth were incestuous; when he enshrined her in idealized heroines like Little Nell and a host of sexless saints, he was trying to disguise those feelings. If he could cast such girls in neutral, innocent roles, applaud their perfection and swathe them

episode and its connection with Dickens' sister Fanny, see pp. 30–34, 121–123, and 131–135.

[8] *The Old Curiosity Shop*, p. 546.

in sentiment, then he could veil, dilute, or repress the sexual nature of their appeal; and in an oedipal attraction, it is precisely this kind of repression which makes forbidden love seem normal and acceptable. If we can assume, for a moment, that such vigorous repression was part of Dickens' makeup, then it helps to explain his strange romantic history: his excessive delight in the childish ways of his first sweetheart, Maria Beadnell; his marriage, on the rebound from Maria, to the childish and incompetent Catherine Hogarth; his turning for support to her younger sisters, first Mary and then Georgina, whose neutral role seems obvious; and his later pursuit of a nineteen-year-old girl, Ellen Ternan, whom he publicly proclaimed to be as innocent, pure and good as his own dear daughters—though he later abandoned this claim, rather quietly, by making her his mistress. Whatever the cause, this was clearly Dickens' burden—this unhealthy predilection for sexless love. It was reflected in his novels by his concentration on the love of fathers and daughters, uncles and nieces, sons and mothers, or brothers and sisters, and by his sentimentalization of adult romance.

Above all, it was reflected by those sweet, good-tempered, virtuous girls who offend so many readers—among them, aptly enough, a young doctor named Sigmund Freud, who made these acute remarks in the days before his great discoveries: "You must have noticed that all our writers have a 'mannerism,' a stereotyped series of motives and arrangements which indicates the limits of their art. . . . To these mannerisms belong, in the case of Dickens, those flawless girls, selfless and good, so good they are quite colorless. . . ."[9] Had Freud returned to "those flawless girls" in later years, he might have traced their lack of color, their notorious *dullness*, to an attempted shrouding of incestuous longing. Without citing the exact neurosis, Aldous Huxley almost says this in a famous essay. He finds the death of Nell insignificant, because Dickens, emotionally blinded by his "pathologically deliberate unawareness," sees only her goodness and innocence, and cuts her off from surrounding reality. This leaves her virtues marooned, and hence meaningless and colorless.[10] Even

[9] See Ernest Jones, *The Life and Work of Sigmund Freud* (New York, 1953), I, 174.

[10] "Vulgarity in Literature," *Music at Night* (London, 1949), pp. 331–336. See also Cockshut's extension of this theory: he holds that Dickens avoids detail, and evades "his own enjoyment of the spectacle of death," because he is forced here to aim at respectability (*op. cit.*, p. 355). This view seems much

in the most detailed scenes, however, Nell's virtues are marooned from her own potential fullness, and taken for her total self. An idealized abstraction, she seems colorless throughout the book, and not merely in her final moments. By that time, Dickens' yearnings have simply reached their peak, and have blinded him to the realities of scene as well as character.

But if such girls are the product of faked feeling, then a flaw exists in Dickens' novels which has nothing to do with changing taste. Human nature has been falsified, not by conscious insincerity or lack of experience, as some critics hold, but by radical distortion at the unconscious level. In creating his Quilps and Bradley Headstones, Dickens could release repressed desires through projection and identification; but in creating his Nells and Florence Dombeys, he came too close, apparently, to the *object* of repressed desire, and was forced to disguise its true appeal. With his male figures, there was a growing awareness of erotic rawness, from the projected lust of Quilp, through the romanticized lust of Copperfield and Pip, to the frank, unmitigated lust of Headstone and John Jasper. But it was one thing to acknowledge conscious feelings, and another to detect and admit their hidden object. Indeed, his Nells were shaped to *prevent* detection. His cloying sentiment was the result, then, of an arrested sensibility, and must be counted a distinct aesthetic flaw. Though it amounts, at times, to no more than a harmless diffusion of vague feeling, it becomes offensive with his flawless, giddy, plain or wizened girls, all sexless danger-figures, or with danger-situations, like the death of young Paul Dombey, who dies with arms around his sister's neck and eyes fixed on his mother, in the world beyond him. When handling such material, Dickens lapses into crude blank verse and overflows with piety; his mood becomes bathetic and incestuous—or as Freud might say, it turns bathetic *because* it is incestuous, because it expresses Dickens' longing for his sister Fanny, and for his wife's sister, Mary Hogarth. One sample, from the death of Dombey, should reinforce the point.

"Now lay me down," he said, "and, Floy, come close to me, and let me see you!"

---

less relevant than his later comment on Arthur Clennam's feelings for the heroine, in *Little Dorrit*: "His affection is neither overtly sexual nor merely fatherly, and the mixture of the two is too alarming to be faced. Therefore a smoke-screen comes down, which is sentimentality" (p. 362). This, it seems to me, is the real nub of the problem through all of Dickens.

Sister and brother wound their arms around each other, and the golden light came streaming in, and fell upon them, locked together.

"How fast the river runs, between its green banks and the rushes, Floy! But it's very near the sea. I hear the waves! They always said so!"

Presently he told her that the motion of the boat upon the stream was lulling him to rest. How green the banks were now, how bright the flowers growing on them, and how tall the rushes! Now the boat was out at sea, but gliding smoothly on. And now there was a shore before him. Who stood on the bank!—

He put his hands together, as he had been used to do at his prayers. He did not remove his arms to do it; but they saw him fold them so, behind her neck.

"Mama is like you, Floy. I know her by the face! But tell them that the print upon the stairs at school is not divine enough. The light about the head is shining on me as I go!"[11]

This passage seems unreadable today, but when Dombey first died, thousands of Victorians read it and wept indiscriminately. Were they too victims of the sister-complex? According to Edgar Johnson, these Victorians were healthier, in their effusions, than we are, in our repulsion from "pure pathos," except in the smallest doses. "It is in fact not we, but Dickens and the Victorians," he says, "who are in the central stream of natural emotion."[12] Yet, as we have seen, there is nothing especially "pure" about Dickens' pathos, whether on psychic or aesthetic grounds. As for the natural feelings of his readers, it is *quality* and not quantity which counts here, and the quality of Victorian sentiment seems neurotic. We can all agree, I think, that the Victorians subdued their sexual feelings in the name of purity; at the same time, they expressed their "higher" feelings freely, in the name of sensibility. When the same object called for both responses, repression met expression, with a sudden rush—which probably released forbidden urges. If this theory is correct, then the lavish flow of tears, over Nell and her counterparts, was a form of cultural neurosis in the audience, a kind of mass indulgence in erotic feeling, in the name of perfect goodness. That the audience came from homes conceived as temples, as sanctified retreats from heartless commerce and declining faith, supports this view. In such holy places, mothers and sisters, cast as saints, were worshiped for their purity, and fathers feared for sternness; there

[11] *Dombey and Son* (Everyman, 1913), pp. 211–212.

[12] *Charles Dickens: His Tragedy and Triumph* (New York, 1952), I, 324–325. Johnson acknowledges the faults of Dickens' style, but claims the Victorians overlooked them in their attraction to "pure pathos."

filial love became intense, at the expense of sex, and crippled sons emerged to cherish childhood in the form of little girls: Ruskin, Dodgson, Dickens, Barrie—the cult called "Love of Little Girls" at Oxford in the 'eighties, where Dowson, in his desire to marry a girl of twelve, quite gave the cult away. For all these Nells, Alices, and Missies were wives, saints, mothers, and sisters, whom one possessed forever in Nature's temple, in sexless country love, away from loveless urban sordidness—but still with faint erotic stirrings, even in second childhood.

In fiction, then, such girls were agents of incestuous release, expressed through sentiment and atoned for by delicious death. Created from private dreams, they helped to crystallize a public myth.[13] It seems significant, in this light, that Nell's popularity declined with the change in sexual mores: the increased frankness, the exposure of hypocrisy, the feminist movement, and, above all, the breakup of the Victorian home, with its hothouse cultivation of domestic feeling and repression. In this changing atmosphere, Nell lost the aura of the domestic goddess writ small. By the turn of the century, she seemed laughable to sophisticated readers; and by 1958, even the popular audience could enjoy that odd performance, Vladimir Nabokov's *Lolita*, which sharply parodies the Nell tradition.[14]

[13] The theory presented here is my own, but I have based it on material drawn from several sources, most notably from Mark Longaker's *Ernest Dowson* (Philadelphia, 1945), pp. 27–28, 73–79, 147–156, for facts about Dowson and the cult at Oxford; from Walter E. Houghton's *The Victorian Frame of Mind, 1830–1870* (New Haven, 1957), for material on the Victorian home; and from Leslie A. Fiedler's "Good Good Girl and Good Bad Boy," *New Leader*, April 14, 1958. pp. 22–25, for scattered insights on the pre-adolescent virgin as archetypal savior: *e.g.*, "It is the unendurable happy ending, as the white slip of a thing too good for this world prepares to leave for the next, while we, abandoned, sob into our handkerchiefs. The Good Good Girl must die, not only that we may weep (and tears are, for the sentimentalist, the way to salvation), but also because there is nothing else for her to do—no course of action which would not sully her." To explain this phenomenon, however, Fiedler resorts to buckshot: "In this single figure are satisfied the middle-class desire to idealize womanhood, the Protestant resolve to celebrate pre-marital chastity, the North American wish to glorify the fair and debase the black, and the sentimental need to deliver childhood from the odor of diapers and the implications of original sin" (p. 22).

[14] For example: in their long trek across country, stopping at motels, and pursued by a mysterious stranger, the narrator and his charge are like Trent and Nell, pursued by Quilp. But Nabokov's hero seeks desperately, and

Still, to give Johnson his due, our present taste reflects a gain in knowledge, not in health. We simply know, more surely than the Victorians, where illness lies, and this gives us an edge when it comes to pathos. In Thomas Mann's "Disorder and Early Sorrow," for instance, the father's joy with his little daughter, his knowledge that his love holds death within it, and his graceful yielding to the future—all this seems truly poignant. Or in Nabokov's *Lolita*, there is genuine pathos in the failure of the hero's morbid love. Through greater awareness, such artists can define the strain of innocence in even the most perverse relations; yet they cannot forge an ideal worth preserving, and perhaps this indicates their moral, rather than emotional, imbalance.

For the nineteenth-century novelists, whatever their faults, could often shape legitimate ideals. Thus Dickens' Russian disciple, Fyodor Dostoevsky, could use Nell (with appropriate ambiguity) to counter mechanistic schemes which reduce men to subhuman status. By sensing the neurotic source of Nell's appeal, he could undercut the half-truths of determinism, and set against them the child's *affective* innocence, her innate capacity and need for love, as living evidence of Christian truth.[15] Dickens too could safely use this contrast in novels where the protagonist is a young boy (whose rougher sex holds off effusion), and where the antagonist is commercial inhumanity. In a single novel, *Hard Times*, where brevity and satire cut down sentiment, a young girl plays this role with some success. Aptly enough, her antagonist is Thomas Gradgrind, whose utilitarian views are matched, in modern times, by Freudian psychology. At his worst, that is, Freud himself implies that fancy is neurotic "nonsense"; poetry, wish-fulfillment; fun, regression; and selflessness, self-flagellation. Like Gradgrind, moreover, he reduces

comically, to amuse his bored "nymphet" with the idle pleasures of the American teen-ager, whereas Nell cares selflessly and tirelessly for her senile grandfather; and Nabokov's strange pursuer is plainly called the hero's "brother," in his lust for little girls, whereas Dickens never sees the sexual link between Quilp and Trent.

[15] I have in mind especially the series of childlike "sisters" in *Crime and Punishment*, whom Dostoevsky uses to invoke the worth and dignity of the individual soul, in opposition to the "louse" view of humanity, and its deterministic variants, upheld by Raskolnikov. Compare this creative use of the novel's defining powers with Nabakov's horror at morality in art, or with Mann's, at the artist's immorality.

parent-child relations to physical and quantitative terms, without any feel for qualitative richness. Yet, to turn the tables now: Dickens, through this novel, can reveal the limits of such thought, and can indicate, in a concrete way, the quality and worth of childhood love. To keep that quality alive, in the face of mechanistic schemes which deny it (*in the name of love!*), is surely one of the functions of great art. In the long run, it is not enough to shape the order and meaning of experience with objective skill; one must also celebrate and define its value. To amend Johnson's formula, then: it is Dickens and his peers, and not ourselves, who are morally creative. If we avoid their affective blindness, we do so at the expense of moral vision.

PAPERS OF THE MICHIGAN ACADEMY OF SCIENCE, ARTS, AND LETTERS
VOL. XLV, 1960 (1959 MEETING)

SUBSTANCE AND SHADOW: THE ORIGINALS OF THE CHARACTERS IN *CAKES AND ALE*

ALLEN B. BROWN

Central Michigan University

WHEN Somerset Maugham's *Cakes and Ale*, or *The Skeleton in the Cupboard*, was published in 1930, only two years after the death of Thomas Hardy, there was a tremendous outcry. "Trampling on Thomas Hardy's grave," "hitting below the shroud," "grave profaned by literary ghoul" were only a few of the milder attacks made on Maugham's *roman à clef*.[1] But Maugham desired more than a *succès du scandale*. He had attempted to make a complete examination from three points of view of the writer's life from puberty to extreme and highly honored old age, contrasting integrity and showmanship and accentuating the part a beautiful and disturbing woman would necessarily play in such an existence.

For his setting he had chosen Whitstable, as he had before, most notably in *Of Human Bondage*, and that part of London he had settled in after finishing his medical studies at St. Thomas's. With his characters he did best, as always, when dealing with those he knew well. "I have always worked from the living model," says Maugham in his spiritual autobiography, *The Summing Up*.[2] In various books he has scattered a few clues as to the identities of his principals in *Cakes and Ale*: Edward Driffield, "the last of the Victorians"—"the greatest novelist of our day"—"an enormous figure. His novels have as good a chance of surviving as any that have been written in the last hundred years"; Alroy Kear, the "timeserver," "snob," "humbug,"[3] the popular, shallow, self-seeking opportunist; the inimitable Rosie Driffield, with fewer morals than a tabby and as inspiring as Helen of Troy. On a cold, gray November afternoon Maugham did not scruple to reveal to me the names of the rest of the literary coterie who

[1] Richard A. Cordell, *W. Somerset Maugham* (New York, 1937), pp. 126–127.
[2] (New York, 1938), p. 68.
[3] *The W. Somerset Maugham Sampler*, ed. Jerome Weidman (Garden City, N. Y., 1943), pp. 24, 40, 24, 13.

hovered like moths around these major flames,[4] but he refused to comment further on Driffield, Kear, and Rosie.

Is Driffield Hardy? This is what we should like to know most, and this may be the very reason that Maugham is most evasive. When we first meet him, Driffield is a man past art, made into a monument by the efforts of his friends, relatively young second wife, and a longevity-loving British public. Amused as we may be by his comment of "rum" upon the occasion of his being awarded the Order of Merit and his rheumy wink to Ashenden (the thinly disguised Maugham), Driffield is the only unsatisfying principal and the only serious defect in *Cakes and Ale*.[5] Maugham admits: "I am conscious that in what I have written of him I have not presented a living man, standing on his feet, rounded, with comprehensible motives and logical activities."[6] But should not he have tried to do so? As it stands, Driffield is "the skeleton in the cupboard," and a little flesh would hardly have been amiss.

The reader of *Cakes and Ale* must make do with tantalizing glimpses. Driffield has been seaman, waiter, cabdriver, and clerk in a booking office. He has had the manifold experiences so useful to the novelist. As a young man he liked to sail, water-color, visit "pubs" and chat about farming, look at country churches and rub brasses. He has a good voice and sings music hall songs while he strums the banjo. His clothes are loud and, like Rosie, he has no conception of taste in home furnishings. He pegs away at his books and writes reviews in spite of or abetted by the confusion in his home. His opinions of Shakespeare and Henry James are not without interest, but the reader is not convinced that these are his own rather than Maugham's. He has more education than morals. Like Strickland, the artist in *The Moon and Sixpence*, shame doesn't seem to be part of his make-up, but unlike Strickland he is as dependent as Dirk Stroeve upon his wife, realizing that her beauty and earthiness are indispensable to his art and his happiness. Mrs. Barton Traf-

[4] At a private interview in Maugham's suite at the Dorchester Hotel, London, November 12, 1951.

[5] Driffield was an easy target for reviewers. V. S. Pritchett stated: Maugham's "artist is too ordinary. We are unable to believe that he wrote anything worth reading." ("Five Established Novelists," *The Spectator*, October 18, 1930, p. 554.) "Maugham does not really convey . . . a sense of the mental agony and desperation felt by Driffield." (*The Saturday Review*, October 4, 1930, p. 409.)

[6] *The W. Somerset Maugham Sampler*, p. 199.

ford's press-agentry makes him distinguished, and her tact produces in him a social brilliance; but this shell of "weary refinement" has lost all claim to manhood. Age and lovelessness gradually wear down his stubbornness, so that at last he parts even with his favorite desk at which he had written so many books. This pathetic figure is noted for his description and his Shakespearean heroines and peasantry, but most interesting is his attempt in a novel to reconstruct what he guessed happened in life. After their child's death, his wife (Rosie) goes out to seek distraction by having dinner with a friend, an actor, and ends by spending the night with him. Art simplified this to the wife's going out and picking up the first likely looking man she meets in Piccadilly Circus. If this was what he really believed, Driffield's opinion of his erring wife was very low indeed and makes us despise him far more than her.

Maugham has always vigorously denied the charge that Driffield was Hardy, claiming that "he was no more in my mind than George Meredith or Anatole France.... I had been struck by the notion that the veneration to which an author full of years and honor is exposed must be irksome to the little alert soul within him that is alive still to the adventures of his fancy.... I had never been so much taken with Hardy's ... books as were most of my contemporaries, and I did not think his English very good.[7] I was never so much interested in him as I was at one time in George Meredith.... I knew little of Hardy's life."[8]

Cordell has ably summed up the similarities and differences of Driffield and Hardy, finding that the similarities are more striking than the differences:

similarities even in personal habits and appearance. Both were of humble origins; both were interested in architecture; both married twice; the themes of their novels were similar; both were accused of coarseness, and their books were at times banned by the libraries; *The Cup of Life* is not unlike *Jude the Obscure*; the reputation of both was augmented by their longevity; both in old age lived in semi-retirement in a beautiful country-house solicitously ministered by a much younger wife; both were earthy and completely free of preciosity or affectation; both were rather slight and frail physically. But there are differences: Driffield ran away to sea when a boy; their tastes in music were violently different; the

[7] But this bears out what Maugham says of Driffield on p. 97 of "Cakes and Ale."

[8] Pref. to *Cakes and Ale*, Collected Edition (London, 1934) and Modern Library Edition (New York, 1950).

Driffield "country" is Kent, not Wessex; no Mrs. Barton Trafford served as publicity director for Hardy, whose reputation was in no way synthetic; Hardy was more poet than novelist; the first Mrs. Hardy was a woman of blameless reputation and refinement.[9]

The completely English Driffield has almost no similarity to Anatole France, but the suggestion that readers should also consider Meredith as well as Hardy may be a significant one; and Joseph Conrad, who made his final home in the same region of England, may have furnished Maugham with some suggestions for his novelist, especially the references to the sea. Meredith also married a flighty woman with many admirers, one who left him but eventually returned; was a well-known literary editor; boarded at The Limes[10]; was prominent in literary coteries and aided by a literary Egeria (Janet Gordon); played well although his ear for music was untrained; received the Order of Merit in 1905; and, unlike Hardy, was likewise refused burial in Westminster Abbey.[11] (See "Cakes and Ale," *The W. Somerset Maugham Sampler*, p. 38.) The first half of Driffield's life is more like Meredith's, who died in 1909, the second half and his writings more like Hardy's.

T. S. Eliot in his penetrating discussion of Thomas Hardy in *After Strange Gods* (New York, 1934, pp. 59–61) repeats five of Maugham's points about Driffield: refusing to cater to the public, carelessness of style, being noted for landscapes and overmelodramatic plots, and "deliberately relieving some emotion of his own at the expense of the reader." Perhaps even more significant is the fact that nothing Eliot says about Hardy contradicts what Maugham says about the character and work of Driffield. Compare "Cakes and Ale," p. 27:

> a lot he wrote seems a bit old-fashioned nowadays. He flourished in a bad period and he was inclined to be long-winded. Most of his plots are melodramatic; but there's one quality you must allow him: beauty.

And p. 97:

> He was long thought to write very bad English, and indeed he gave you the impression of writing with the stub of a blunt pencil; his style was laboured, an uneasy mixture of the classical and the slangy, and his dialogue was such as could never have issued from the mouth of a human being. Toward the end of his

---

[9] Cordell, pp. 125–126.

[10] The Driffield home in Blackstable is called Lime Cottage.

[11] *The Dictionary of National Biography 1901–11* (New York, 1912), pp. 604–616.

> career, when he dictated his books, his style, acquiring a conversational ease, became flowing and limpid;[12] and then the critics, going back to the novels of his maturity, found that their English had a nervous, racy vigour that eminently suited the matter. His prime belonged to a period when the purple patch was in vogue and there are descriptive passages in his works that have found their way into all the anthologies of English prose. His pieces on the sea, and spring in the Kentish woods, and sunset on the lower reaches of the Thames are famous.

and p. 210 where the death of the child parallels that of Little Father Time in *Jude the Obscure.*

We are more certain that the smooth and sycophantic Alroy Kear stands for Hugh Walpole. Cordell (p. 128) states that Walpole was furious. Walpole himself has recorded: "Read on with increasing horror. Unmistakable portrait of myself."[13] Alroy Kear has almost no talent, but unflagging industry and an eye for the main chance seem to have done more for him than wrestling with the muse could have. He is confident that some day he will be hailed as one of the "greats" of English fiction. He has a closed mind and fixed point of view and adapts everything, willy-nilly, to that point of view. He reveals to Ashenden his plan for a critical biography of Driffield, intimate, homey details, "allusive and charming," combined with "sympathetic, searching . . . subtle" and exhaustive criticism. He hopes to be able to combine "reserve and delicacy" with "manly frankness."[14] And he hopes to be able to do all this for someone he has scarcely known, someone he respects only professionally, by means of others' notes and patiently gathered Driffieldiana.

Maugham wrote Hugh Walpole an apologetic letter denying all[15] and later stated that the figure is a composite of all he found detestable and ridiculous in opportunistic contemporaries: "I took the appearance from one writer, the obsession with good society from another, the heartiness from a third, the pride in athletic prowess from a fourth, and a great deal from myself."[16] But note what he says about Hugh Walpole in *A Writer's Notebook* (pp. 348–349): "he has pretensions to be a man of letters and it is a mortification to him that critical opinion will not give him what he considers his due. . . . I have little doubt that he would have given all his popularity to gain

[12] Especially true of Meredith.
[13] Rupert Hart-Davis, *Hugh Walpole* (New York, 1952), p. 316.
[14] *Cakes and Ale*, ch. 11.
[15] Hart-Davis, pp. 316–317.
[16] Orig. pref. to *Cakes and Ale*.

the esteem of the intelligentsia. He knocked humbly at their doors and besought them to let him in, and it was a bitterness to him that they only laughed."[17] In a new introduction to the Modern Library edition of *Cakes and Ale* (New York, 1950), Maugham admits that Kear is modeled on Walpole but that he had tried to disguise this, as he had no wish to hurt Hugh's feelings.

When Somerset Maugham granted me an interview at the Dorchester Hotel in London, November 12, 1951, he had no compunction in admitting that the completely obnoxious Barton Traffords—Mrs. Trafford, entirely without a skeleton, so sweet and soft as to resemble a jello or swansdown, is peculiarly despicable—were the Sidney Colvins. Sir Sidney (1845–1927), Slade Professor of Fine Art at Cambridge, a close friend of Robert Louis Stevenson, the editor of the Edinburgh edition of Stevenson and his letters, was a well-known critic of art and literature. *Walter Savage Landor* (1881) and *John Keats* (1917) are his best-known works.

Snobbish and spiteful Allgood Newton he identified as Sir Edmund Gosse—then the "best-known critic in England," charming to authors when he met them but "no one could with more point tell a malicious story about a friend"[18]—who had completely patronized him from earliest days.[19] The fact that Gosse was an excellent Scandinavian scholar and authority on Ibsen, Donne, Gray, Jeremy Taylor, Sir Thomas Browne, Congreve, and Swinburne, and close friend of Stevenson and Henry James was, of course, ignored.

The drunken flash-in-the-pan poet Jasper Gibbons was perhaps fairer game. Stephen Phillips' (1868–1915) *Poems* (1897) was far too highly acclaimed, but his *Paolo and Francesca* and *Christ in Hades* are well worth reading.[20]

Maugham's weak defense he gave me was that all these have been dead for many years. (Note the identical number of syllables in

[17] W. Somerset Maugham, *A Writer's Notebook* (Garden City, N. Y., 1949).

[18] *The W. Somerset Maugham Sampler,* p. 169.

[19] *A Writer's Notebook,* p. 261.

[20] Fred Bason, the author of *Fred Bason's Diaries* (1950, 1952) and an early bibliography of W. S. Maugham (1931), an amusing little "cockney sparrow," as he called himself, lived around the corner from me in Walworth, London. A friend of Somerset Maugham's nearly his whole life, one who claimed that he had met most of the literary and theatrical greats of London in his professional capacity of autograph hound, Bason declared that John Drinkwater, the well-known poet-dramatist, considered Gibbons a cruel caricature of himself.

most of these names.) Does it not seem likely that Maugham, who has always worked so closely from life,[21] should also have patterned his major characters with definite individuals in mind, especially if one could feel no pain and the other was beneath his contempt? When Maugham graciously received me in his suite that November afternoon, he had just gotten up from his postprandial nap. His rather disheveled appearance, coupled with the fact that he had forgotten to button his fly and his suspenders were showing, did not, to paraphrase a speech in the recent hit play *Separate Tables*, make him appear "a crashing bore, a dirty old man, or a wicked liar," but, instead, carried conviction.

There were some people in Whitstable who told me when I was there early in 1952 that the real life prototype of Rosie, whom Maugham had long had in mind,[22] is common knowledge; but they would give me no hint as to whom she was. The prototype of both Rosie and the odious Mildred of *Of Human Bondage* is probably revealed on page 167 of *The Summing Up*:

> I was at the time much taken with a young person of extravagant tastes and the gratification of my desires was frustrated by the attentions of more opulent admirers who were able to provide the luxuries that her frivolous soul hankered after. I had nothing much to offer but a serious disposition and a sense of humour. I determined to write a book that would enable me to earn three or four hundred pounds with which I could hold my own with my rivals.

Rosie even says, "I don't mind,"[23] as Mildred often did. Rosie, as Cordell has pointed out (p. 128), and R. H. Ward has likewise noted,[24] is really the most important character outside of Maugham (Ashenden) himself. Weak as she is, so easily trampled into the mire by any brute or worn as any man's boutonniere, she is the middle-aged Maugham's ideal, a pure "white rose," "as good as she was beautiful."[25] It was one he had long held. An early short novel, "Daisy," in *Orientations* (1899) has the same basic person and situation.

One of the pleasanter characters in *Cakes and Ale* is Maugham's irrepressible London landlady, Mrs. Foreman. In *A Writer's Note-*

[21] Maugham defends his lifelong practice at length in his preface to *East and West* (New York, 1934), pp. xvi–xviii.

[22] Pref. to *Cakes and Ale*.

[23] *The W. Somerset Maugham Sampler*, p. 71.

[24] *William Somerset Maugham* (London, 1937), p. 160.

[25] *The W. Somerset Maugham Sampler*, pp. 200, 201.

book (p. 10) he admits, "I did no more than suggest her many excellences."

Maugham's old friend, the doctor's son, who is so summarily disposed of as a has-been at the beginning of Chapter 24, although at that time he still was to give many years of outstanding community service, is the only character of *Cakes and Ale*, except Maugham himself, that I have met in the flesh. This harmless old man and his charming family gave me a most pleasant Sunday afternoon. The old doctor, a collector of Nelsoniana, was less bitter than his family; but he wanted me to know that Maugham was never much good at games. (Ashenden claimed to have been good at football, on page 78.)

The uncle, aunt, and their maid have changed little from *Of Human Bondage* and are drawn straight from life. They are a little older and seen, perhaps, a little more mildly.

Hugh Walpole is thought to have returned the attack in *Captain Nicholas* (1934),[26] and Rupert Hart-Davis (p. 371) states that Archie Bertrand [Willie Maugham] in *John Cornelius* (1937) is "clearly based on Somerset Maugham." Certainly an author called Somerset Ball, whose novels are "complete shams," is mentioned in *Captain Nicholas*. Captain Nicholas with his little daughter is reminiscent of Maugham's Captain Nichols in *The Moon and Sixpence* and is very similar to the "hero" in Elinor Mordaunt's libelous *Gin and Bitters* by "A. Riposte."[27]

When Maugham wrote *Cakes and Ale* did he try his best to give a true picture of the artist's life, with all its trials and tribulations, or did he take this golden opportunity to humiliate everyone whom he thought had not sufficiently admired him in the past? Still, as Maugham is fond of remarking, "there is no kick in the milk of human kindness."

[26] Fred B. Millet et al., *Contemporary British Literature*, 3d ed. (New York, 1935), p. 357.

[27] See "Mr. Maugham Issues Writ for Libel against Mrs. E. Mordaunt," *The Times*, October 6, 1931.

Papers of the Michigan Academy of Science, Arts, and Letters
Vol. XLV, 1960 (1959 Meeting)

THE BUSINESS NOVEL

STEWART H. BENEDICT

Michigan College of Mining and Technology

SINCE the end of World War II, and increasingly within the past five years, American readers have been confronted by a new fictional phenomenon—the business novel. This novel of corporate business does not attempt to contrast particular businesses, but rather to treat elements common to all at the managerial level, such as the problems of business ethics, of selection and advancement, and of personal adjustment to the economic milieu. This definition, then, excludes any business novel which is historical, like Caldwell's *Dynasty of Death*, futuristic, like Rand's *Atlas Shrugged*, devoted to the operational details of a special type of business, like Wakeman's *The Hucksters*; which deals with the struggle of an executive with a basically nonbusiness problem, like Harris's *The Golden Jungle*; or which treats the subject as a source of humor, like Grisman's *Early to Rise.*

Since Twain's *The Gilded Age*, most novels about business—except for the proletarian novels of the 1930's—have dealt sympathetically with the "big businessman." Howells draws Silas Lapham as a warm and sincere human; even such ardent socialists as Upton Sinclair and Frank Norris present business leaders whose cultural interests are broad and whose personal refinement cannot be questioned;[1] Sinclair Lewis portrays William W. Eathorne[2] and Sam Dodsworth as intelligent, sincere, and sensitive to people.[3]

It is true, of course, that the hyper-aggressive *nouveaux riches* like Hunter in Anderson's *Poor White*, Dryfoos in Howells' *A Hazard of New Fortunes*, and Kynance in *Dodsworth* have not been gently handled, but it is not until Dos Passos and Steinbeck that we find all businessmen shown as unrelievedly vicious and villainous. The recent rise of the business novel, then, represents, in part, some kind of reaction.

[1] See especially *The Jungle*, Chapter XXIV, and *The Pit, passim.*

[2] Zenith's leading banker, in *Babbitt*.

[3] See also Lewis's lesser-known book, *Work of Art*, virtually a hymn of praise to the type of businessman who has a sense of service.

But the school of historians spearheaded by Allan Nevins has certainly contributed to its development. Beginning with his monumental study of John D. Rockefeller,[4] Nevins took the lead in urging intellectuals to reconsider the historical role and the personal outlook of the entrepreneur and the corporate executive. In assessing Rockefeller, Nevins called him "a great innovator," a man of "vision" who brought order to an industry that was "lawless . . . and chaotic," and who succeeded because of his "organizing genius . . . tenacity of purpose . . . keenness of mind, and . . . firmness of character."[5]

Published in 1940, Nevin's work prompted a number of prominent historians to question the fairness of the legend of the robber barons and to carry on what amounts to a collective re-evaluation of American economic history and the men who made it. *Harvard Studies in Business History*, for instance, a series published under the aegis of the Harvard School of Business Administration, has become the foremost spokesman for the Nevins point of view. Also significant is the increasingly strong influence of the New Conservatism.

Frederic Wakeman's *The Hucksters*, published just at the end of World War II, may also have served as a stimulus. This book implied what Herbert Spencer had suggested in the nineteenth century, namely that the American businessman in his responsibility is like the soldier: the job of each calls for courage and aggressiveness and involves the dangerous, the unexpected; failure in either job means disgrace or disaster. The novel itself is hardly a landmark in American literature, but it may well have been very influential.

The business novel, as it has developed, has begun to assume a standard pattern. Each work presents at least one of three stereotyped characters. The first of these is the older man, often of upper-class background, who has devoted his life to the corporation which he has built or expanded. His attitude toward the corporation is pure mysticism of the type that Veblen called "medieval"[6] and that was first enunciated by Andrew Undershaft in *Major Barbara*: the business does not belong to him; he belongs to it. For Avery Bullard in *Executive Suite* the factory is "a living, breathing organism."[7]

[4] Allan Nevins, *John D. Rockefeller* (Scribner's, 1940).

[5] *Ibid.*, Vol. II, pp. 707–14.

[6] See Chapter IV, "Business Principles," in *The Theory of Business Enterprise*.

[7] Cameron Hawley, *Executive Suite* (Houghton Mifflin, 1952), p. 3.

Cash McCall, in the novel of that name, says of his father, "Wallpaper was his religion and the plant was his temple, his shrine, his place of worship," and he defends this attitude, saying, "If the high priests lose their faith, the temple walls start to crumble."[8] Ralph Hopkins of *The Man in the Gray Flannel Suit* bursts out to his young assistant, who wants time to spend with his family, "This world has been built by men like me! To really do a job, you have to live it, body and soul!"[9]

In addition to his absorption in business affairs, this character has some outside interest which suggests leisure or refinement of taste, or both. Often this is listening to classical music or reading serious literature, but it may be only having acquired a familiarity with exotic herbs or vintage wines. Conrad Selby in *The Big Store* discusses the relative merits of the Koussevitzky and Walter interpretations of Brahms' "Academic Festival Overture"; Robert Kilgore in *The Empire—and Martin Brill* had read Gibbon and Voltaire by the time he was nine; Cash McCall has a collection of East Indian art and demonstrates an expert's knowledge of Buddhist theology.

The second stereotyped character is the protagonist. This young man is frequently from the upper middle class, but with sufficient breeding or innate good taste to be respectful, if not actually diffident, toward those to the dollar born. Paul Blaze in *The Big Store* realizes that he does not want to defeat the Selby family by wresting control of the store from them. He feels that such people ". . . had to be preserved as ideals. They represented what all others like himself wanted to attain—dignity, respectability, integrity. Qualities like that had to be eternal. . . ."[10]

The protagonist is intelligent, highly imaginative, and combines great drive with a consuming interest in business and the company. His aggressiveness, however, is not so overpowering as to cause him to lose faith in the popular American version of the Christian ethic. In *The Durable Fire* Stephen Lowry regretfully brings about the disgrace and firing of his friend and fellow executive Jim Peale when he learns that Peale has embezzled $50,000 from Continental Industries Corporation. And so sensitive to this ethical lapse on the

[8] Hawley, *Cash McCall* (Pocket Books, 1955), p. 400.

[9] Sloan Wilson, *The Man in the Gray Flannel Suit* (Pocket Books, 1955), p. 262.

[10] Oscar Schisgall, *The Big Store* (Pocket Books, 1955), p. 183.

part of a trusted employee is the corporation's president that, after firing the embezzler, he resigns himself.

In addition, the hero never lets himself forget that he lives in a world of people and it is they, not money or power, that matter. Sam Glidden in *A Man of Affairs*, having been momentarily blind to his company loyalty and community responsibility, comes to his senses and says, "I was willing to betray a trust. A hell of a lot of trust from a lot of different people. I thought the money would make up for it. But I've thought about the money . . . and I feel . . . like a heel."[11] In *Executive Suite*, Don Walling states, ". . . we have an obligation to our stockholders—but it's a bigger obligation than just paying dividends. We have to keep this company *alive*. That's the important thing—and a company is like a man. No man can work for money alone. It isn't enough. You starve his soul when you try it. . . ."[12]

The third stereotype is the villain, or, in some of these novels, the antagonist. He is usually middle class, sometimes lower middle class, and he too is exceedingly aggressive. In the novel in which he is the villain he bears a close resemblance to the popular image of Louis Wolfson—a sort of pecuniary vampire devoid of all humanity. In the novel in which he is merely the antagonist, he is either too aggressive or too coldly calculating and his motive is self-advancement rather than a religious attachment to the company or to the ideal of service. His preoccupation with himself and his own goals makes him insensitive to others, and he is generally depicted as tense and unhappy. John Tyson in *The Magnate* comes to grief because, as another character describes him, "He would attempt to use the corporation, his position, to further purely personal ambitions and not even legitimate ambitions . . . [and] he will use any power he may have to gratify certain inward emotional needs of his own."[13] In *The Empire—and Martin Brill*, the chief statistician, who conceives a plan to increase company profits by bribing public officials, is disgraced, while a junior executive who attempts to advance by claiming credit for the work of others is permanently denied admittance to management. Loren Shaw in *Executive Suite*, an inhibited, neurotic perfectionist who has planned since his college days to become a cor-

[11] John MacDonald, *A Man of Affairs* (Dell, 1957), p. 188.
[12] Hawley, *Executive Suite*, p. 333.
[13] John Harriman, *The Magnate*, (Wyn, 1946), p. 215.

poration president, is stopped just short of the goal because it becomes clear that his chief interest in business is profits, not people, or even pride in the company's product. Least subtle of all is the scene in *A Man of Affairs* in which Mike Dean is struck down by a coronary occlusion as he shouts at the hero, "By God, I despise your kind.... You pollyanna boys want to go around thinking the business world is honorable... and decent.... Listen to me. There's no more morality in industry than there is in that pack of barracuda out there...."[14]

The women in these novels are for the most part uninteresting fictional clichés: the loyal helpmeet, the bored dipsomaniac, the bored nymphomaniac, the clubwoman, the social climber, and so on. The only exceptions occur in two novels of Howard Swiggett, *The Power and the Prize* and *The Durable Fire.* The heroines of both these works are Europeans. Both, too, are interesting creations, but in each novel the author leaves no doubt that their personalities hinder, rather than help, their husbands in the business world.

The plot of the business novel is of necessity undeviating. The problem presented is that of whether or not the protagonist will get the promotion he wants. Since he has the personal attributes he does and since his rival is what he is, there is really no problem at all. The only possible variation on this plot, namely that the protagonist is chosen for a top executive position but rejects it, has been used by Sloan Wilson in *The Man in the Gray Flannel Suit.* Another alternative, that of a qualified protagonist's being passed over for promotion, involves an attitude toward the business community which most of these authors do not permit themselves to entertain.[15]

What criticism, then, can be made of business novels as a type? First, few of them have been written by men who are primarily writers. The majority of those who have ventured into this genre either are or have been businessmen themselves, most frequently business executives.[16] The implications of this, as far as both craftmanship and outlook are concerned, is obvious.

[14] MacDonald, *op. cit.*, p. 189.

[15] John P. Marquand's *Sincerely, Willis Wayde,* which details the rise of a man without the personal qualities that a man "ought to have" to succeed, lacks the properly reverent tone toward business, and probably cannot be considered a business novel in the sense here intended.

[16] Among them Cameron Hawley, Howard Swiggett, George DeMare, John Harriman, and Oscar Schisgall.

Both surprising and disappointing, however, especially in view of the authors' backgrounds, is the fact that these novels are, as a class, extremely romantic, with only superficial resemblances to the realities of the world of management. We can see this romanticism most strikingly in the writers' insistence that the Christian ethic plays a vital role in the operation of the American corporation. Almost no objective study of executive leadership confirms this assertion, and many indicate that it is simply untrue. In his essay "Self-Development for Business Leadership,"[17] George S. Dively, President of the Harris-Seybold Company, states that a leader should have "character," but nowhere mentions ethics of any sort. Herbert Simon, in his book *Administrative Behavior*, says that the only possible criterion for management is the realization of the objectives of the particular business. To quote his very apt summary, ". . . a physician treating a patient does not take time to determine what difference the life or death of his patient will make to the community."[18] Finally, the most extreme position is that of Mills in *White Collar*; he writes that "men in power [in business and industry] do not expect that things can be done legitimately."[19]

Equally unrealistic is the management structure of the typical corporation pictured in the business novel. It appears to have its roots in Nietzsche, for virtually every one of these fictional companies has at its head a man who runs his organization by himself, with little interference from his stockholders and with little authority delegated to his subordinates. Once again the analysts of management differ sharply. In *The Organization Man* Whyte notes that, although the average executive dislikes "the group way" and resents the time spent on "committeemanship," "he is . . . very, very good at team playing—he wouldn't have risen if he wasn't."[20] Robert Gordon, discussing the changing role of the chief executive in his book *Business Leadership in the Large Corporation*, mentions "the diffu-

[17] Included in Marvin Bower, *The Development of Executive Leadership* (Harvard, 1941).

[18] Herbert A. Simon, *Administrative Behavior* (Macmillan, 1947), p. 83.

[19] C. Wright Mills, *White Collar* (Oxford, 1951), p. 95.

[20] William H. Whyte, Jr., *The Organization Man* (Simon and Schuster, 1956), p. 169.

sion of decision-making in the large firm" and says that, because of it, "the chief executive must above all be a coordinator."[21]

Finally, the intellectual level of the fictional executive is far higher than that of his real counterpart. The business novel presents corporation managers who are virtual geniuses, but the students of management take a different view. Whyte writes that "... brilliance and the kind of well-roundedness management asks are a contradiction in terms,"[22] and quotes a president whose organization used to look for brilliance in management trainees but now tries to hire "well-rounded" people with "character." Bender, in *The Technique of Executive Leadership*, presents a list of the traits common to successful executives; the six characteristics listed include drive, sustained interest, and willingness to work hard, but the ability to think is absent. Lastly, in *White Collar*, Mills presents his finding that executives "... are often personally not very bright."[23]

The business novel, indeed, contains so many inherent limitations and circumscriptions that it can never become much more than eulogistic pamphleteering. Because, however, the type offers a wide variety of settings and is essentially romantic escapism in the trappings of verisimilitude, it can be expected to persist in our literature for a long time and eventually to take its place beside the historical novel as a major branch of second-rate fiction.

[21] Robert A. Gordon, *Business Leadership in the Large Corporation* (Brookings, 1945), p. 108.

[22] Whyte, *op. cit.*, p. 233.

[23] Mills, p. 95.

Papers of the Michigan Academy of Science, Arts, and Letters
Vol. XLV, 1960 (1959 Meeting)

ART

MICHIGAN ARTISTS AND THEIR WORK: A SURVEY OF THE FINE ARTS EXHIBITION AT THE MEETING OF THE MICHIGAN ACADEMY OF SCIENCE, ARTS, AND LETTERS, 1959

LILLIAN M. PEAR

This year's Fine Arts exhibition, which presents simultaneously realistic, surrealistic, romantic, and abstract points of view, makes eminently clear that there are certain intangible, significant, and individual contributions in each category of style. The exhibition showed not only a healthy variety but a quality which the following black and white reproductions can only suggest. There were variations of Dove, who succeeded in turning poetic insights into striking examples of private symbolism. There were paintings that sought nuance, combination, and a blending of colors—*rapport de tons.* Some artists apparently have inherited the legacy of Leger in an intricate maze of spatial recession. Others followed somewhat the style of De Kooning in cutting subject matter apart, reversing, exchanging, and otherwise manipulating their subjects. There were also expressions of Picasso's early cubist still life. Others reflected the fireworks, the pitfalls, the surprises and delights of Jackson Pollock. It is evident that painters in Michigan may look with equal interest at what is being done in California, New York, or Rome.

May Jane Bigler's *Orange Shot*, done in broad brush strokes, is almost a textile design. One is reminded of the Taliesin fabric designs of Frank Lloyd Wright. It is also reminiscent of Picasso's early still life painting. Ruth M. Smith has caught the spirit of carnival colors, the clamor, the din in her interlocking lines suggesting elephant tusks and clowns in *Carnival.* In his abstract *Composition*, Richard Royal builds up prismatic forms in strong primary colors and black accents. In *Color Harmony* there is an entirely opposite approach with emphasis on the nuance of analogous harmony. The calligraphic painting of Jackson Pollock and its influence is evident in

John Davies' *Tam O'Shanter's Ride,* but the artist has added his own highly personal attributes.

Realism is handled by Elva Lahti MacMullen with impressionistic overtones. One can almost smell the dust in the *Toreador and Bull.* Pauline Angle's *Bouquet* is a palette painting consisting of a mixture of impressionistic and abstracted forms. James Calder's *Assumption College* is a glowing, richly colored canvas of Romantic Realism. In contrast Jean Paul Slusser's *Image of Our Time* represents the artist's sensitive approach to color and has a Cézanne quality in its positions of the planes. John Pappas' expressionistic *Flowing Winter* depicts all the dark dreariness of a winter's day. Some artists who do very successful abstractions depart occasionally from that style and present work which has a strong impressionistic and traditional connotation such as Edgar Yaeger's *Landscape in Catalonia.*

Louise Jansson Nobili's *Procession* uses blue as a spiritual symbol; the figures become almost unnecessary to the design. Her *Pueblo Market* is strongly reminiscent of Marin's *Corn Dance of the San Domingo Indians in New Mexico* (1929). *Red Still Life* by Stefan Davidek is an expression of a remembered vision of red glow on a hillside. Barbara Johnson in *High Noon at the Shore* shows Roualt's influence; the prow of boats have a stained-glass effect in the forms and planes of the picture space. Eleanor Smart suggests the abstracted patterns of Georgia O'Keefe in *Shore Activity,* and Elizabeth Dulmage's *Building Designs* is rightly named for its spatial design and reflects a departure from this artist's usually more realistic approach. William Lewis reflects an oriental influence in his pen and wash drawing of *Camp at Culpepper.*

These painters of the Academy's 1959 Fine Arts show, indeed, display something of the variety, the chaos, and the vitality of the twentieth century, a volcanic world for which the language of expressionism, whether Faulkner's, Steinbeck's or Picasso's, exists to express the human predicament within an atomic age: "But within this violent landscape like mounds of earth created by the flood, then by-passed by it, there are islands remote, serene with a mythical pastoral calm."[1] It is quite apparent, as Professor Paul Love told the Fine Arts Section this year in his paper "Crystallization of Modern Art," that there "are two streams flowing simultaneously,

[1] Wright Morris, "The Violent Land," *Magazine of Art* (Mar., 1952), p. 102.

the one a conclusion of the Renaissance tradition, more and more sluggishly—the other, a movement into the unknown more and more rapidly."

Nevertheless, "ever and ever is man's search for wholeness, what he is becomes a battlefield, an inner contradiction breaking out periodically on one hand in moral platitudes, escape into verbalization, and on the other violent and arbitrary action."[2]

In Europe, as is here, there are those who, for a variety of reasons, have been watching for signs of a returning realism and have been eager to spot and encourage such a trend should it appear. The direct influence of Klee, Kandinsky, and Mondrian, which was so prevalent after the war, seems to have subsided. But there is a great variety of style, since all of them have their imitators and the paintings of different schools are everywhere. With modern jet travel, Americans can now go in great numbers to feast on architecture and paintings that have withstood all onslaughts of time and criticism. Today, as never before, because old masters are practically off the market place, nineteenth and twentieth century Impressionisms are bringing fabulous prices.

The Fine Arts exhibition demonstrates that there need be no war against the past, that no one need be belligerently modern, that the artist is free to draw from all sources whatever is valid. Henry McBride stated in 1953 that the fashion of the day was abstract, but, on the other hand, Andrew Wyeth, stark realist and academician, quietly sells not only to his neighbors but to every museum.

The visitors to the 1959 Fine Arts exhibition had ample opportunity to discover that there is a time element in the enjoyment of modern painting—time spent by the eye in unraveling the accumulated color, form, and light—and that it can become an endless delight. There were, here and there, however, some signs of an eventual crisis in modern art. The problem of Jackson Pollock's legacy is that some artists have a belief that self-destruction, rather than releasing a freedom, is necessary for a work of art. Action painting should not become a cliché, and, on the other hand, a painting which can be all things to all people ceases to be a communication at all. It is not a question of living in one's day or age; it rather seems there has been too much searching for space in the laboratories and at Cape Canaveral. There are other parts in this violent land

[2] Glen Olds, *Saturday Review of Literature,* (Feb. 21, 1959), p. 28.

that are not fractured segments; there are those islands which hold significant art rather than signaletic art. The 1950's present the artist with two alternatives: "he may either view the eternal world, and if he wishes make a significant comment on it—or he must retreat into an inner world, consolidate himself, and everlastingly emerge with new symbols."[3] The Michigan Academy exhibitions can well serve as a citadel to which both roads must inevitably lead, since the artist must ever have a choice if he is to survive.

There are several specific characteristics which set the exhibition of the Fine Arts Section (which includes musicians in its membership) apart from other shows in Michigan: (1) it presents work by painters, sculptors, weavers, ceramists, architects, silver craftsmen, and photographers in one over-all exhibit; (2) it presents workshops, films, lectures by faculty members of the various universities and art schools, colleges, and museums in which the members may participate without any added fees; (3) its membership is composed of professors, teachers, and professional and amateur artists; (4) it offers a medal annually to the person, artist, or otherwise, who has made an outstanding contribution to the arts in Michigan; (5) it never adheres to one style, embracing Impressionism, Cubism, Expressionism, Realism, and Abstraction, so long as the work has stature, sincerity, and quality in execution; (6) it makes purchase awards and bestows them to smaller museums in Michigan for a future collection of Michigan art.

Since the exhibition and opening tea occurred on Good Friday this year, it seemed to this writer that a work which was completed shortly before Easter should introduce the photographs representative of the 1959 show.

CATALOGUE*

(NFS means Not for Sale)

Paintings, Oil

1. Pauline Angle, Bouquet	$150
2. William Bostick, Mill Town	200
3. James Calder,† Assumption University	300

[3] Paul Love, "Crystallization of Modern Art." Paper presented to Fine Arts Section, 1959. East Lansing, Michigan, 1959.

* The exhibit was hung in Kresge Art Center, East Lansing, March 27–April 10, 1959.

† Honorable mention.

4.	Stefan Davidek, Red Still Life	$150
5.	John Davies, Tam O'Shanter's Ride	150
6.	Stuart Hodge, Rectilinear	300
7.	Stuart Hodge, Afterglow	150
8.	Barbara Johnson, High Noon at the Shore	100
9.	Maude Kronquist, The Arena	50
10.	Allen Leepa, Composition	—
11.	Eva Lahti MacMullen, Encounter	150
12.	David Mitchell, Landscape	300
13.	Maxine Murphy, View of Lake Michigan	—
14.	Maxine Murphy, French Beach Sept. 20	75
15.	Louise Jansson Nobili, Procession	400
16.	Harry Packman, End of Summer	300
17.	Harry Packman,‡ Minstrels	750
18.	Guy Palazzola,‡ Corridor of Fossils	700
19.	John Pappas, Flowing Winter	300
20.	Richard Royal, Color Symphony	125
21.	Richard Royal, Composition	250
22.	Eleanor Smart, Shore Activity	150
23.	Ruth M. Smith, Solar Flight	150
24.	Ruth M. Smith, Carnival	150
25.	Edgar Yaeger, Landscape in Catalonia	150

Water Colors

26.	Mary Jane Bigler, Orange Shot	200
27.	Ray J. De Francesco, St. John, Virgin Islands	135
28.	Elizabeth Dulmage, Building Designs	75
29.	William Lewis, Camp at Culpepper	NFS
30.	Louise Jansson Nobili,† Pueblo Market	195
31.	Jean Paul Slusser, Image of Our Time	100
32.	Sally Swiss, Bargain Days	75

Graphics

33.	Frank Cassara, Triptych (Color Etching)	39
34.	William T. Woodward, Agiochook (Wood Block)	45

Photography

35.	William R. Crise, Wee Wayward Weed	—
36.	Maurice H. Smith, Contacts	25

Sculpture

37.	Howard Church, Centurion	NFS
38.	Howard Church, Prophet	NFS
39.	Walter Midener, Shepherd Boy	NFS

‡ Award $50.

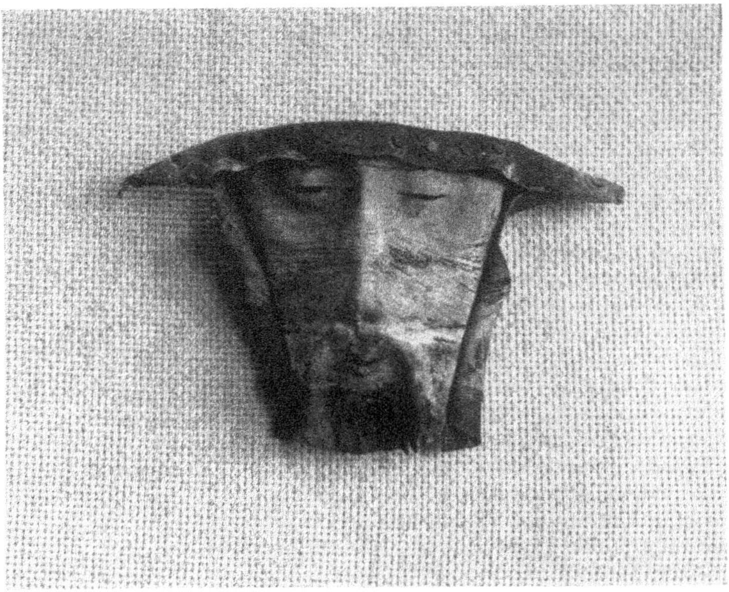

FIG. 1.
Howard V. Church, *The Centurion.* Lead Sculpture, 1959.

JURY: Mr. Harold Zigla, South Bend Art Association.
DONORS: Lewis Art Supply, Jonathan and Dorothy Taylor, Mrs. Sadie McBrearty, Mrs. B. G. Buell, Mrs. Wayne Cowell, Mr. Schemeyer, Mrs. Frances Rinkle, and Mr. H. Davidson.

C. HOWARD CHURCH, East Lansing. Born 1904, South Sioux City, Nebraska. EDUCATION: B.F.A., 1935, Art Institute of Chicago. B.A., 1938, University of Chicago (Phi Beta Kappa). M.A., 1939, Ohio State University; 1940 additional study. POSITIONS HELD: Head of Art Department, Washburn University, Director, Mulvane Art Museum, Topeka, Kansas, 1940–45; Head of Art Department, Michigan State University, 1945. MEMBER OF: Michigan Academy Science, Arts, and Letters since 1950, Vice-Chairman, Fine Arts Section, 1953–56, Chairman, 1956–59; American Association of University Professors; College Art Association; Midwestern College Art Association. Represented in private collections.

WILLIAM A. BOSTICK, Detroit, Born in Detroit. EDUCATION: Printing management, Carnegie Institute of Technology. GRADUATE WORK at: Cranbook Academy of Art; Detroit Society of Arts and Crafts; Berlitz School of Languages; M.A., Wayne University. POSITIONS HELD: Typographer, Printing Designer, Commercial Artist, Printing Plant Superintendent, City Supervisor of Printing, illustrator, and author. Has exhibited in Michigan Artists Exhibition

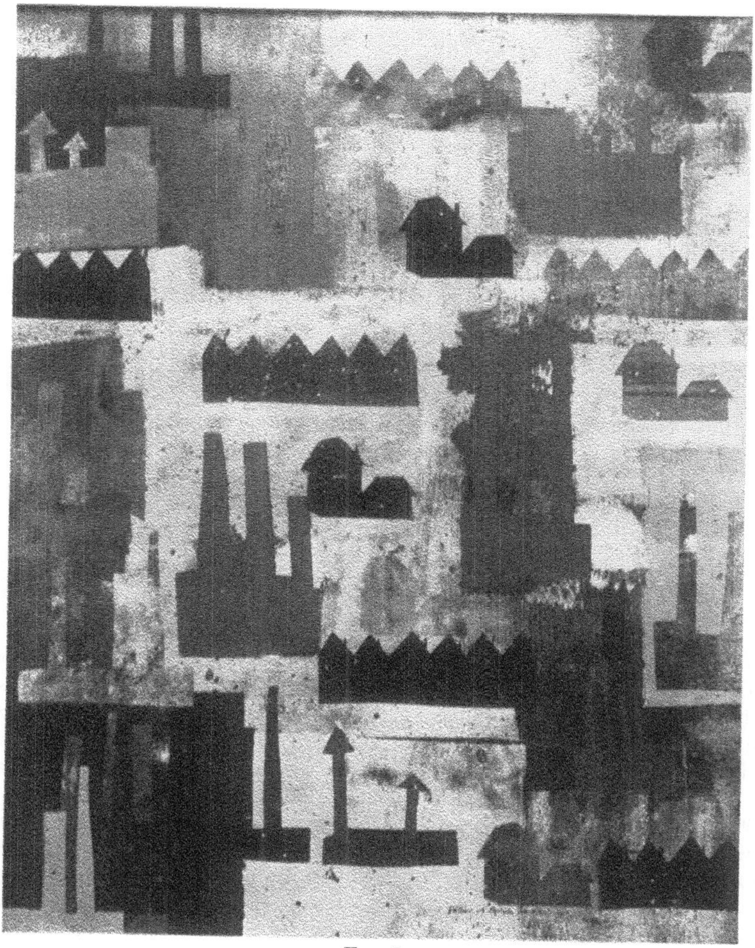

FIG. 2.
William A. Bostick, *Mill Town.* Oil, 1959.

since 1937, winning two prize awards in this period. Has exhibited ceramics and textiles in Michigan Craftsman Show. Has held one-man shows in Detroit, Norfolk, Virginia, Youngstown, and Canton, Ohio. MEMBER OF: American Institute of Graphic Arts, Scarab Club, Past President of Torch Club of Detroit, Michigan Water Color Society, Past President of the Midwest Museums Conference, Book Club of Detroit. POSITION: At present Secretary of the Detroit Institute of Arts since 1946.

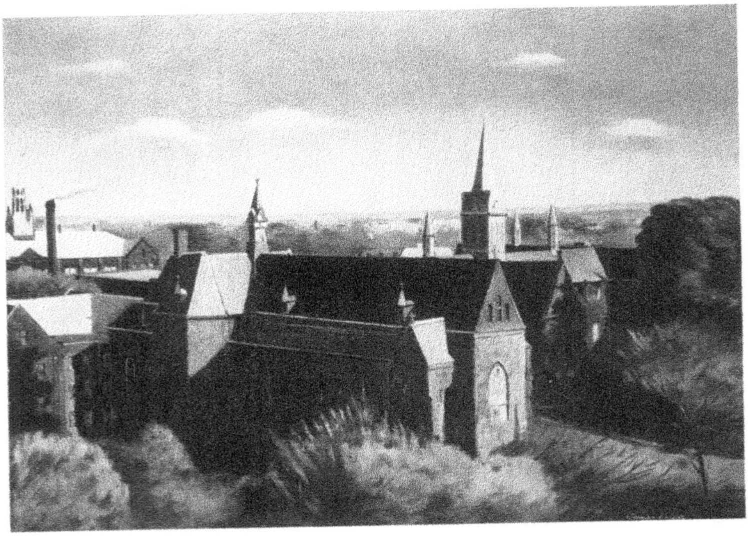

FIG. 3.
James Calder, *Assumption College*. Oil, 1959, Honorable Mention.

JAMES CALDER, Detroit. Born in Detroit, 1907. EDUCATION: Graduate of Arts and Crafts School, Detroit, 1931; student of John Carroll. AWARDS: Halpert Memorial Scholarship, 1930; Modern Art Prize, Michigan Artists Exhibition, 1932; Alvan Macauley Prize, 1936; Anna Scripps Whitcomb Prize, 1951; Museum Purchase Prize, 1954; Mrs. Henry Wineman Prize, 1954; Mr. and Mrs. Anthony Maiullo Prize, 1955, 1957; Garelicks Gallery Purchase Prize, 1957; Elmer Lakatos Memorial Prize, 1959; Third Prize, Michigan State Fair 1956. MURAL COMMISSIONS: St. Clair, Michigan, Post Office, 1938; Grand Ledge, Michigan, Post Office, 1938; Rogers City, Michigan, Post Office, 1939. EXHIBITIONS: New York World's Fair; Golden Gate Exhibition; Pennsylvania Academy of Fine Arts; National Gallery of Canada; Corcoran Gallery, Washington, D. C.; A.C.A. Gallery, New York City; Magnificent Mile Art Festival, Chicago; Provincetown Arts Festival, 1958; Michigan Academy Science, Arts, and Letters, 1959. ONE-MAN SHOWS: J. L. Hudson Gallery, Detroit, 1952; Detroit Artists Market, 1954; Women's City Club, 1958. In many private collections of nationally known collectors.

STEFAN DAVIDEK, Flint. Born in Flint, 1924. EDUCATION: Flint Institute of Art, Art Students League, New York City, Cranbrook Academy of Art. EXHIBITIONS: Twenty-five and Under Show, New York City, Flint Institute Annuals, Michigan Artists Show, the past ten years, Grand Rapids Art Gallery, South Bend, Indiana, San Francisco. POSITION: Painter and decorator, Davidek Co., Flint.

RAY J. DE FRANCESCO, Mt. Clemens. Born in Detroit, 1924. EDUCATION: Institute of Design, Chicago; Michigan State University, B.A.; Wayne State

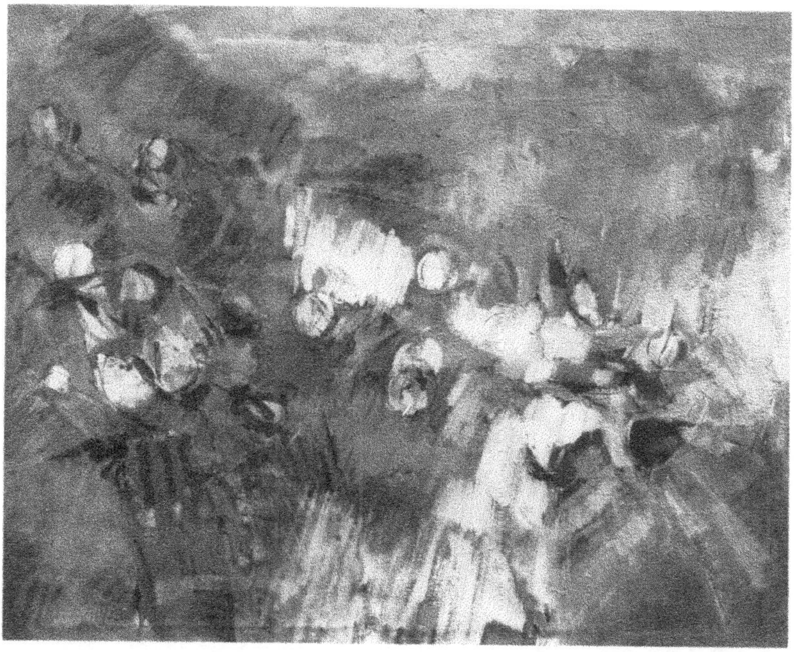

FIG. 4.
Stefan Davidek, *Red Still Life.* Oil, 1959 Award.

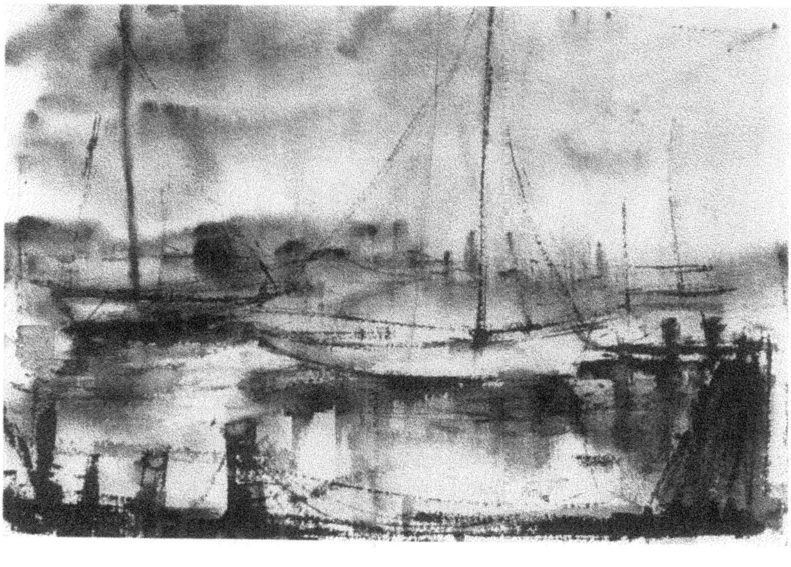

FIG. 5.
Ray J. De Francesco, *St. John, Virgin Islands.* 1959.

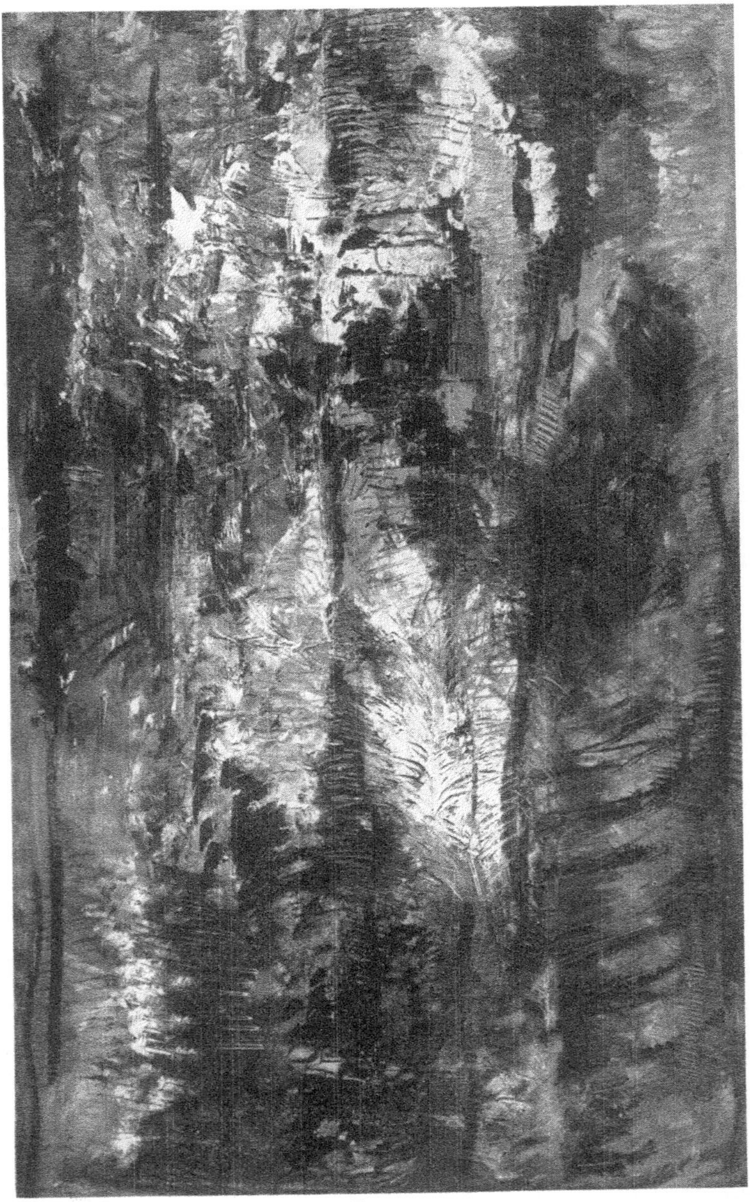

FIG. 6.
Stuart Hodge, *After Glow*. Oil, 1959.

University (graduate work). EXHIBITIONS: Michigan Artist Annuals, Michigan Water Color Society Annuals, Little Gallery, Birmingham, Summer Art School, Leland, Michigan. POSITION: Industrial Designer, Chrysler Corporation. Designs, interiors, furnishings, furniture.

STUART HODGE, East Lansing. Born and raised in Worcester, Massachusetts. EDUCATION: B.S. in Education, Massachusetts School of Art, Boston, 1938; B.F.A., Yale University, 1941; M.F.A., Cranbrook Academy of Art, 1949; Ph.D., State University of Iowa, 1957. Studied at Sorbonne and Ecole des Beaux Arts, 1945. EXHIBITIONS: Painted and exhibited as resident artist, Paris, 1951–53; One-Man Show at Galerie Huit, Paris, 1953; has exhibited in general shows at Paris; St. Louis, Missouri, and Worcester, Massachusetts; many group shows and in the Missouri Annual for many years; represented in permanent collections, Springfield, Maryland Art Museum. POSITIONS: Professor of Art, Principia College, Ilsah, Illinois; Art Instructor, Choate School for Boys, 1941–42; Professor of Painting, Michigan State University.

DAVID MITCHELL, Detroit. Born 1918, Detroit, Michigan. EDUCATION: Denison University, State University of Iowa, Cranbrook Academy of Art. AWARDS: Detroit Institute of Arts, Newberry Water Color Prize, Owen Skelton Prize, Cranbrook Academy Prize, Dr. Morgan Prize. Fulbright Grant, Rome, Italy, 1953–54. POSITION: Instructor in Painting, Wayne State University, 1948–59.

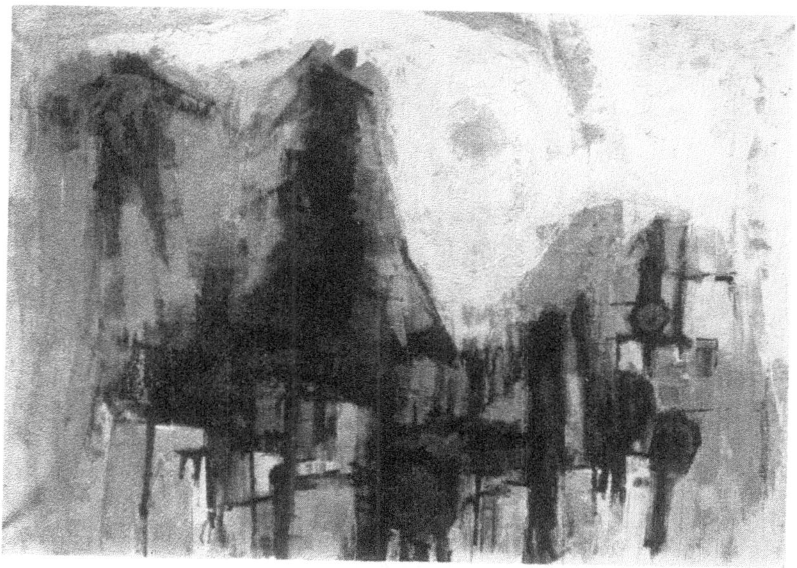

FIG. 7.
David Mitchell, *Landscape*. Oil, 1959.

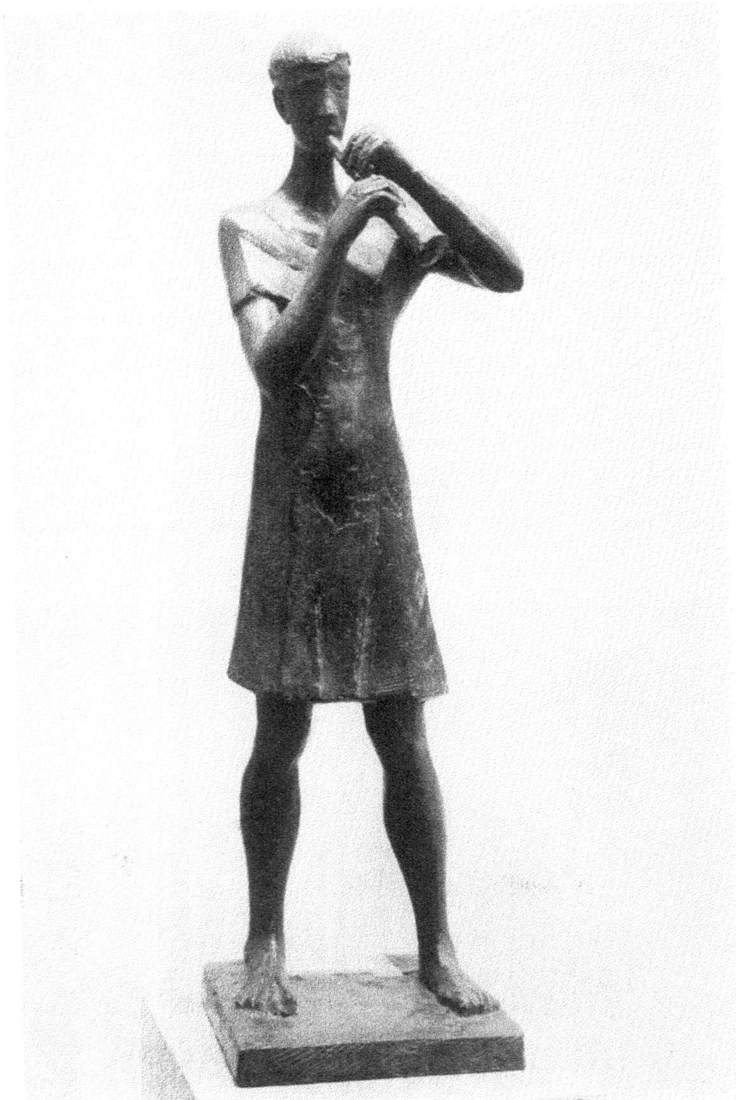

Fig. 8.
Walter Midener, *Shepherd Boy*. Sculpture, 1959.

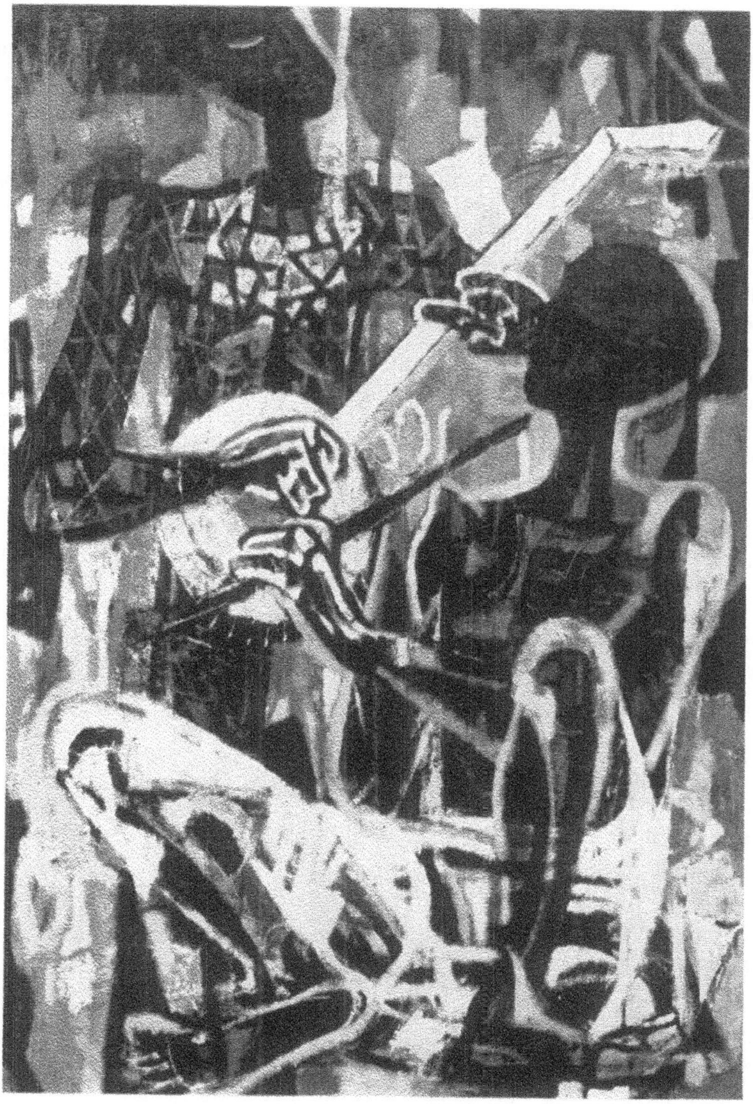

FIG. 9.
Harry Packman, *Minstrels*. Oil, 1959, Award.

WALTER MIDENER, Detroit. Born, 1912, Germany, emigrated to United States, 1938; American citizen since 1943. EDUCATION: Humanistic Gymnasium —Arbitur, 1932; Berlin Academy of Fine and Applied Arts, 1932–36; Teachers Professor Gerstel, Hitzberger, Diederich; Wayne State University, M.A., 1950. POSITIONS HELD: Instructor, Wood and Stone Carving, Henry Street Settlement, New York City, 1939–41; Head of Sculpture Department, Arts and Crafts School, Detroit since 1946; Assistant Director, Arts and Crafts School since 1958; Visiting Instructor, Cleveland Institute of Arts, 1955. PRIZES: Museum Purchase Prize, Detroit Institute of Arts, 1950; Founder's Prize, Michigan Artists Show, 1952; Lewers and Cook Prize, Sculpture, Honolulu, 1943; Tiffany Fellowship, 1940; Lewis Art Supply Co. Prize, Detroit, 1954; Winkleman Foundation Prize, 1955. EXHIBITIONS: Michigan Artists Show, 1946–57, every year; Museum of Modern Art Traveling Show, 1950; Pennsylvania Academy, 1941–48–50–53–58; Third International Sculpture Show, Philadelphia, 1949; Whitney Museum Sculpture, Metropolitan Museum of Art, New York City, 1951; John Herron Institute, Indianapolis, 1951; Detroit Artist Market One-Man Show, 1953; Barone Gallery, New York City, One-Man Show, 1955; represented in several museums and private collections.

HARRY PACKMAN, Detroit. Born in St. Louis, Missouri, 1913. EDUCATION: Cass Technical School, Detroit; Arts and Crafts School, Detroit, under Sarkes Sarkisian. EXHIBITIONS: Pepsi Cola National Exhibition, 1947; Butler Art Institute several times—won first prize in one show; Michigan Artists Show

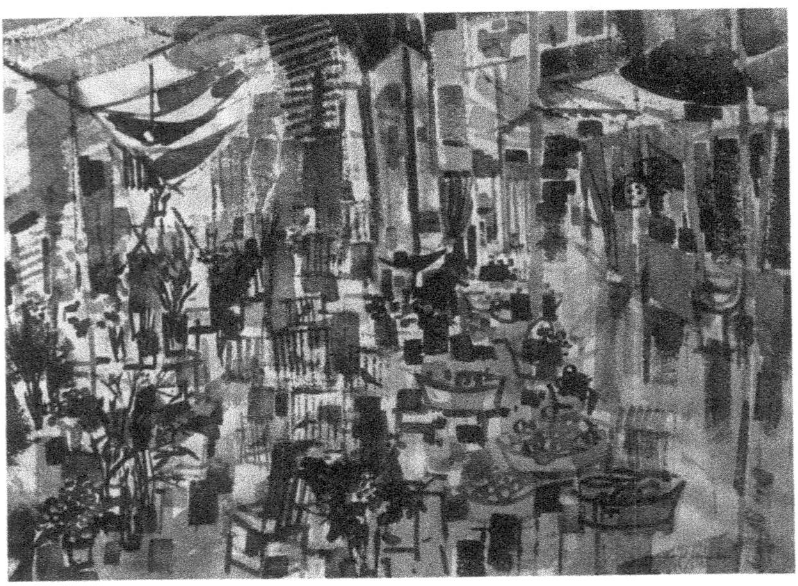

FIG. 10.
Louise Jansson Nobili, *Pueblo Market*. Honorable Mention, 1959.

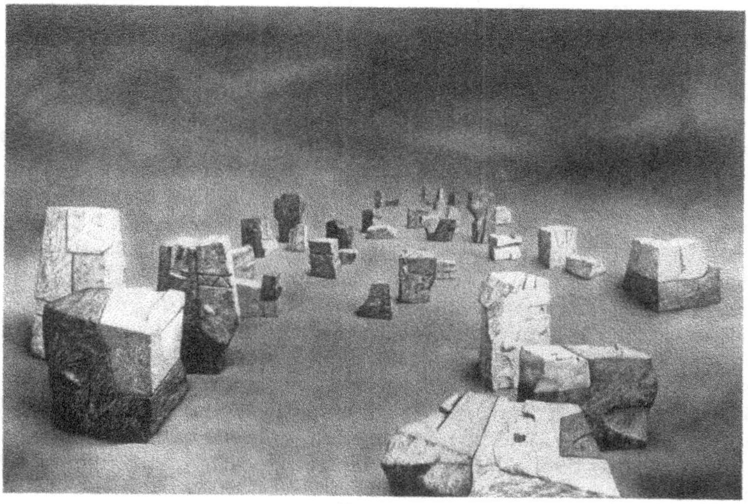

FIG. 11.
Guy Palazzola, *Corridor of Fossils.* Oil, 1959, Award.

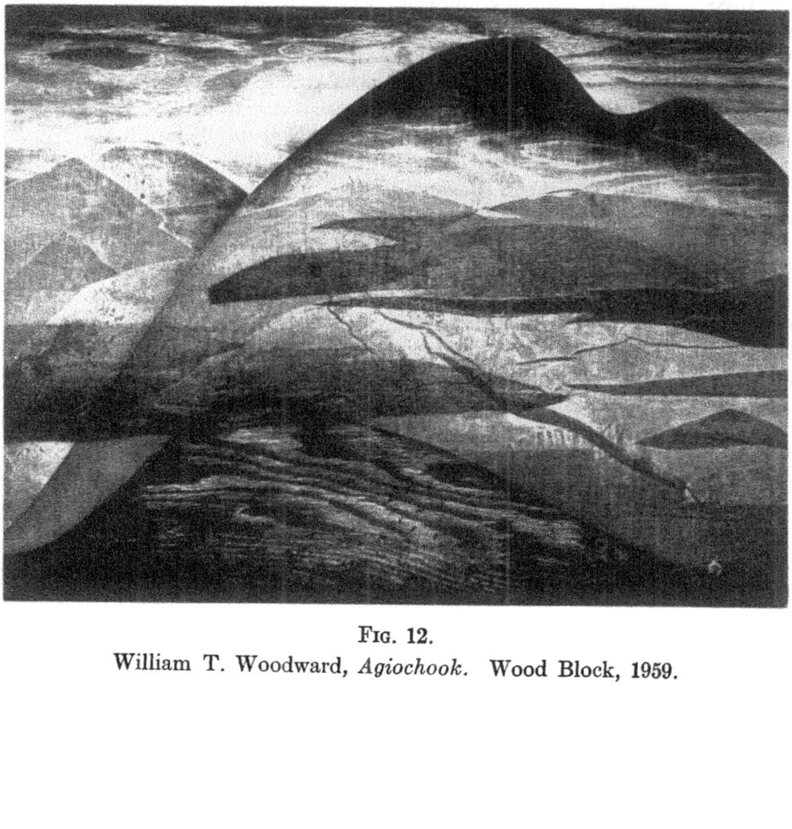

FIG. 12.
William T. Woodward, *Agiochook.* Wood Block, 1959.

several times; Two One-Man Shows Garelichs Gallery, Detroit; Detroit Artist Market; A.C.A. Gallery, New York City Group Shows.

LOUISE JANSSON NOBILI, Detroit. Born in Detroit. EDUCATION: B.F.A., Wayne State University, 1939; M.A., Wayne State University, 1941; studied at University of Wisconsin, summers, 1939–40–41; Chicago School of Design with Moholy Nagy; Chicago Art Institute, 1943–44. POSITION: Professor, Art Department, Wayne State University. REPRESENTED IN COLLECTIONS: Detroit Institute of Arts, Detroit Women's City Club, Art Institute of Dearborn, Sewanhaka High School, Floral Park, New York, Wayne State University, Women's City Club, Detroit, Art Museum, University of Michigan, Ford Times Collection of American Art, Dearborn, Ford Estate, Buhl Building, Detroit, many nationally known private collections. EXHIBITIONS: One-Man Show, Detroit Artists Market, 1945–46–49–54–57; National Water Color, Chicago Art Institute, 1948; Beloit College, 1950; Detroit Scarab Club annually; Ecclesiastic Art Exhibition, 1954–56; Denison College, 1955; Brooklyn Museum, 1953; Michigan State Fair annually; Pennsylvania Academy and Detroit Institute of Arts, 1958; Contemporary American Painters, 1958; Europe National Provincetown Arts Festival, 1958; Forsythe Gallery, Ann Arbor, One-Man Show, 1958.

GUY PALAZZOLA, Ann Arbor. Born, 1916, Kalamazoo, Michigan. EDUCATION: Scholarship for three years, Arts and Crafts School, Detroit, 1937–40; Studied with John Carroll and Sarkis; graduate of Wayne State University. POSITIONS: Instructor, painting, drawing, design, Arts and Crafts School, 1944–57; Assistant Director, Arts and Crafts School, 1950–57; Assistant Professor of Art, University of Michigan, 1957. Has exhibited nationally and internationally since 1938. AWARDS WON: Friends of Modern Art, Detroit, 1939; Sarah Sheridan Award, Detroit, 1940, 1956; Museum Purchase Prize, Detroit, 1949; Owen R. Skelton Prize, Detroit, 1950; Founders Award, Detroit, 1951; Butler Art Institute, 1955, Third Award; Purchase Award, Butler, 1956; Michigan State Fair, 1955, Third Award; Michigan Regional, 1955, Third Award. Chosen by *Magazine of Art* one of twenty-seven young American painters. American Federation of Arts Traveling Show, 1957; Keeler Award, best painting in Western Michigan Show, 1959; work included in many private and nationally known collections.

WILLIAM T. WOODWARD, Grosse Ile. Born in Cincinnati, Ohio, 1913. EDUCATION: University of Wisconsin, B.S., 1936; Wayne State University, M.A., 1941. EXHIBITIONS: Wayne State University, Michigan Water Color Society, Wichita Kansas Graphic Art Association, Wayne County Graphic Society. Graphic illustrations. PUBLICATIONS: F. Clever Bald, *Michigan in Four Centuries,* Harper; Lucy and Sidney Corbett, *Pot Shots from Grosse Isle Kitchen,* Harper; *French Cooking in Old Detroit,* Wayne University Press; *Perennial Harvest* by Philip H. Smith, Harper. POSITION: Associate Professor of Art, Wayne State University.

Printed and bound by CPI Group (UK) Ltd, Croydon, CR0 4YY
16/09/2026
14963492-0004